Biostatistics Decoded

Biostatistics Decoded

A. Gouveia Oliveira

Department of Pharmacy
Universidade Federal do Rio Grande do Norte
Natal, RN, Brazil

Library of Congress Cataloging-in-Publication Data

Oliveira, A. Gouveia de.
  Biostatistics decoded / A. Gouveia Oliveira.
    p. ; cm.
  Includes bibliographical references and index.
  ISBN 978-1-119-95337-1 (pbk.)
  I. Title.
  [DNLM: 1.   Biostatistics. WA 950]
  R853.S7
  610.72'7–dc23
                          2013013976

A catalogue record for this book is available from the British Library.

ISBN: 978-1-119-95337-1 (pbk)

Set in 10/12pt Times by Thomson Digital, Noida, India
Printed and bound in Singapore by Markono Print Media Pte Ltd

1   2013

Contents

Preface   **ix**

1 Introduction   **1**
1.1   The object of biostatistics   1
1.2   Defining the population   4
1.3   Study design   4
1.4   Sampling   6
1.5   Inferences from samples   9

2 Basic concepts   **15**
2.1   Data reduction   15
2.2   Scales of measurement   15
2.3   Tabulations of data   18
2.4   Central tendency measures   19
2.5   Measures of dispersion   20
2.6   Compressing data   21
2.7   The standard deviation   23
2.8   The $n-1$ divisor   24
2.9   Properties of means and variances   26
2.10   Common frequency distributions   28
2.11   The normal distribution   29
2.12   The central limit theorem   31
2.13   Properties of the normal distribution   31
2.14   Statistical tables   34

3 Statistical inference   **37**
3.1   Sampling distributions   37
3.2   The normal distribution of sample means   39
3.3   The standard error of the mean   40
3.4   The value of the standard error   42
3.5   Inferences from means   43
3.6   Confidence intervals   45
3.7   The case of small samples   46
3.8   Student's t distribution   48
3.9   Statistical tables of the t distribution   51

3.10   Estimation with binary variables   53
3.11   The binomial distribution   54
3.12   Inferences from proportions   55
3.13   Statistical tables of the binomial distribution   58
3.14   Sample size requirements   59

4   Descriptive studies   **63**
4.1   Classification of descriptive studies   63
4.2   Probability sampling   64
4.3   Simple random sampling   66
4.4   Replacement in sampling   67
4.5   Stratified sampling   70
4.6   Multistage sampling   74
4.7   Prevalence studies   77
4.8   Incidence studies   78
4.9   The person-years method   80
4.10   Non-probability sampling in descriptive studies   81
4.11   Standardization   82

5   Analytical studies   **87**
5.1   Design of analytical studies   87
5.2   Non-probability sampling in analytical studies   91
5.3   The investigation of associations   92
5.4   Comparison of two means   93
5.5   Comparison of two means from small samples   96
5.6   Comparison of two proportions   98
5.7   Relative risks and odds ratios   100
5.8   Attributable risk   102
5.9   Logits and log odds ratios   104

6   Statistical tests   **107**
6.1   The null hypothesis   107
6.2   The z-test   108
6.3   The p-value   111
6.4   Student's t-test   112
6.5   The binomial test   115
6.6   The chi-square test   116
6.7   Degrees of freedom   121
6.8   The table of the chi-square distribution   122
6.9   Analysis of variance   123
6.10   Statistical tables of the F distribution   129

7   Issues with statistical tests   **131**
7.1   One-sided tests   131
7.2   Power of a statistical test   135
7.3   Sample size estimation   136

7.4	Multiple comparisons	139
7.5	Scale transformation	142
7.6	Non-parametric tests	143

8   Longitudinal studies   **147**
8.1	Repeated measurements	147
8.2	The paired Student's *t*-test	147
8.3	McNemar's test	150
8.4	Analysis of events	151
8.5	The actuarial method	151
8.6	The Kaplan–Meier method	155
8.7	The logrank test	159
8.8	The adjusted logrank test	161
8.9	The Poisson distribution	163
8.10	The incidence rate ratio	166

9   Statistical modeling   **169**
9.1	Linear regression	169
9.2	The least squares method	171
9.3	Linear regression estimates	174
9.4	Regression and correlation	179
9.5	The *F*-test in linear regression	180
9.6	Interpretation of regression analysis results	183
9.7	Multiple regression	185
9.8	Regression diagnostics	188
9.9	Selection of predictor variables	192
9.10	Regression, *t*-test, and anova	194
9.11	Interaction	196
9.12	Nonlinear regression	199
9.13	Logistic regression	201
9.14	The method of maximum likelihood	204
9.15	Estimation of the logistic regression model	206
9.16	The likelihood ratio test	208
9.17	Interpreting the results of logistic regression	209
9.18	Regression coefficients and odds ratios	211
9.19	Applications of logistic regression	211
9.20	The ROC curve	213
9.21	Model validation	216
9.22	The Cox proportional hazards model	219
9.23	Assumptions of the Cox model	223
9.24	Interpretation of Cox regression	225

10   Measurement   **229**
10.1	Construction of clinical questionnaires	229
10.2	Factor analysis	230
10.3	Interpretation of factor analysis	234

10.4	Factor rotation	236
10.5	Factor scores	238
10.6	Reliability	239
10.7	Concordance	245
10.8	Validity	251

11   Experimental studies   **253**

11.1	The purpose of experimental studies	253
11.2	The clinical trial population	255
11.3	The efficacy criteria	256
11.4	Non-comparative clinical trials	258
11.5	Controlled clinical trials	261
11.6	Classical designs	262
11.7	The control group	266
11.8	Blinding	267
11.9	Randomization	268
11.10	The size of a clinical trial	272
11.11	Non-inferiority clinical trials	277
11.12	Adaptive clinical trials	284
11.13	Group sequential plans	286
11.14	The alpha spending function	288
11.15	The clinical trial protocol	291
11.16	The data record	292

12   The analysis of experimental studies   **295**

12.1	General analysis plan	295
12.2	Data preparation	296
12.3	Study populations	297
12.4	Primary efficacy analysis	301
12.5	Analysis of multiple endpoints	303
12.6	Secondary analyses	307
12.7	Safety analysis	308

13   Meta-analysis of clinical trials   **311**

13.1	Purpose of meta-analysis	311
13.2	Measures of treatment effect	312
13.3	The inverse variance method	313
13.4	The random effects model	316
13.5	Heterogeneity	317
13.6	Publication bias	319
13.7	Presentation of results	322

Further reading   **325**

Index   **327**

Preface

The purpose of the book is to present statistical theory and biostatistical methods and applications through a different approach than the one usually adopted by conventional statistical texts.

First, the book integrates topics that, typically, are dealt with in separate books, that is, it covers sampling methods, study design, and statistical methods in a single book and is organized in short chapters structured according to the areas of application of the statistical methods and relating the methods with the corresponding study designs.

Second, and probably the most appealing aspect of the book, biostatistics are presented in a strictly non-mathematical approach, emphasizing the rationale of statistical theory and methods rather than mathematical proofs and formalisms. Illustrations, working examples, computer simulations, and geometrical approaches, rather than mathematical expressions and formulas, are used throughout the book to explain every statistical method.

Third, the topics selected for this book cover most needs of clinical researchers, regarding both study designs and statistical methods, considering the contents of the current scientific literature. The reader will find an explanation of every statistical method, from simple interval estimation and standard statistical tests to advanced methods such as multiple regression, survival analysis, factor analysis, and meta-analysis.

Fourth, the presentation of statistical theory is gradually built upon very simple basic concepts, such as the properties of means and variances, the properties of the normal distribution, and the central limit theorem. This will allow the reader to understand the conditions required for the application and the limitations of each method.

Therefore, this book will satisfy most needs of clinical researchers and medical professionals, offering in a single volume a clear and simple explanation of over 90% of the statistical methods they are likely to find in scientific publications or are likely to need in the course of their own research.

In addition, the book is written according to two skill levels, one for readers who are interested only in understanding the methods and results presented in scientific papers, and one for readers who also wish to know how calculations are done. Even for these, no mathematical skills are required beyond the basic arithmetic operations and an understanding of what square roots and logarithms are.

In conclusion, this book attempts to translate basic, intermediate, and even some advanced statistical concepts into a language and an approach with which health professionals feel comfortable. The topics have been selected according to their relevance for the medical professions and are introduced from a non-mathematical perspective in a sequence that makes sense to clinicians.

All the datasets used for illustration are from my own former research work. Examples of computer outputs and many graphs were produced using Stata (Stata Corporation, College Station, TX, USA).

Lastly, a word of appreciation to the many friends who have offered me continual encouragement and support throughout this project, and in particular to Ana Cristina and my sons Miguel and Ivan, to whom I dedicate this book.

1

Introduction

1.1 The object of biostatistics

Biostatistics is a science that allows us to make abstractions from instantiated facts, therefore helping us to improve our knowledge and understanding of the real world. Most people are aware that biostatistics is concerned with the development of methods and of analytical techniques that are applied to establish facts, such as the proportion of individuals in the general population who have a particular disease. The majority of people are probably also aware that another important application of biostatistics is the identification of relationships between facts, for example, between some characteristic of individuals and the occurrence of disease. Consequently, biostatistics allows us to establish the facts and the relationships among them, that is, the basic building blocks of knowledge.

Therefore, it can be said that it is generally recognized that biostatistics plays an important role in increasing our knowledge of medical science, and health professionals in their routine clinical practice make extensive use of statistical information to reason about patients. When considering the likelihood of a given disease, clinicians need to have a lot of information about the clinical picture, diagnostic methods, treatment, prognosis, and prevention of that particular disease. More specifically, physicians use statistical information like the proportion of people in the general population who have that disease (the disease prevalence) or the annual rate of disease episodes (the incidence), the frequency of symptoms and physical signs in that disease (the clinical picture), the proportion of patients that have abnormalities in selected diagnostic tests (the test sensitivity), and the accuracy of each diagnostic test (the test specificity).

However, clinical practice is not just involved in understanding the cause of the patients' complaints. Clinical practice is largely involved in taking action to prevent, correct, remedy, or cure diseases. But before each action is taken, a decision must be made as to whether an action is required and which action will benefit the patient

---

Biostatistics Decoded, First Edition. A. Gouveia Oliveira.
© 2013 John Wiley & Sons, Ltd. Published 2013 by John Wiley & Sons, Ltd.

most. This, of course, is the most difficult part of clinical practice simply because people can make decisions about alternative actions only if they can predict the likely outcome of each action. In other words, to be able to make decisions about the care of a patient, a clinician needs to be able to predict the future. And it is precisely here that the central role of biostatistics in clinical practice resides.

Actually, biostatistics is the science that allows us to predict the future. How is this accomplished? Simply by assuming that, for any given individual, the expectation is that his or her features and behavior are the same, on average, as those in the **population** to which the individual belongs. Therefore, once we know the average features of a given population, we are able make a reasonable prediction of the features of each individual belonging to that population.

Let us take a further look at how biostatistics allows us to predict the future using, as an example, personal data from a nationwide survey of some 45 000 people in the population. The survey estimated that 27% of the population suffers from chronic venous insufficiency (CVI) of the lower limbs. With this information we can predict, for each member of the population, knowing nothing else, that this person has a 27% chance of suffering from CVI. We can refine our prediction about that person if we know more about the population. Figure 1.1 shows the prevalence of CVI by gender and by age group. With this information we can predict, for example, for a 30-year-old women, that she has a 40% chance of having CVI and that in, say, 30 years she will have a 60% chance of suffering from CVI.

Health professionals constantly use statistical information to make predictions that allow them to make good decisions. Some examples of such statistical information are how the population responds to existing treatment options (treatment efficacy), the proportion of patients that will experience adverse reactions to treatments (treatment safety), the proportion of patients that will relapse after a

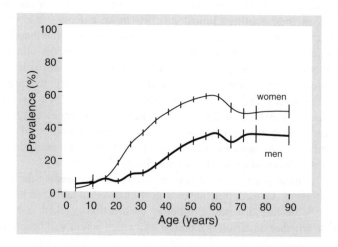

Figure 1.1   Using statistics for predictions. Age- and gender-specific prevalence rates of chronic venous insufficiency.

successful course of treatment (prognosis), and how the patient population feels about possible alternatives to care (treatment effectiveness).

Therefore, the key to prediction is knowing about individual characteristics and disease and treatment outcomes in the population. So we need to study, measure, and evaluate populations. This is a sensible conclusion, but not so easily accomplished. The first difficulty is that, in practice, most populations of interest to medical research have no material existence. One reason for this is that patient populations are very dynamic entities. For example, the population of patients with acute myocardial infarction, with flu, or with bacterial pneumonia is changing at every instant, because new cases are entering the population all the time, while patients resolving the episode or dying from it are leaving the population. Therefore, at any given instant there is one population of patients, but in practice there is no possible way to identify each and every member of the population.

The other reason is that the definitions of populations are based on medical concepts, and usually there is no absolute way of determining whether a given person truly belongs to the population. Therefore, for any individual there is virtually always some uncertainty about whether he or she really belongs to the population. For example, imagine a population like the diabetes mellitus population. Which clinical criteria will identify a person with this disease? There is a choice among a fasting blood glucose level measured once or several times with some interval in between, a single casual blood glucose level, and an oral glucose tolerance test. However, all these methods are known to have false positives and false negatives. Therefore, there is always some amount of uncertainty as to whether each individual actually has or does not have the disease.

The point here is that, in practice, there is no way we can identify and evaluate all the individuals that belong to a given population. This is why we said that populations have no actual physical existence and are only conceptual. So, if we cannot study the whole population, what can we do? Well, the most we can do is to study, measure, and evaluate a **sample** of the population. We then may use the observations we made in the sample to estimate what the population is like. This is what biostatistics is about, **sampling**. Biostatistics studies the sampling process and the phenomena associated with sampling, and by doing so it gives us a method for studying populations which are immaterial. Knowledge of the features and behavior of a conceptual population allows us to predict the features and future behavior of an individual patient belonging to that population and, thus, makes it possible for the health professional to make informed decisions.

Biostatistics is involved not only in helping to create knowledge and to make individual predictions, but also in measurement. Material things have weight and volume and are usually measured with laboratory equipment, but what about things that we know exist but have no weight, no volume, and cannot be seen? Pain, for example. Well, one important area of research in biostatistics is on methods for the development and evaluation of instruments to measure virtually anything we can think of. This includes not just things that we know exist but are not directly observable, like pain or anxiety, but also things that are only conceptual and have no real existence in the physical world, such as the quality of life.

In summary, biostatistics is concerned with the measurement of characteristics (that may not even exist in our material world) in populations (that are virtual) to enable us to predict the future.

1.2 Defining the population

Whenever we want to study a population, we need three basic things: a definition of the population, a study design, and a sampling method.

In clinical research, the population is almost always defined in terms of a recognized disease or condition and the definition of the population is, therefore, mostly a clinical issue. It is of critical importance, however, that the criteria enabling one to assign a given individual to the population being studied are precisely defined. In other words, it is paramount to distinguish the **conceptual definition** of the population from the **operational definition**, and in any study it is necessary to establish the two definitions.

For example, a population definition such as 'arterial hypertension' corresponds to a conceptual definition and its importance is that it allows one to immediately grasp the scope of the study. However, in operational terms this definition is worthless, since two investigators trying to identify subjects eligible for that population might obtain different results simply by using their own criteria for diagnosing hypertension.

On the other hand, if it had been agreed that the population included every person with a sitting blood pressure above 140/90 mmHg after 5 minutes' rest, assessed with a digital sphygmomanometer placed on the right brachial artery, on the average of two measurements made on three different occasions, there would be no ambiguity in identifying those subjects who actually belong to that population.

We can delineate the main properties of a good definition as follows. First, **recognition**, or the property of the definition to refer to a clinical condition recognizable by the medical and scientific communities. Second, **relevance**, that is, the definition should identify a population which is relevant from the clinical standpoint. Finally, **attributability**, or the ability of the definition to allow one to decide unambiguously whether or not a given individual belongs to the population under study.

1.3 Study design

Once we have defined accurately the population of interest, we need to decide which study design we will use. The study design is related to the specific aim of the investigation, and we will go over this subject later on. For the moment, let us just say that, among the large diversity of study designs, we can make a straightforward classification of study types based on a simple notion: in clinical research, the ultimate purpose of an investigation is to establish a **cause–effect relationship**. This goal is actually implicit in any investigation – if we can discover what causes an illness or a symptom, then possibly we will find a way to solve or prevent that

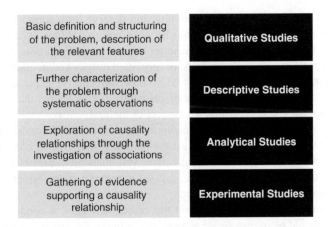

Basic definition and structuring of the problem, description of the relevant features	**Qualitative Studies**
Further characterization of the problem through systematic observations	**Descriptive Studies**
Exploration of causality relationships through the investigation of associations	**Analytical Studies**
Gathering of evidence supporting a causality relationship	**Experimental Studies**

Figure 1.2   The path to causality. Types of research studies.

illness or symptom – and it is the interest of the investigator on the management of a disease that drives him or her to start a study. The different types of clinical studies that we use are, in fact, successive steps in the way of establishing a causality relationship (Figure 1.2).

The first step on the way to causality is, then, to gather whatever information we can on the subject. We investigate patients presenting that problem, we analyze blood and tissue samples, we do x-rays, CT scans, ultrasound examinations, and whatever we believe will provide information on the patient's condition. Then we interpret and summarize the data and present it to others, usually in the form of a **clinical case** or a **case series**. These could be called **qualitative studies**, and the purpose of these investigations is mainly to define the more general aspects of the problem, such as the clinical picture and evolution, the scope of the disease process, and which organ systems appear to be involved.

Once we have some leads on a clinical problem, the next step will naturally be to gather information in a systematic way. The initial qualitative studies have enabled us to focus the problem, and have provided enough information to let us define the population of interest. We can now conduct studies based on the systematic observation of a larger number of people affected by the condition. We call these studies **descriptive studies**. As a result of these studies, we will eventually have an almost complete description of the condition, including its prevalence in the general population, the frequency of the various symptoms and signs, its outcome, and so forth.

Now that we have a picture of the problem, the next question we ask ourselves is what its causes are. This leads us to the question of how we establish causality. This is a complex issue, but it is generally accepted that three conditions must be met for establishing a cause–effect relationship. First, there must be evidence that a strong **association** exists between a stimulus and an observed response, where a stimulus may be something like an intervention or exposure to some product or environment. Second, there must be an **order factor**, that is, there must be evidence that the stimulus preceded the response.

These two conditions are quite obvious and deserve little comment. It is quite clear that if a cause–effect relationship between any two things exists, then those things must be associated and the cause must precede the response. Now the last condition, although also quite obvious, is very often overlooked in clinical research. The third condition is that there must be no other **alternative explanation** for the response, within a reasonable degree of plausibility. This means that in scientific research we must always consider all aspects of the problem and thoroughly search for an explanation of the observed response, other than the stimulus under study, often called a contaminant of the investigation. Only after careful consideration of all the possible contaminants, and after systematically excluding them as responsible for the observed response, can we presume a cause–effect relationship with a reasonable degree of confidence.

Therefore, if we wish to understand the causes of a clinical condition, the next logical step would be to investigate associations between the disease and a number of stimuli. We do this with **analytical studies**, which are also called by a variety of other names, including association studies. The same as descriptive studies mentioned above, analytical studies are **observational studies**, that is, no intervention is carried out on the study subjects. The main purpose of these studies is to identify which factors are related to a disease, to its features, or to its outcome, because those factors will be candidates for further evaluation by experimental studies.

Experimental studies are widely recognized as the most reliable means of establishing a causality relationship. However, in special situations where an association is so strong that it is hard to give an alternative and plausible explanation for the observed effect, many scientists accept the establishment of causality based only on analytical studies.

Experimental studies are designed to verify simultaneously the three conditions for causality. In those studies, we apply an intervention and measure any response occurring after the intervention to establish the order condition. To demonstrate that there is an association between the intervention and the response, we compare the observed responses to those obtained in controls that were not exposed to the intervention. Finally, to avoid any contamination of the experiment, we conduct it under highly controlled conditions. If a response is observed, and if we can exclude contamination of the experiment, then theoretically we can establish causality with reasonable confidence. Some analytical studies are **interventional studies** but they are not experimental studies, in the sense that they will not be able to ascertain the three conditions for causality. We will see later on this book that establishing causality is not a simple task, as there are always many factors external to the experiment that might explain the observed response. Some of them are, precisely, the methods used to analyze the study data.

1.4 Sampling

The third thing to consider when planning a research study is the sampling method. Sampling is such a central issue in biostatistics that an entire chapter of this book is devoted to discussing it. This is necessary for two main reasons: first, because an understanding of the statistical methods requires a clear understanding of the sampling phenomena; second, because most people do not understand at all the purpose of sampling.

Sampling is a relatively recent addition to statistics. For almost two centuries, statistical science was concerned only with **census**, the study of entire populations. Nearly a century ago, however, people realized that populations could be studied more easily, faster, and more economically if observations were used from only a small part of the population, a sample of the population, instead of the whole population. The basic idea was that, provided a sufficient number of observations were made, the patterns of interest in the population would be reproduced in the sample. The measurements made in the sample would then mirror the measurements in the population.

This approach to sampling had, as a primary objective, to obtain a miniature version of the population. The assumption was that the observations made in the sample would reflect the structure of the population. This is very much like going to a store and asking for a sample taken at random from a piece of cloth. Later, by inspecting the sample, one would remember what the whole piece was like. By looking at the colors and patterns of the sample, one would know what the colors and patterns were in the whole piece (Figure 1.3).

Now, if the original piece of cloth had large, repetitive patterns but the sample was only a tiny piece, by looking at the sample one would not be able to tell exactly what the original piece was like. This is because not every pattern and color would be present in the sample, and the sample would be said not to be representative of the original cloth. Conversely, if the sample was large enough to contain all the patterns and colors present in the piece, the sample would be said to be representative (Figure 1.4).

This is very much the reasoning behind the classical approach to sampling. The concept of **representativeness** of a sample was tightly linked to its size: large samples tend to be representative, small samples give unreliable results because

Population

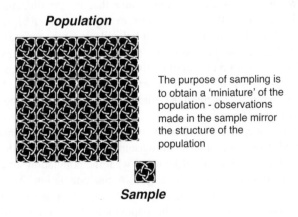

The purpose of sampling is to obtain a 'miniature' of the population - observations made in the sample mirror the structure of the population

Sample

Figure 1.3   Classical view of the purpose of sampling.

non-
representative   *representative*   *excessive*

The concept of
representativeness is
closely related to the
sample size

Figure 1.4   Relationship between representativeness and sample size in the classic view of sampling. The concept of representativeness is closely related to the sample size.

they are not representative of the population. The fragility of this approach, however, is its lack of objectivity in the definition of an adequate sample size.

Some people might say that the sample size should be in proportion to the total population. If so, this would mean that an investigation on the prevalence of, say, chronic heart failure in Norway would require a much smaller sample than the same investigation in Germany. This makes little sense. Now suppose we want to investigate patients with chronic heart failure. Would a sample of 100 patients with chronic heart failure be representative? What about 400 patients? Or do we need 1000 patients? In each case, the sample size is always an almost insignificant fraction of the whole population, since in mainland Portugal, for example, the estimates are that about 300 000 people suffer heart failure.

If it does not make much sense to think that the ideal sample size is a certain proportion of the population, even more so because in many situations the population size is not even known, would a representative sample then be the one that contains all the patterns that exist in the population? If so, how many people will we have to sample to make sure that all possible patterns in the population also exist in the sample? For example, some findings typical of chronic heart failure, like an S3-gallop and alveolar edema, are present in only 2 or 3% of the patients, and the combination of these two findings (assuming they are independent of each other) should exist in only 1 out of 2500 patients. Does this mean that no study of chronic heart failure with less than 2500 patients should be considered representative? And what happens when the structure of the population is unknown?

The problem of the lack of objectivity in defining sample representativeness can be circumvented if we adopt a different reasoning when dealing with samples. Let us accept that we have no means of knowing what the population structure truly is, and all we can possibly have is a sample of the population. Then, a realistic procedure would be to look at the sample and, by inspecting its structure, formulate a hypothesis about the structure of the population. The structure of the sample constrains the hypothesis to be consistent with the observations.

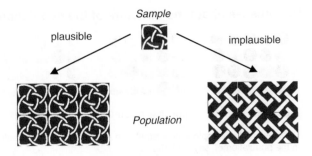

The purpose of sampling is the evaluation of the plausibility of a hypothesis about the structure of the population, considering the structure of a limited number of observations

Figure 1.5   Modern view of the purpose of sampling. The purpose of sampling is the evaluation of the plausibility of a hypothesis about the structure of the population, considering the structure of a limited number of observations.

Taking the above example on the samples of cloth, the situation now is as if we were given a sample of cloth and asked what the whole piece would be like. If the sample were large, we probably would have no difficulty answering that question. But if the sample were small, something could also be said about the piece. For example, if the sample contained only red circles over a yellow background, one could say that the sample probably did not come from a Persian carpet. In other words, by inspecting the sample one could say that it was consistent with a number of pieces of cloth but not with other pieces (Figure 1.5).

Therefore, the purpose of sampling is to provide a means of evaluating the plausibility of several hypotheses about the structure of the population, through a limited number of observations and assuming that the structure of the population must be consistent with the structure of the sample. One immediate implication of this approach is that there are no sample size requirements in order to achieve representativeness.

Let us verify the truth of this statement and see if this approach to sampling is still valid in the extreme situation of a sample size of one. We know that with the first approach we would discard such a sample as non-representative. Will we reach the same conclusion with the current approach?

1.5 Inferences from samples

Imagine a swimming pool full of small balls. The color of the balls is the attribute we wish to study, and we know that it can take only one of two possible values: black and white. The problem at hand is to find the proportion of black balls in the population of balls inside the swimming pool. So we take a single ball out of the pool and imagine that such a ball happened to be black (Figure 1.6). What can we say about the proportion of black balls in the population?

Studied attribute: color
Attribute values: black and white.
Sample size: 1
Sampling result: ⬤

Hypotheses about the structure of the population:

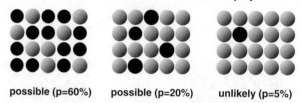

possible (p=60%)   possible (p=20%)   unlikely (p=5%)

Conclusion: this sample is consistent with a proportion of black balls
in the population of, say, 5 to 100%.

Figure 1.6   Interpretation of the result of sampling.

We could start by saying that it is perfectly possible that the population consists 100% of black balls. We could also say that it is also quite plausible that the proportion of black balls is, say, 80% because then it would be quite natural that, by taking a single ball at random from the pool, we would get a black ball. However, if the proportion of black balls in the population is very small, say less than 5%, we would expect to get a white ball, rather than a black ball. In other words, a sample made up of a black ball is not very consistent with the hypothesis of a population with less than 5% of black balls. On the other hand, if the proportion of black balls in the population is between 5 and 100%, the result of the sampling is quite plausible. Consequently, we would conclude that the sample was consistent with a proportion of black balls in the swimming pool between 5 and 100%. The inference we would make from that sample would be to estimate as such the proportion of black balls, with a high degree of confidence.

You can say that this whole thing is nonsense, because such a conclusion is completely worthless. Of course it is, but that is because we did not bother spending a lot of effort in doing the study. If we wanted a more interesting conclusion, we would have to work harder and collect some more information about the population. That is, we would have to make some more observations to increase sample size.

Before going into this, think for a moment about the previous study. There are three important things to note. First, this approach to sampling still works in the extreme situation of a sample size of one, while that is not true for the classical approach. Second, the conclusion was correct (remember, it was said that one was very confident that the proportion of black balls in the population was a number between 5 and 100%). The problem with the conclusion, better said with the study, was that it lacked **precision**. Third, the inference procedure described here is valid only for random samples of the population, otherwise the conclusions may be completely wrong. Suppose that the proportion of black balls in the population is minimal, but because their color attracts our attention we are much more likely to select a flashy black ball than a boring white one. We would then make the same

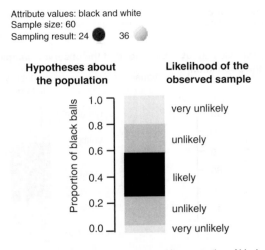

Attribute values: black and white
Sample size: 60
Sampling result: 24 ● 36

**Hypotheses about
the population**

**Likelihood of the
observed sample**

Proportion of black balls

1.0 — very unlikely

0.8 —

0.6 — unlikely

0.4 — likely

0.2 — unlikely

0.0 — very unlikely

Conclusion: this sample is consistent with a proportion of black
balls in the population of 0.25 to 0.55.

Figure 1.7   Interpretation of the result of sampling.

reasoning as before and reach the same conclusion, but we would be completely
wrong because the sample was **biased** toward the black balls.

Suppose now that we decide to take a random sample of 60 balls, and that we
have 24 black balls and 36 white balls (Figure 1.7). The proportion of black balls in
the sample is, therefore, 40%. What can we say about the proportion of black balls
in the population? Well, we can say that if the proportion is below, say, 25%, there
should not be so many black balls in a sample of 60. Conversely, we can also say
that if the proportion is above, say, 55%, there should be more black balls in the
sample. Therefore, we would be confident in concluding that the proportion of
black balls in the swimming pool must be somewhere between 25 and 55%. This is
a more interesting result than the previous one because it has more precision; that is,
the range of possibilities is narrower than before. If we need more precision, all we
have to do is to increase the sample size.

Let us return to the situation of a sample size of one and suppose that we want to
estimate another characteristic of the balls in the population, for example, the
average weight. This characteristic, or **attribute**, has an important difference from
the color attribute, because weight can take many different values, not just two.

Let us see if we can apply the same reasoning in the case of attributes taking
many different values. To do so, we take a ball at random and measure its weight.
Let us say that we get a weight of 60 grams. What can we conclude about the
average weight in the population?

Now the answer is not so simple. If we knew that the balls were all about the
same weight, we could say that the average weight in the population should be a
value between, say, 50 and 70 grams. If it were below or above those limits, it
would be unlikely that a ball sampled at random would weigh 60 grams.

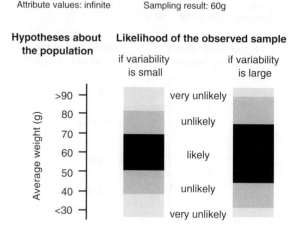

Figure 1.8   Interpretation of the result of sampling.

However, if we knew that the balls varied greatly in weight, we would say that the average weight in the population should be a value between, say, 40 and 80 grams (Figure 1.8). The problem here, because now we are studying an attribute that may take many values, is that for making inferences about the population we also need information about the amount of variation of that attribute in the population. It thus appears that this approach does not work well in this extreme situation. Or does it?

Suppose we take a second random observation and now have a sample of two. The second ball weighs 58 grams, and so we are compelled to believe that balls in this population are relatively homogeneous regarding weight. In this case, we could say that we were quite confident that the average weight of balls in the population was between 50 and 70 grams. If the average weight were under 50 grams, it would be unlikely that we would have two balls with 58 and 60 grams in the sample; and similarly if the average weight were above 70 grams. So this approach works properly with a sample size of two, but is this situation extreme? Yes it is, because in this case we need to estimate not one but two characteristics of the population, the average weight and its variation, and it is only normal that it is now required to have at least two observations.

In summary, in order that the modern approach to sampling be valid, sampling must be at random. The representativeness of a sample is primarily determined by the sampling method used, not by the sample size. Sample size determines only the precision of the population estimates obtained with the sample.

Now, if sample size has no relationship to representativeness, does this mean that sample size has no influence at all on the validity of the estimates? No it does not. Sample size is of importance to validity because large sample sizes offer protection against accidental errors during sample selection and data gathering, which might have an impact on our estimates. Examples of such errors are selecting an individual who does not actually belong to the population under study, measurement errors, transcription errors, and missing values.

Where do we go from here? Well, we have already eliminated a lot of subjectivity by putting the notion of sample representativeness within a convenient framework. Now we must try to eliminate the remaining subjectivity in two other statements. First, we need to find a way to determine, objectively and reliably, the limits for population proportions and averages that are consistent with the samples. Second, we need to be more specific when we say that we are confident about those limits. Terms like confident, very confident, or quite confident lack objectivity, so it would be very useful if we could express quantitatively our degree of confidence in the estimates. But before going into this, we have to review some basic statistical concepts.

2

Basic concepts

2.1 Data reduction

The intermediate result of any clinical investigation is typically a large set of numeric and coded data. If we take a look at the data, we will immediately realize that it is virtually meaningless to us. Contrary to the written word, which we can read, abstract, and understand immediately, we have no such ability when it comes to a list of numbers. So, in order to understand the information contained in such lists of numbers we need to compress the data into just a few numbers, trying to lose as little information as we can.

One commonly used method of data compression is to average all observations, by summing all the values of an attribute and dividing the total by the number of observations. The **average**, or **mean**, gives us an immediate grasp of the order of magnitude of the values, but unfortunately its use is limited to values measured on a numeric scale. Such an approach would not work with an attribute measured in categories, such as profession.

The first thing we must do when we evaluate the results of a clinical study is, therefore, to abstract the data. To do that, we must first identify the **scale of measurement** used with each attribute in the dataset, and then we must decide which is the best method for summarizing the data.

2.2 Scales of measurement

We may measure patient characteristics, or attributes, with many scales, but these will usually fall into one of four types.

The simplest scale is the **binary scale**, which has only two values. Patient gender (female, male) is an example of an attribute measured in a binary scale. Everything that has a yes/no answer (e.g., obesity, previous myocardial infarction,

Biostatistics Decoded, First Edition. A. Gouveia Oliveira.
© 2013 John Wiley & Sons, Ltd. Published 2013 by John Wiley & Sons, Ltd.

family history of hypertension, etc.) is being measured in a binary scale. Very often the values of a binary scale are not numbers but terms, and this is why the binary scale is also a **nominal scale**. However, the values of any binary attribute can readily be converted to 0 and 1. For example, the attribute gender with values female and male can be converted to the attribute female gender with values 0 meaning no and 1 meaning yes.

Next in complexity is the **categorical scale**. This is simply a nominal scale with more than two values. In common with the binary scale, the values in the categorical scale are usually terms, not numbers, and the order of those terms is arbitrary: the first term in the list of values is not necessarily smaller than the second. Arithmetic operations with categorical scales are meaningless, even if the values are numeric. Examples of attributes measured on a categorical scale are profession, race, and education.

It is important to note that in a given person an attribute can have only a single value. However, sometimes we see categorical attributes that seem to take several values for the same person. Consider, for example, an attribute called cardiovascular risk factors with values arterial hypertension, hypercholesterolemia, diabetes mellitus, obesity, and tabagism. Obviously, a person can have more than one risk factor and this attribute is called a **multi-valued attribute**. This attribute, however, is just a compact presentation of a set of related attributes grouped under a heading that is commonly used in data forms. For analysis, these attributes must be converted into binary attributes. In the example, cardiovascular risk factors is the heading, while arterial hypertension, hypercholesterolemia, diabetes mellitus, obesity, and tabagism are binary variables that take the values 0 and 1.

When values can be ordered, we have an **ordinal scale**. In the particular case when all the consecutive values in the scale are at the same distance, we call that an interval scale (Figure 2.1). An example of an ordinal scale is the staging of a tumor (stage I, II, III, IV). There is a natural order of the values, since stage II is more invasive than stage I and less than stage III. However, one cannot say that the difference, either biological or clinical, between stage I and stage II is larger or smaller than the difference between stage II and stage III. This is an important thing to remember about ordinal scales: differences between values are meaningless.

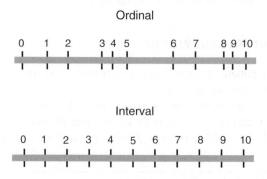

Figure 2.1   Difference between an ordinal and interval scale.

Item list

Compared to last month, today you feel:
- ○ much worse
- ○ worse
- ○ no change
- ○ better
- ○ much better

Likert scale:

Today I feel much better than last month:

Strongly disagree				Strongly agree
1	2	3	4	5

Visual analogic scale:

How do you feel today? Place a vertical mark in the line below to indicate how you feel today

Worst ├──────────────────────┤ Best
imaginable   imaginable

Figure 2.2   Examples of commonly used ordinal scales.

Attributes measured in ordinal scales are often found in clinical research. Figure 2.2 shows three examples of ordinal scales: the item list, where the subjects select the item that more closely corresponds to their opinion, the Likert scale, where the subjects read a statement and indicate their degree of agreement, and the visual analogic scale where the subjects mark on a 100 mm line the point that they feel corresponds to their assessment of their current state. Psychometric, attitudinal, quality of life, and, in general, many questionnaires commonly used in clinical research have an ordinal scale of measurement.

Interval scales are very common in research and in everyday life. Examples of attributes measured in interval scales are age, height, blood pressure, most clinical laboratory results, and so forth. Some interval-measured attributes are **continuous**, for example, height, while others are not continuous and they are called **discrete**. Examples of discrete attributes are **counts**, like the leukocyte count. Because all values in the scale are at the same distance from each other, we can perform arithmetic operations on them. We can say that the difference between, say, 17 and 25 is of the same magnitude than between, say, 136 and 144.

If an interval scale has a meaningful zero, it is called a **ratio scale**. Examples of ratio scales are height and weight. An example of an interval scale that is not a ratio scale is the Celsius scale, where zero does not represent the absence of temperature, but rather the value that was by convention given to the temperature of thawing ice. In ratio scales, not only are sums and subtractions possible, but also multiplication and division, although the latter two operations are meaningless in non-ratio scales. For example, we can say that a weight of 21 g is half of 42 g, and a height of 81 cm is three times 27 cm, but we cannot say that a temperature of 40 °C is twice as warm as 20 °C. In practical data analysis, however, we often make no distinction between interval and ratio scales, or continuous and discrete variables.

It is important to note that the simple inspection of the values of an attribute does not allow us to decide which scale was used. We must know what was being measured. Take, for example, the results of a written examination presented on a scale of 0 to 100, all questions having the same score. Now, what kind of scale was used? Well, that depends on what was being measured. If the teacher was measuring the number of correct answers, than it must be an interval scale: a student with a score of, say, 68 had twice the number of correct answers than a student with a score of 34. However, if the teacher was trying to evaluate the knowledge of the students, then it is probably an ordinal scale. The difference in knowledge between a student with 30 and a student with 40 is normally much smaller than the difference in knowledge between a student with 85 and a student with 95.

2.3 Tabulations of data

Now that we can identify the scales of measurement of each attribute, let us see how we can compress the data. One simple method is the tabulation of the data, whereby we make a list of all the different values found in the dataset and, in front of each one, we write down the number of times it occurred. This is called the **absolute frequency** of each value. In order to improve readability, it is customary to also write down the number of occurrences of each value as a percentage of the total number of values, the **relative frequency**.

When we look at a table, such as the ones shown in Figure 2.3, we are evaluating how the individual values are distributed in our sample. Such a display of data is called a **frequency distribution**.

Tabulations of data with absolute and relative frequencies are the best way of presenting binary and categorical data. Tables are a very compact means of data presentation, and tabulation does not involve any significant loss of information.

Binary data

Gender	n	%
male	15	25.0
female	45	75.0
Total	60	100.0

Categorical data

Location	n	%
fundus	3	3.5
body	26	30.6
antrum	43	50.6
pylorus	13	15.3
Total	85	100.0

Figure 2.3   Tabulation of nominal data.

Interval data with grouping

Age (years)	n	%	cumulative
20-29	3	5.0	5.0
30-39	6	10.0	15.0
40-49	12	20.0	35.0
50-59	12	20.0	55.0
60-69	27	45.0	100.0
Total	60	100.0	100.0

Figure 2.4   Tabulation of ordinal and interval data.

We can also use tables for ordinal and interval data, provided the number of different values is not too large. In those tables we present the values in ascending order and write down the absolute and relative frequencies, as we did with binary and categorical data. For each value we can also add the **cumulative frequency**, or the percentage of values in the dataset that are equal to or smaller than that value. If the number of values is large, then it is probably better to group the values into larger intervals, as in Figure 2.4, but this will lead to some loss of information.

If an attribute has been measured on an ordinal or interval scale, there are other ways of presenting the data that are much more compact than tables, although generally there is a greater loss of information. These are commonly called central tendency measures, and the ones most used are the mean, the median, and the mode.

2.4 Central tendency measures

The **mean** is a very common measure of central tendency. We use the notion of mean extensively in everyday life, so it is not surprising that the mean plays an extremely important role in statistics. Furthermore, being a sum of values, the mean is a mathematical quantity and therefore amenable to mathematical processing. This is the other reason why it is such a popular measure in statistics.

As a measure of central tendency, however, the mean is valid only when the values are symmetrically distributed about its value. This is not the case with a number of attributes we study in biology and medicine – they often have a large number of small values and a few very large values. In this case, the arithmetic mean is not a good measure of central tendency, since a large number of values will be on one side of the mean and a small number of values will be on the other side. A better measure of central tendency is, in these cases, the median.

The **median** is the quantity that divides the sample into two groups with an equal number of observations: one group has all the values smaller than that quantity, and the other group has all the values greater than that quantity. The median, therefore, is a quantity that has a straightforward interpretation: half the observations are smaller than that quantity. Actually, the interpretation of the median is exactly the same as the mean when the values are symmetrically distributed about the mean and, in this case, the mean and median will have the same value. With asymmetric distributions such as the ones noted above, however, the median will be smaller than the mean.

One problem with the median is that it is not a mathematical result. To calculate the median, we must count the number of observations, as we do to compute the mean. Then we must sort all the values in ascending or descending order, divide the number of observations by 2, and round the result to the nearest integer. Then we take this result, go to the observation that occupies that position in the sorted order, and obtain the value of that observation. The value is the median value. Further, if the number of observations is even, then we must take the value of the observation that has a rank in the sorted order equal to the division of the number of observations by 2, then add that value to the value of the next observation in the sorted order, and divide the result by 2 to finally obtain the median value.

The median, therefore, requires an algorithm for its computation. This makes it much less amenable to mathematical treatment than the mean and, consequently, less useful. In many situations, however, the median is a much better measure of central tendency than the mean. For example, attributes that are measured on ordinal scales – recall that with ordinal scales sums and differences are meaningless – should almost always be summarized by the median, not the mean. One possible exception to this rule is when an ordinal scale has so many distinct values, say, more than 50, that we can assume that we are measuring a continuous attribute with a somewhat faulty instrument in which the measurement error varies slightly across the range of values, as if we were measuring lengths with a metric tape in which the marks were erased in some sections so we have to take an approximate reading in those sections. In such a case, it would appear that the attribute had been measured in an ordinal scale while it has actually been measured in an interval scale. This is why we often see data obtained with some clinical questionnaires presented and analyzed as if it were interval data.

The last central tendency measure is the mode. The **mode** is simply the most common value of an attribute. It has the advantage over the other measures of central tendency in that it can be used with all types of scales of measurement, including categorical scales. The mode, however, has many disadvantages and this is why it is seldom used in clinical research. One important problem with the mode is that there is no guarantee that it is a unique value. The most frequent values of an attribute may occur with the same frequency, and then we will have several modes. In addition, in very small samples, each value may occur only once and we will have as many modes as values. No further mention will be made to the mode throughout this book.

2.5 Measures of dispersion

Central tendency measures are a drastic method of data abstraction, whereby a large number of values are condensed into just one quantity, but at the cost of a severe loss of information. Consider a list of numbers: one obvious characteristic of the list will be the degree of heterogeneity of the values, and it would be important to have at least some information about that heterogeneity. In other words, simple inspection of the mean will not tell us anything about whether all the values are very close to the mean or whether they are largely spread about the mean. In order to keep that information, and in the spirit of central tendency measures, **measures of dispersion** have been developed.

As with the central tendency measures, there are a number of available measures of dispersion, each one having some useful properties and some short-comings. As noted above, the basic idea when selecting a data reduction measure must always be to lose as little information as possible.

One possible way of expressing the degree of dispersion of individual values could be to write down the **limits**, that is, the minimum and maximum values. One good thing about this approach is that it is easy to interpret. If the two values are similar then the dispersion is small, otherwise it is large. There are a few bad things, though. First, we will have to deal with two quantities, which is not very practical. Second, the limits are very unstable, in the sense that if one adds a dozen observations to a study, it will most probably be necessary to redefine the limits. This is because, as one adds more observations, the more extreme values will have a greater chance of appearing.

The first problem can be solved by using the difference between the maximum and minimum values, a quantity commonly called the **range**, but this will not solve the problem of instability.

The latter problem can be minimized if, instead of using the minimum and maximum values to describe the dispersion, we use the lower and upper **quartiles**. The lower quartile (also called the 25th percentile) is the value below which is one-quarter, or 25%, of all the values in the dataset. The upper quartile (or 75th percentile) is the value below which lie three-quarters, or 75%, of the values in the dataset (note, incidentally, that the median is the same as the 50th percentile). The advantage of the quartiles over the limits is that they are more stable because the addition of one or two extreme values to the dataset will probably not change the quartiles.

However, we still have the problem of having to deal with two values, which is certainly not as practical and easy to remember and reason with if we had just one value. One way around this could be to use the difference, the upper quartile minus the lower quartile, to describe the dispersion. This is called the **interquartile range**, but the interpretation of this value is not straightforward: it is not amenable to mathematical treatment and therefore it is not a very popular measure, except perhaps in epidemiology.

2.6 Compressing data

Before moving on to more adequate measures of dispersion, let us look at an illustration of the measures of dispersion presented in the previous sections. As noted earlier, the central idea is to describe a whole set of values, losing as little information as possible. Can this actually be achieved? This example will show that it does.

For this illustration we used a patient attribute with complete random values and created a graph with the frequency distribution of those values. We are in a position

Original distribution

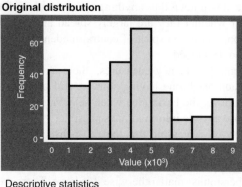

Descriptive statistics

observations:	309
median:	3927
range:	24–8947
percentile 25:	2037
percentile 75:	5012

Reconstructed distribution

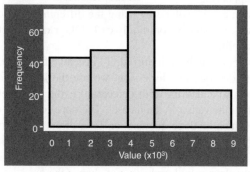

Figure 2.5   Data abstraction by descriptive statistics.

of assuring that the values were completely random because that attribute was simply the patient chart number.

The result was the graph shown in the top half of Figure 2.5. This type of graph is called a histogram. The **histogram** is a bar chart with the bars adjacent to each other, which is an adequate way of designing a bar graph for grouped continuous variables.

The dataset included 309 patients, so we had 309 observations on that attribute. We computed a set of descriptive statistics, namely, the median, range, and lower and upper quartiles, whose values are also shown in the figure. Then we used only those values to create the graph in the bottom half of Figure 2.5, that is, what we could say about the frequency distribution if all we had were those descriptive statistics.

Now when we compare the two graphs we will see that, although the dataset was compressed from 309 to 6 values (a 2% compression rate), the loss of information was actually relatively small. This example shows us that describing the data by summary measures is an efficient and adequate method of summarizing information.

2.7 The standard deviation

Let us now consider other measures of dispersion. Another possible measure could be the average of the deviations of all individual values about the mean or, in other words, the average of the differences between each value and the mean of the distribution. This would be an interesting measure, being both a single value and easy to interpret, since it was an average. Unfortunately, it would not work because the differences from the mean in those values smaller than the mean are positive, and the differences in those values greater than the mean are negative. The result, if the values were symmetrically distributed about the mean, would always be close to zero regardless of the magnitude of the dispersion.

Actually, what we want is the average of the size of the differences between the individual values and the mean. We do not really care about the direction (or sign) of those differences. Therefore, we could use instead the average of the absolute value of the differences between each value and the mean. This quantity is called the **mean deviation**. It satisfies the desired properties of a summary measure: single value, stability, and interpretability. The mean deviation is easy to interpret because it is an average, and people are used to dealing with averages. If we were told that the mean of some patient attribute is 256 mmol/L and the mean deviation is 32 mmol/L, we could immediately figure out that about half the values were in the interval 224 to 288 mmol/L, that is, $256 - 32$ to $256 + 32$.

There is a small problem, however. The mean deviation uses absolute values, and absolute values are quantities that are difficult to manipulate mathematically. Actually, they pose so many problems that it is standard mathematical practice to square a value when one wants the sign removed. Let us apply that method to the mean deviation. Instead of using the absolute value of the differences about the mean, let us square those differences and average the results. We will get a quantity that is also a measure of dispersion. This quantity is called the **variance**. The way to compute the variance is, therefore, first to find the mean, then subtract each value from the mean, square the result, and add all those values. The resulting quantity is called the sum of squares about the mean, or just the **sum of squares**. Finally, we divide the sum of squares by the number of observations to get the variance.

Because the differences are squared, the variance is also expressed as a square of the attribute's units, something strange, like $mmol^2/L^2$. To put things right we have to convert these awkward units into the original units by taking the square root of the variance. This new result is also a measure of dispersion and is called the **standard deviation**.

As a measure of dispersion, the standard deviation is single valued and stable, but what can be said about its interpretability? Let us see: the standard deviation is the square root of the average of the squares of the differences between individual values and the mean. It is not easy to understand what this quantity really represents. However, the standard deviation is the most popular of all measures of dispersion. Why is that?

One important reason is that the standard deviation has a large number of interesting mathematical properties, but this probably will not be very reassuring. The other important reason is that, actually, the standard deviation has a straightforward interpretation, very much along the lines given earlier to the value of the mean deviation. Unfortunately, we cannot go into this right now, because in order to understand what it might mean, we first have to go over some facts. But before we proceed to the next section, let us review the methods of data abstraction:

- Recall that the idea of using summary statistics is to display the data in an easy-to-grasp format while losing as little information as possible.
- Begin by understanding what scale of measurement was used with each attribute.
- If the scale is binary or categorical, the appropriate method is tabulation, and both the absolute and relative frequencies should always be displayed.
- If the scale is ordinal, the mean and standard deviation should not be presented, which would be wrong because arithmetic operations are not allowed with ordinal scales; instead, present the median and one or more of the other measures of dispersion. In the medical literature, the limits seem to be most popular.
- If the scale is interval, the mean and the standard deviation should be presented unless the distribution is very asymmetrical about the mean. In this case, the median and the limits may provide a better description of the data.

A final word is in order on the standard deviation. In most statistical packages the standard deviation is computed differently from the way explained above, in that the sum of squares is not divided by the number of observations (n), but by the number of observations minus one ($n - 1$). In some packages and hand calculators, we have the option of selecting either method. The standard deviation with the n divisor is usually called the **sample standard deviation**, while the standard deviation with the $n - 1$ divisor is called the **population standard deviation**. As a rule, you should use the $n - 1$ divisor throughout.

2.8 The $n - 1$ divisor

The reason why we use the $n - 1$ divisor instead of the n divisor for the sum of squares when we calculate the variance and the standard deviation is because, when we present those quantities, we are implicitly trying to give an estimate of their value in the population. Now, since we use the data from our sample to calculate the variance, the resulting value will always be smaller than the value of the variance in the population. We say that our result is biased toward a smaller value. What is the explanation for that bias?

Remember that the variance is the average of the squared differences between individual values and the mean. If we calculated the variance by subtracting the individual values from the true mean (the population mean), the result would be unbiased.

This is not what we do, however. We subtract the individual values from the mean computed from our sample and, because of this, the mean occupies a central position among all the data values. Since the **sample mean** is the quantity closest to all the values in the dataset, individual values are more likely to be closer to the sample mean than to the **population mean**. Therefore, the value of the **sample variance** tends to be smaller than the value of the **population variance**.

It is an easy mathematical exercise to demonstrate that dividing the sum of squares by $n - 1$ instead of n provides an adequate correction of that bias. However, the same thing can be shown by the small experiment illustrated in Figures 2.6 and 2.7.

Using a computer's random number generator, we obtained random samples of a variable with variance equal to 1. This is what is called the population variance of that variable. Starting with samples of size 2, we obtained 10 000 random samples and computed their sample variances using the n divisor. Next, we computed the average of those 10 000 sample variances and retained the result. We then repeated the procedure with samples of size 3, 4, 5, and so on up to 100.

The plot of the averaged value of sample variances against sample size is represented by the solid line in Figure 2.6. It can be clearly seen that, regardless of the sample size, the variance calculated with the n divisor is on average less than the population variance, and the deviation from the true variance increases as the sample size decreases.

Average value of the sample variance, in 10,000 random samples of size 2 to 100, from an attribute with variance 1.

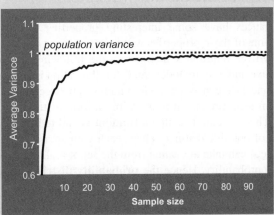

Figure 2.6   The n *divisor of the sum of squares.*

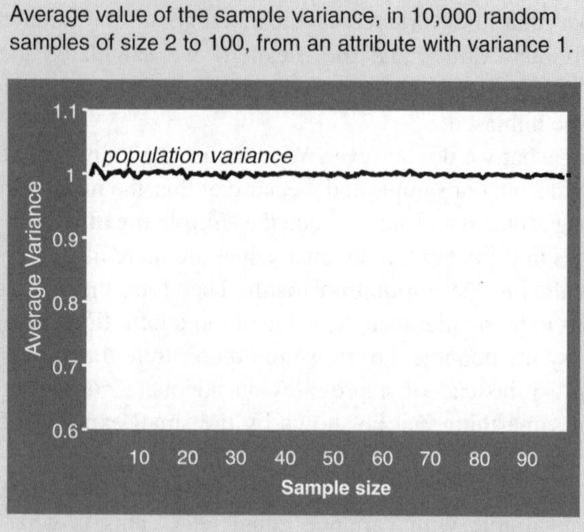

Average value of the sample variance, in 10,000 random
samples of size 2 to 100, from an attribute with variance 1.

Figure 2.7   The n − *1 divisor of the sum of squares.*

Now let us repeat the procedure, exactly as before, but this time using the $n-1$ divisor. The plot of the average sample variance against sample size is shown in Figure 2.7. The solid line is now exactly over 1, the value of the population variance, for all sample sizes.

This experience clearly illustrates that, contrary to the sample variance using the n divisor, the sample variance using the $n-1$ divisor is an unbiased estimator of the population variance.

2.9 Properties of means and variances

Means and variances have some interesting properties that deserve mention. Knowledge of some of these properties will help you when you are analyzing your data, and they will be required sometimes in the following sections. Regardless, they all are intuitive and easy to understand. An illustration is also provided.

With a computer, we generated random numbers between 0 and 1, representing observations from a continuous attribute with uniform distribution, which we will call variable A. This attribute is called a **random variable** because it can take any value from a set of possible distinct values, each with an associated probability. In this case, variable A can take any value from the set of real numbers between 0 and 1, all with equal probability. Hence the **probability distribution** of variable A is called the uniform distribution.

A second variable, called variable B, with uniform distribution but with values between 0 and 2, was also generated. The histograms are shown in Figure 2.8. Let us now see what happens to the mean and variance when we perform arithmetic operations on a random variable.

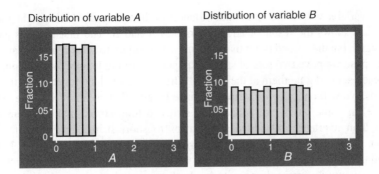

Figure 2.8   Two random variables with uniform distribution.

When a constant amount is added to, or subtracted from, the values of a random variable, the mean will, respectively, increase or decrease by that amount but the variance will not change. This is illustrated in Figure 2.9, left graph, which shows the distribution of variable A plus 2. This result is obvious, because, as all values are increased (or decreased) by the same amount, the mean will also increase (or decrease) by that amount and the distance of each value to the mean will thus remain the same, keeping the variance unchanged.

When a constant amount multiplies, or divides, the values of a random variable, the mean will be, respectively, multiplied or divided by that amount, and the variance will be, respectively, multiplied or divided by the square of that amount. Therefore, the standard deviation will be multiplied or divided by the same amount. Figure 2.9, middle graph, shows the distribution of A multiplied by 2. As an example, consider the attribute height with mean 1.7 meters and standard deviation 0.6 meters. If we want to convert the height to centimeters, we multiply all values by 100. Now the mean will of course be 170 cm and the standard deviation 60 cm. Thus, the mean was multiplied by 100 and the standard deviation also by 100 (and, therefore, the variance was multiplied by 100^2).

When observations from two independent random variables are added or subtracted, the mean of the resulting variable will be the sum or the subtraction, respectively, of the means of the two variables. In both cases, however, the variance of

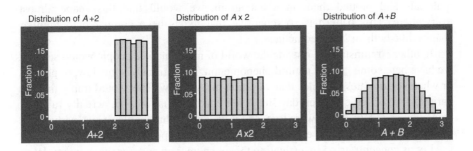

Figure 2.9   Properties of means and variances.

the new variable will be the sum of the variances of the two variables. The right graph in Figure 2.9 shows the result of adding variables A and B. The first result is easy to understand, but the second is not that evident, so we will try to show it by an example.

Suppose we have two sets of strips of paper of varying length. We take one strip from each set and glue them at their ends. When we have glued together all the pairs of strips, we will end up with strips that have lengths that are more variable. This is because, in some cases, we added long strips to long strips, making them much longer than average, and added short strips to short strips, making them much smaller than average. Therefore, the variation of strip length increased. Now, if instead of adding the two strips of paper we cut a variable amount from each strip, we will eventually make large cuts in short strips and small cuts in large strips, again increasing variation.

Note that this result will not hold if the variables are not **independent**, that is, if they are correlated. Taking the example above, if we decided to always make large cuts in long strips and small cuts in short strips, we would end up with a smaller variance. If we did it the other way around, the final variance would be much larger than the sum of the two variances.

2.10 Common frequency distributions

In the discussion on measures of dispersion, it was mentioned that many attributes measured in interval scales had their values symmetrically distributed about the mean. Let us investigate some patient attributes to evaluate how common this pattern is.

Figure 2.10 presents several histograms showing the frequency distribution of several commonly assessed clinical laboratory variables measured in interval scales, obtained from a sample of over 400 patients with hypertension.

Notice not only that all distributions are approximately symmetrical about the mean, but also that the very shape of the histograms is strikingly similar. In all of them, the frequency of the individual values is highest near the middle, declining smoothly from there and at the same rate on each side.

Actually, if we went around taking some kind of interval-based measurements (e.g., length, weight, concentration) from samples of any type of biological materials and plotted them in a histogram, we would find this shape almost everywhere. This pattern is so repetitive that it has been compared to familiar shapes, like bells or Napoleonic hats.

In other circumstances, outside the world of mathematics, people would say that we have here some kind of natural phenomenon. It seems as if some law, of physics or whatever, dictates the rules that variation must follow. This would imply that the variation we observe in everyday life is not chaotic in nature, but actually ruled by some universal law. If this were true, and if we knew what that law says, perhaps we could understand why, and especially how, variation appears.

Let us concentrate on investigating this phenomenon and, for the moment, let us not think about how we can make use of it. That will be evident in the next sections.

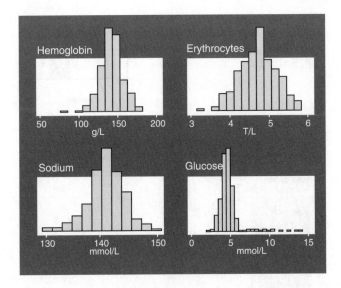

Figure 2.10   Frequency distributions of some biological variables.

So, what would be the nature of that law and is it known already? Yes it is, and it is actually very easy to understand how it works. Let us conduct a little experiment to see if we can create something whose values have a bell-shaped distribution pattern.

2.11 The normal distribution

Consider some attribute that may take only two values, say 1 and 2, and that those values occur with equal frequency. Technically speaking, we say a **random variable** taking values 1 and 2 with equal probability; this is the **probability distribution** for that variable (see Figure 2.11, upper part). Consider also four variables that behave exactly like this, that is, they have the same probability distribution. Now let us create a fifth variable that is the result of adding all four variables together. Can we predict what will be the probability distribution of this variable?

We can, and the result is also presented in Figure 2.11. We simply write down all the possible combinations of values of the four equal variables and see in each case what the value of the fifth variable is. If all four variables have value 1, the fifth variable will have value 4. If three variables have value 1 and one has value 2, the fifth variable will have value 5. This may occur in four different ways – either the first variable had the value 2, or the second, or the third, or the fourth. If two variables have the value 1 and two have the value 2, then the sum will be 6, and this may occur in six different ways. If one variable has value 1 and three have value 2, the result will be 7 and this may occur in four different ways. Finally, if all four variables have value 2 the result will be 8 and this can occur in only one way.

Probability distribution of identical random variables *A, B, C* and *D*

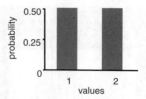

Values of variable $Z = A+B+C+D$

Value	Is the result of				number of possibilities	%
	A	B	C	D		
4	1	1	1	1	1	6.25
5	1	1	1	2	4	25.0
	1	1	2	1		
	1	2	1	1		
	2	1	1	1		
6	1	1	2	2	6	37.5
	1	2	1	2		
	1	2	2	1		
	2	1	1	2		
	2	1	2	1		
	2	2	1	1		
7	1	2	2	2	4	25.0
	2	1	2	2		
	2	2	1	2		
	2	2	2	1		
8	2	2	2	2	1	6.25
Total					16	100.0

Probability distribution of random variable $Z = A+B+C+D$

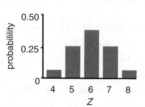

Figure 2.11   The origin of the normal distribution.

So, of the 16 different possible ways or combinations, in one the value of the fifth variable is 4, in four it is 5, in six it is 6, in four it is 7, and in one it is 8. If now we graph the relative frequency of each of these results, we obtain the graph shown in the lower part of Figure 2.11. This is the graph of the probability distribution of the fifth variable. Do you recognize the bell shape?

If we repeat the experiment with not two but a much larger number of variables, the variable that results from adding all those variables will have not just five different values but many more. Consequently, the graph will be smoother and more bell shaped. The same will happen if we add variables taking more than two values.

If we have a very large number of variables, then the variable resulting from adding those variables will take an infinite number of values and the graph of its probability distribution will be a perfectly smooth curve. This curve is called the **normal curve**. It is also called the **Gaussian curve** from the German mathematician Karl Gauss who described it.

This is the explanation of why so many biological attributes, or variables, have a normal distribution. It is reasonable to think that many biological attributes are the result of a large number of independent factors added together, each one making a small contribution to the final value of the attribute. If the normal curve arises whenever we add a large number of identically distributed variables, then it should not be surprising that those biological variables will have a normal distribution.

However, it must be said that it is actually of no great importance to us, as far as biostatistics is concerned, whether biological attributes have a normal distribution or not. The important fact is that the normal distribution arises from the sum of variables with identical distribution. We will see shortly how the whole theory of sampling was built upon this simple fact.

2.12 The central limit theorem

What was presented in the previous section is known as the **central limit theorem**. This theorem simply states that the sum of a large number of independent variables with identical distribution has a normal distribution. The central limit theorem plays a major role in statistical theory, and the following experiment illustrates how the theorem operates.

With a computer, we generated random numbers between 0 and 1, obtaining observations from two continuous variables with the same distribution. The variables had a uniform distribution, which is a distribution where all values occur with exactly the same probability.

Then, we created a new variable by adding the values of those two variables and plotted a histogram of the frequency distribution of the new variable. The procedure was repeated with three, four, and five identical uniform variables. The frequency distributions of the resulting variables are presented in Figure 2.12.

Notice that, the more variables we add together, the more the shape of the frequency distribution approaches the normal curve. The fit is already fair for the sum of four variables. This result is a consequence of the central limit theorem.

2.13 Properties of the normal distribution

The normal distribution has many interesting properties, but we will present just a few of them. They are very simple to understand and, occasionally, we will have to call on them further on in this book.

First property: The normal curve is a function solely of the mean and the variance. In other words, given only a mean and a variance of a normal

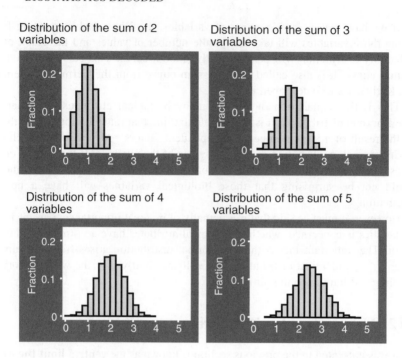

Figure 2.12   Frequency distribution of sums of identical variables with uniform distribution.

distribution, we can find all the values of the distribution and plot its curve using the equation of the normal curve (technically, that equation is called the **probability density function**). This means that in normally distributed attributes we can describe its distribution entirely by using only the mean and the variance (or equivalently the standard deviation). This is the reason why the mean and the variance are called the **parameters** of the normal distribution, and what makes these two summary measures so important. It also means that if two normally distributed variables have the same variance, then the shape of their distribution will be the same; if they have the same mean, their position on the horizontal axis will be the same.

Second property: The sum or difference of normally distributed independent variables will result in a new variable with a normal distribution. According to the properties of means and variances, the mean of the new variable will be, respectively, the sum or difference of the means of the two variables, and its variance will be the sum of the variances of the two variables (Figure 2.13).

Third property: The sum, or difference, of a constant to a normally distributed variable will result in a new variable with a normal distribution. According to the properties of means and variances, the constant will be added to or subtracted from its mean, and its variance will not change (Figure 2.13).

Fourth property: The multiplication, or division, of the values of a normally distributed variable by a constant will result in a new variable with a normal

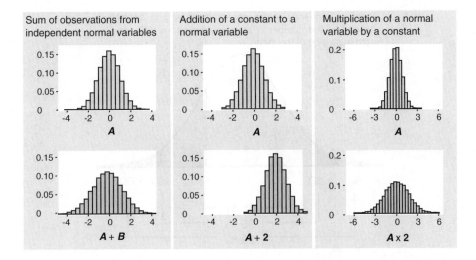

Figure 2.13   Properties of the normal distribution.

distribution. Because of the properties of means and variances, its mean will be multiplied, or divided, by that constant and its variance will be multiplied, or divided, by the square of that constant (Figure 2.13).

Fifth property: In all normally distributed variables, irrespective of their means and variances, we can say that two-thirds of the observations have a value lying in the interval defined by the mean minus one standard deviation to the mean plus one standard deviation (Figure 2.14). Similarly, we can say that 95% of the observations have a value lying in the interval defined by the mean minus two standard deviations to the mean plus two standard deviations. The relative frequency of the observations with values between the mean minus three standard deviations and the mean plus three standard deviations is about 99%, and so on. This means that if we know that an attribute, for example, height, has a normal distribution with a population mean of 170 cm and a standard deviation of 20 cm, then we also know that the height of about 66% of the population is 150 to 190 cm, and the height of 95% of the population is 130 to 210 cm.

Recall what was said earlier, when we first discussed the standard deviation: that its interpretation was easy but not evident at that time. Now we can see how to interpret this measure of dispersion. In normally distributed attributes, the standard deviation and the mean define intervals corresponding to a fixed proportion of the observations. This is why summary statistics are sometimes presented in the form of mean ± standard deviation (e.g., 170 ± 20), which, incidentally, is not appropriate and may confuse those who are not familiar with the notation.

We also have ways of calculating the relative frequency of values of a normal variable lying in whatever interval we specify. This can be done manually with statistical tables, or by using statistical software.

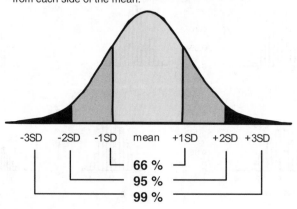

Relative frequency of the observations lying in intervals defined by the number of standard deviations (SD) away from each side of the mean.

Figure 2.14   Relationship between the area under the normal curve and the standard deviation.

2.14 Statistical tables

Given any variable with a normal distribution, we can estimate the proportion of observations with values lying within a specified interval by consulting a statistical table of the normal distribution. The table displays the proportion of observations on a variable with a normal distribution that exceed a specified value.

From the properties of the normal distribution, we know that the proportion of observations between intervals defined in terms of the distance to the mean measured in number of standard deviations is the same for all normal distributions. Therefore, **statistical tables** have been created where these proportions are tabulated in relation to the values of a measure of the distance from the mean, expressed by the number of standard deviations (up to two decimal places). This measure is called the **standard normal deviate**. Hence, a value of, say, 2.3 for the standard normal deviate means a distance from the mean equal to the value of 2.3 standard deviations.

For example, if an attribute has mean 12 and standard deviation 4, then the value 18 is 1.5 standard deviations above the mean. We reached this conclusion simply by first subtracting the value of the mean from 18 and then dividing the result by the standard deviation. If now we look at the statistical table for the number 1.5, we can see that the corresponding proportion is 0.0668, that is, 6.68% of the observations on that variable exceed the value 18.

In order to obtain a more compact display, these tables are typically presented as tables with two entries. One entry is the value of the standard normal deviate up to the first decimal place, and the other entry is the second decimal place. The desired proportion is read at the intersection of the two quantities.

Table A1   Areas in the tail of the normal distribution

z	0.00	0.01	0.02	0.03	0.04	0.05	0.06	0.07	0.08	0.09
0.0	0.5000	0.4960	0.4920	0.4880	0.4840	0.4801	0.4761	0.4721	0.4681	0.4641
0.1	0.4602	0.4562	0.4522	0.4483	0.4443	0.4404	0.4364	0.4325	0.4286	0.4247
0.2	0.4207	0.4168	0.4129	0.4090	0.4052	0.4013	0.3974	0.3936	0.3897	0.3859
0.3	0.3821	0.3783	0.3745	0.3707	0.3669	0.3632	0.3594	0.3557	0.3520	0.3483
0.4	0.3446	0.3409	0.3372	0.3336	0.3300	0.3264	0.3238	0.3192	0.3156	0.3121
0.5	0.3085	0.3050	0.3015	0.2981	0.2946	0.2912	0.2877	0.2843	0.2810	0.2776
0.6	0.2743	0.2709	0.2676	0.2643	0.2611	0.2578	0.2546	0.2514	0.2483	0.2451

Figure 2.15   Statistical table of the normal distribution.

Figure 2.15 shows a fragment of a statistical table of the normal distribution. The graph on the upper left tells us that the tabulated values represent the proportion of observations exceeding a given value. Note that tables may differ from the one presented, some tables displaying instead the proportion of observations smaller than a given value.

Continuing with the same variable, if we wanted to know the proportion of observations greater than, say, 13.84 we would first calculate the difference to the mean, which is $13.84 - 12 = 1.84$, which means that value is at 0.46 standard deviations $(= 1.84/4)$ above the mean. Then, we would find in the first column of the table the value 0.4, and look in the first row for the second decimal place (0.06). At the intersection of the two we would read off the value 0.3238, which means that 32.38% of the observations are more than 0.46 standard deviations above the mean and, thus, exceed the value 13.84.

Because the normal distribution is symmetrical about the mean, the proportion of observations that are more than 0.46 standard deviations above the mean is the same as the proportion of observations that are more than 0.46 standard deviations below the mean. Therefore, there really is no need to tabulate all the values of the standard normal deviate and most tables actually only present proportions for values above the mean.

A more common use of statistical tables, as we will see later on, is to work the other way around, that is, to find the value that is exceeded by a pre-specified proportion of observations. For example, suppose the mean body height in a population is 160 cm and the standard deviation is 12 cm and that we wish to find how tall are the tallest 5% of the population. First we look in the statistical table for the tabulated value of 5% and we find the value 0.0505, which is very near 5%. Then, we sum the z-values of the corresponding column and row (see Figure 2.16). The sum is 1.64, meaning that the height of the tallest 5% is more than 1.64

Table A1   Areas in the tail of the normal distribution

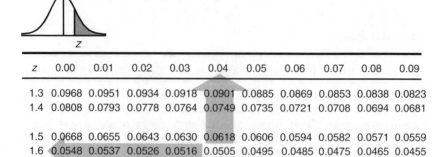

z	0.00	0.01	0.02	0.03	0.04	0.05	0.06	0.07	0.08	0.09
1.3	0.0968	0.0951	0.0934	0.0918	0.0901	0.0885	0.0869	0.0853	0.0838	0.0823
1.4	0.0808	0.0793	0.0778	0.0764	0.0749	0.0735	0.0721	0.0708	0.0694	0.0681
1.5	0.0668	0.0655	0.0643	0.0630	0.0618	0.0606	0.0594	0.0582	0.0571	0.0559
1.6	0.0548	0.0537	0.0526	0.0516	0.0505	0.0495	0.0485	0.0475	0.0465	0.0455
1.7	0.0446	0.0436	0.0427	0.0418	0.0409	0.0401	0.0392	0.0384	0.0375	0.0367
1.8	0.0359	0.0351	0.0344	0.0336	0.0329	0.0322	0.0314	0.0307	0.0301	0.0294
1.9	0.0287	0.0281	0.0274	0.0268	0.0262	0.0256	0.0250	0.0244	0.0239	0.0233

Figure 2.16   Statistical table of the normal distribution.

standard deviations higher than the mean height. Then, the height of the tallest 5% is over $160\,\text{cm} + 1.64 \times 12\,\text{cm} = 179.68\,\text{cm}$.

In statistical notation, the proportion of observations in a probability distribution exceeding a given value is named α, and the notation for the corresponding number of standard deviations away from the mean is z_α. In this example, $\alpha = 0.05$ and $z_\alpha = 1.64$. If we wanted to know the height of the shortest 5%, then $\alpha = 0.05$ and $z_\alpha = -1.64$ and the height would be $160\,\text{cm} - 1.64 \times 12\,\text{cm} = 140.32\,\text{cm}$.

Sometimes we want to find the values exceeded, in both directions, by a given proportion of the population, for example, by 15% of the population. We first have to halve this proportion because we are looking for the height of the tallest 7.5% and of the shortest 7.5%. So we search for the tabulated value $\alpha = 0.075$ and we find 0.0749, corresponding to $z_{\alpha/2} = 1.44$. Therefore, the $\alpha = 15\%$ extreme heights are more than 1.44 standard deviations on either side of the mean, that is, $160 \pm 17.28\,\text{cm}$.

3

Statistical inference

3.1 Sampling distributions

Now that we have all the basic information we need, we can return to the discussion on sampling and on how we can speak about the characteristics of a population after inspecting a random sample from it.

For the moment, let us restrict our discussion to the situation of interval variables. Therefore, consider that we have a number of observations on an interval-scaled attribute, body weight for example, from a random sample of a defined population. We have the individual measurements, and we have collapsed them into two summary statistics, the mean and the standard deviation.

The first thing we must discuss is: exactly what do we want to know about the population? Recall from the published studies you have read that the authors seem to be interested only in the mean. This may look confusing to us, because we know that the variation is also an important element in the description of an attribute, but you probably have never read a paper where the authors were interested in estimating the variance in the population. Why, then, all this focus on the mean?

One answer is because we want to be able to make predictions on individual patients, and our best guess as to which value of an attribute a patient might have will be the most frequent value in the population to which the patient belongs. That value is, of course, the population mean.

Another reason is because we know that biological attributes are virtually always influenced by a number of factors that contribute to their variation. Therefore, the mean may be seen as the true value of an attribute, and the variation as a sign of the presence of factors of variation influencing that attribute. We will see later on in this book that there are methods that allow us to identify those factors and even estimate how much they contribute to the final value of the attribute. For now, we will concentrate on estimating the mean value of an attribute in the population.

Biostatistics Decoded, First Edition. A. Gouveia Oliveira.
© 2013 John Wiley & Sons, Ltd. Published 2013 by John Wiley & Sons, Ltd.

Let us consider the real situation: all we have is a mean obtained from a random sample of the population. Now, in considering sample means, it is not reasonable to expect the sample mean to have exactly the same value as the population mean, other than by an amazing coincidence. It is also easy to realize that, if we take two random samples with the same sample size from the same population, their means will not have the same value. In order for them to be the same, the sum of all the observations would have to be exactly the same in both samples, since the mean is the sum of all observations divided by the sample size. This, again, would be an amazing coincidence.

There is another explanation for **sampling variation**: we can view a sampling process as a measurement of the value of the population mean using a method that has a random measurement error. Because of this measurement error, the reading we have, that is, the sample mean, does not correspond to the exact value of the population mean and, because the error is random, each time we try to measure the population mean we get a different value.

Consequently, the sample means behave as a random variable. If this is so, then the sample means must have a probability distribution. The question now is: What is the probability distribution of the sample means? Is it a known distribution? Can this distribution be described by parameters?

In order to investigate these questions, we could perform a simple experiment on sampling. With the help of a computer's random number generator, we created samples of observations on a random variable. For this experiment we created samples of 60 observations from a variable with a uniform distribution. Then, we calculated the means of these samples and plotted them on a histogram. The result is shown in Figure 3.1.

The result is very interesting. The shape of the distribution of the sample means looks very similar to the shape of the normal distribution. Actually, it can be demonstrated that the distribution of sample means is the normal distribution, whatever the distribution of the variable under study might be, on the condition that

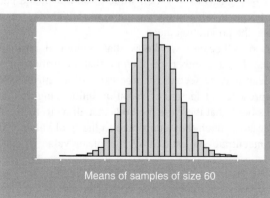

Distribution of the means of samples of 60 observations
from a random variable with uniform distribution

Means of samples of size 60

Figure 3.1   Distribution of sample means of large samples.

Distribution of the means of samples of 2 observations
from a random variable with normal distribution

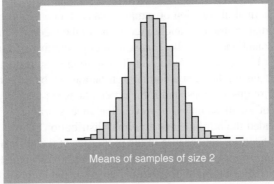

Means of samples of size 2

Figure 3.2   Distribution of sample means of small samples of attributes with a normal distribution.

the observations are mutually independent and the sample size is large. How large samples must be is not absolutely defined, but most statisticians would agree that with a sample size above 40 we may be confident that the distribution of sample means will be very nearly normal.

In the case of small samples, however, the means will also have a normal distribution, provided the attribute has a normal distribution. This is illustrated in Figure 3.2, showing the distribution of samples of size 2 from a random variable with a normal distribution.

If the distribution of the variable is not normal, or if its distribution is unknown, we cannot assume that the sample means have a normal distribution. In these cases, we simply do not know how sample means behave.

3.2 The normal distribution of sample means

The reason for the pattern of variation of sample means observed in the previous section can be easily understood.

We know that a mean is calculated by summing a number of observations on a variable and dividing the result by the number of observations. Normally, we look at the values of an attribute as observations from a single variable. However, we could very well view each single value as an observation from a distinct variable, with all variables having an identical distribution. For example, suppose we have a sample of size 100. We can think that we have 100 independent observations from a single random variable, or we can think that we have single observations on 100 variables, all of them with identical distribution. What would be the consequences of that?

In this perspective, a sample mean would correspond to the sum of a large number of observations from variables with identical distribution, each observation being divided by a constant amount which is the sample size. Under these circumstances, the central limit theorem applies and, therefore, we must conclude that the sample means have a normal distribution, regardless of the distribution of the attribute being studied.

Because the normal distribution of sample means is a consequence of the central limit theorem, certain restrictions apply. According to the theorem, this result is valid only under two conditions. First, there must be a large number of variables. Second, the variables must be mutually independent. Transposing these restrictions to the case of sample means, this implies that a normal distribution can be expected only if there is a large number of observations, and if the observations are mutually independent.

In the case of small samples, however, the means will also have a normal distribution provided the attribute has a normal distribution. This is not because of the central limit theorem, but because of the properties of the normal distribution. If the means are sums of observations on identical normally distributed variables, then the sample means have a normal distribution whatever the number of observations, that is, the sample size.

3.3 The standard error of the mean

We now know that the means of large samples may be defined as observations from a random variable with a normal distribution. We also know that a normal distribution is completely characterized by its mean and variance. The next step in the investigation of sampling distributions, therefore, must be to find out whether the mean and variance of that variable can be determined.

We can conduct an experiment simulating a sampling procedure. With the help of a computer's random number generator, we can create a random variable with normal distribution with mean 0 and variance 1. Incidentally, this is called a **standard normal variable**. Then, we obtain a large number of random samples of size 4 and calculate the means of those samples. Next, we calculate the mean and standard deviation of the sample means. We repeat the procedure with samples of size 9, 16, and 25. The results of the experiment are shown in Figure 3.3.

As expected, as the variable we used had a normal distribution, the sample means also have a normal distribution. We can see that the average value of the sample means is, in all cases, the same value as the population mean, that is, 0. However, the standard deviations of the values of sample means are not the same in all four experiments. In samples of size 4 the standard error is 0.50, in samples of size 9 it is 0.33, in samples of size 16 it is 0.25, and in samples of size 25 it is 0.20.

If we look closer at these results, we will realize that those values have something in common. Thus, 0.50 is 1 divided by 2, 0.33 is 1 divided by 3, 0.25 is 1 divided by 4, and 0.20 is 1 divided by 5. Now, can you see the relation between the divisors and the sample size, that is, 2 and 4, 3 and 9, 4 and 16, and 5 and 25? The divisors are the square root of the sample size and 1 is the value of the population standard deviation. This means that the standard deviation of the sample means is equal to the population standard deviation divided by the square root of the sample size. Therefore, there is a fixed relationship between the standard deviation of the

Distribution of the means of samples from a normal
variable with mean 0 and standard deviation 1

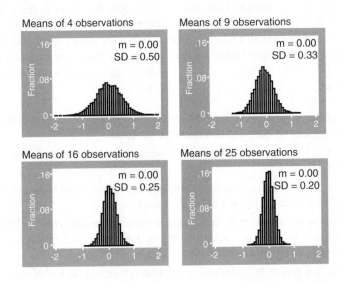

Figure 3.3   Distribution of sample means of different sample sizes.

sample means of an attribute and the standard deviation of that attribute, where the former is equal to the latter divided by the square root of the sample size.

In the next section we will present an explanation for this relationship, but for now let us consolidate some of the concepts we have discussed so far.

The standard deviation of the sample means has its own name of **standard error of the mean** or, simply, standard error. If the standard error is equal to the population standard deviation divided by the square root of the sample size, then the variance of the sample means is equal to the population variance divided by the sample size.

Now we can begin to see why people tend to get confused with statistics. We have been talking about different means and different standard deviations, and students often become disoriented with so many measures. Let us review the meaning of each one of those measures.

There is the **sample mean**, which is not equal in value to the **population mean**. Sample means have a frequency distribution whose mean has the same value as the population mean. Authors of statistics books often, but not always, refer to the value of the sample mean by the letter m, and to the value of the population mean by the letter μ, the letter 'm' in Greek.

Next, there is the **sample standard deviation**, which is not equal in value to the **population standard deviation**. Sample means have a distribution whose standard deviation, also known as **standard error**, is different from the sample standard deviation and from the population standard deviation. In statistics books we will often find the sample standard deviation represented by the letter s, and the population standard deviation by the letter σ, which is the letter 's' in Greek. There is no specific notation for the standard error.

Then there is the **sample variance**, which is also not equal to the **population variance**. These quantities are usually represented by the symbols s^2 and σ^2, respectively. Sample means also have variance, which is the square of the standard error, but the **variance of sample means** has neither a specific name, nor a specific notation.

From all of the above, we can conclude the following about sampling distributions:

- Sample means have a normal distribution, regardless of the distribution of the attribute, but on the condition that they are large.
- Small samples have a normal distribution only if the attribute has a normal distribution.
- The mean of the sample means is the same as the population mean, regardless of the distribution of the variable or the sample size.
- The standard deviation of the sample means, or standard error, is equal to the population standard deviation divided by the square root of the sample size, regardless of the distribution of the variable or the sample size.
- The above results are valid only if the observations in the sample are mutually independent.

3.4 The value of the standard error

Let us continue to view, as in Section 3.2, the mean value of a large sample as the sum of single observations from a large number of independent, identically distributed variables.

As we saw before, according to this view the means of large samples are a random variable that results from the sum of a large number of identically distributed independent variables. Therefore, according to the central limit theorem, the sample means must have a normal distribution.

The mean and variance of each of these identical variables are, of course, the same as the population mean and variance, respectively μ and σ^2. When we calculate sample means we sum all observations and divide the result by the sample size. This is exactly the same as if, before we summed all the observations, we divided each one by the sample size. If we represent the sample mean by m, each observation by x, and the sample size by N, what was just said can be represented by

$$m = \frac{x_1 + x_2 + \cdots + x_n}{N} = \frac{x_1}{N} + \frac{x_2}{N} + \cdots + \frac{x_n}{N}$$

This is the same as if every one of the identical variables was divided by a constant amount equal to the sample size. From the properties of means, we know that if we divide a variable by a constant, its mean will be divided by the same constant. Therefore, the mean of each x/N is equal to the population mean divided by N, that is, μ/N.

Now, from the properties of means we know that if we add independent variables, the mean of the resulting variable will be the sum of the means of the independent variables. Sample means result from adding together N variables, each one having a mean equal to μ/N. Therefore, the mean of the resulting variable will be $N \times \mu/N = \mu$, the population mean. The conclusion, then, is that the distribution of sample means m has a mean equal to the population mean μ.

Now, what about the variance of sample means? We saw above that, to obtain a sample mean, we divide every single identical variable x by a constant, the sample size N. Therefore, according to the properties of variances, the variance of each identical variable x/N will be equal to the population variance σ^2 divided by the square of the sample size, that is, σ^2/N^2. Sample means result from adding together all the x. Consequently, the variance of the sample mean is equal to the sum of the variances of all observations, that is, N times the population variance divided by the square of the sample size, or $N \times \sigma^2/N^2$. This is equivalent to σ^2/N, that is, the variance of sample means is equal to the population variance divided by the sample size. Therefore, the standard deviation of sample means (the standard error of the mean) equals the population standard deviation divided by the square root of the sample size.

Do not forget that these properties of means and variances only apply in the case of independent variables. Therefore, the results presented above will also only be valid if the sample consists of mutually independent observations. On the other hand, these results have nothing to do with the central limit theorem and, therefore, there are no restrictions related to the normality of the distribution or to the sample size. Actually, whatever the distribution of the attribute and the sample size might be, the mean of the sample means will always be the same as the population mean, and the standard error will always be the same as the population standard deviation divided by the square root of the sample size, provided that the observations are independent. The problem is that, in the case of small samples from an attribute with unknown distribution, we cannot assume that the sample means will have a normal distribution. Therefore, knowledge of the mean and of the standard error will not be sufficient to completely characterize the distribution of sample means.

3.5 Inferences from means

We are now in a position to make inferences about the true value of the population mean by inspecting the value of a sample mean.

Consider again the distribution of means from a sample of given size (Figure 3.4). It is evident from the shape of the distribution that most samples

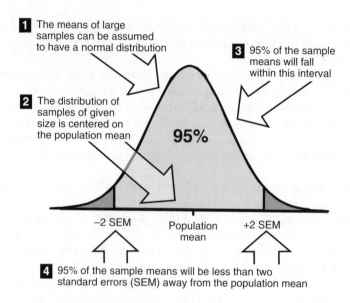

1 The means of large samples can be assumed to have a normal distribution

3 95% of the sample means will fall within this interval

2 The distribution of samples of given size is centered on the population mean

95%

−2 SEM   Population mean   +2 SEM

4 95% of the sample means will be less than two standard errors (SEM) away from the population mean

Figure 3.4   Steps in the inference of the value of the population mean from the value of a single sample mean.

will have a mean value that is close to the population mean, while a few will have values quite different from that one. If the sample means have a normal distribution, because of the properties of the normal distribution we know that about 95% of the samples will have means that are less than two standard errors away, on either side, from the population mean. For the same reason, we know that 99% will have means that are less than three standard errors on either side, and so forth.

Therefore, given a single sample from the population, we can state that, with a probability of 95%, the difference of the sample mean from the population mean is neither smaller nor greater than two standard errors in either direction. Along similar lines, we can state that, with a probability of 99%, the difference is neither smaller nor greater than three standard errors in either direction. And so forth.

For example, assume you have measured the total serum cholesterol in a random sample of 100 patients with coronary heart disease. The sample mean of total cholesterol was 210 mg/dL and the standard deviation was 20 mg/dL. The observations were independent and the sample size is large enough so that we can accept that the sample mean comes from a normal distribution. Therefore, we know with a probability of 95% that the interval from 210 mg/dL minus two standard errors, to 210 mg/dL plus two standard errors, will include the value of the population mean.

This is progress, but still we cannot place accurately the limits of that interval. To do so, we need to know the value of the standard error. Of course, we can never know exactly what the true value of the standard error is, but perhaps we have the means to estimate it from our data.

3.6 Confidence intervals

We have seen that the standard error is equal to the population standard deviation divided by the square root of the sample size. We know the square root of the sample size, but not the population standard deviation. However, the sample standard deviation (s) is an estimate of the population standard deviation (σ). Therefore, we could obtain an estimate of the standard error, simply by dividing the sample standard deviation by the square root of the sample size.

The question is: How good would that estimate be? We know that the estimate is unbiased, because the sample standard deviation is an unbiased estimate of the population standard deviation, provided that the $n - 1$ divisor is used. The question, then, is mostly about the accuracy of the estimate.

If the sample size is large, the accuracy of the estimate of the population standard deviation is high and, therefore, so is the estimate of the standard error. Furthermore, because when we calculate the standard error we divide the sample standard deviation by a large number, the already small difference between the sample and the true standard deviations will be divided by a large number. Thus, we may be confident that the estimate of the standard error will be very close to the true value.

Returning to the example given in the previous section, our estimate of the standard error would be 2 mg/dL, that is, the sample standard deviation, 20 mg/dL, divided by 10, the square root of 100 observations. Two standard errors equal 4 mg/dL. Therefore, we can say with a probability of 95% that the difference between the sample mean and the population mean is neither smaller nor larger than 4 mg/dL. In other words, we can say with a probability of 95% that the interval 210 minus 4 mg/dL, to 210 plus 4 mg/dL, or 206 to 214 mg/dL, contains the true value of the population mean.

The situation is thus the same as was described in the section on sampling. We can look at our observations and make plausibility judgments about the structure of the population. In this example, we see that the sample mean was 210 mg/dL and the sample standard deviation was 20 mg/dL. This result is quite plausible if the population mean is a quantity between 206 and 214 mg/dL, but very unlikely (although not impossible) if the population mean is smaller or higher than those limits. We would conclude that, based on our observations, we are very confident that the population mean lies somewhere between those limits. Technically speaking, we could say that, based on the sampling results, we are 95% confident that the true value of the population mean is between 206 and 214 mg/dL. This is why these limits are called 95% **confidence limits**, and the corresponding interval is called the 95% **confidence interval**.

In the discussion on sampling, it was also said that sample size is mostly related to the precision of the estimate, rather than to the representativeness of the sample. As we saw above, because the standard error equals the standard deviation divided by the sample size, then as the sample size increases, the standard error will decrease and the confidence interval will narrow. Consequently, as the sample size increases, we narrow the range of plausible values for the population mean. Alternatively, we may say that our estimate becomes increasingly accurate.

We can find confidence limits for any probability. To do so, we have to know how many standard errors on each side of the mean define the desired proportion of sample means. We can get this from statistical tables. For example, if we wanted to define a 72% confidence interval, the statistical tables say that, in a normal distribution, 1.08 standard deviations on each side of the mean delimit 72% of the observations. The 72% confidence limits could be found by multiplying the standard error by 1.08 and then adding and subtracting the result from the sample mean.

If we look at the statistical tables, we will see that 95% of the observations are delimited precisely by 1.96 standard deviations, not by 2.00 as has been said. This was for simplicity's sake, but from now on we will always use the exact value of 1.96 standard deviations.

Finally, where does that magical 95% number come from? Actually, it is just a convention based on an ancient statement that an event occurring only 5% of the time could be regarded as very uncommon. Thus, most statisticians agree that it is reasonably safe to reject a hypothesis about the population if, were that hypothesis true, the results from one's observations had a probability of occurrence of less than 5%.

3.7 The case of small samples

The approach for making inferences from means when all we have is a small sample is basically the same as described for large samples in the previous section. However, some important differences need to be understood and accounted for.

We have seen that the construction of confidence limits for the population mean of an attribute is based on the assumption of the normal distribution of sample means. This assumption is always verified for large samples, as long as the scale of measurement is interval and the observations are independent. We have also seen that for small samples this assumption is valid only if an attribute has a normal distribution. Therefore, if the distribution of an attribute is unknown, or is known not to have a normal distribution, we have no possibility of determining confidence intervals in small samples. The only solution in these cases is to increase the sample size.

The other problem with small samples is that the sample standard deviation is a less accurate estimator of the population standard deviation than it is in large samples. Because the sample standard deviation is obtained with a small number of observations, its variation from sample to sample is greater than in large samples, and increases as the sample size decreases.

This is illustrated in Figure 3.5, which presents the results of an experiment where we generated with a computer several random samples of size 5, 10, 15, and 20 from a standard normal variable. The histograms show the distribution of the sample standard deviations. Clearly, the smaller the sample size, the larger the dispersion of values of the sample standard deviations. Therefore, the standard deviation obtained in small samples is more likely to be significantly different from the population standard deviation than in the case of large samples.

Distribution of the sample standard deviation in small
samples from a standard normal variable

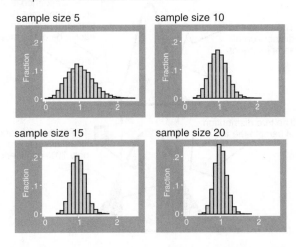

Figure 3.5   Relationship between the sample size and the spread of the sample standard deviation.

Furthermore, in small samples, the divisor for the calculation of the standard error estimate is also a small number. Therefore, we do not have the assurance that we have in large samples: that any differences between the sample and the population standard deviations will make an insignificant contribution to the value of the standard error estimate. These two factors contribute to the much lower accuracy of the estimate of the standard error from small samples than is the case with large samples.

The consequences of the lower accuracy of the standard error estimator are especially important when, by chance, a sample standard deviation is significantly lower than the population standard deviation. In this case, the method for drawing confidence limits as used in the case of large samples will inevitably produce the wrong conclusions.

This situation, and its consequences, are illustrated in Figure 3.6. If the sample standard deviation is significantly smaller than the population standard deviation, the estimate of the standard error will also be smaller than the true standard error. Therefore, our estimate of the distribution of sample means will not correspond to the true distribution, its shape being less spread on both sides of the mean. When we draw the 95% confidence limits as the values that are 1.96 standard errors from each side of the mean, in the true distribution those values actually encompass a lower proportion of observations. Contrary to our beliefs, the limits we find do not correspond to 95% confidence limits but to a lower, and actually unknown, confidence level.

In the opposite case, that is, when we obtain a sample standard deviation that is larger than the population standard deviation, the 95% confidence interval will be wider than it is in reality. However, this situation is not as serious as the former

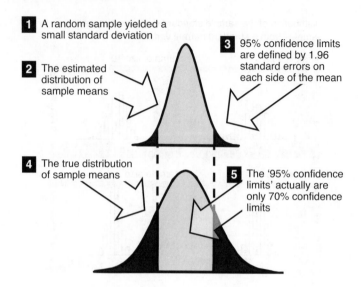

1 A random sample yielded a small standard deviation

2 The estimated distribution of sample means

3 95% confidence limits are defined by 1.96 standard errors on each side of the mean

4 The true distribution of sample means

5 The '95% confidence limits' actually are only 70% confidence limits

Figure 3.6   The error in the determination of confidence limits from standard error estimates obtained from sample standard deviations in the case of small samples.

because the probability that the 95% confidence interval we constructed contains the true value of the population mean will actually be 95% or greater. Therefore, we will err on the conservative side, which is not usually seen as being of serious importance.

To sum up, this is what can be said about the distribution of means of small samples:

- The distribution of sample means is normal if the variable has a normal distribution.
- As in large samples, the mean of the distribution of sample means is equal to the population mean.
- As in large samples, the standard error of the distribution of sample means is equal to the population standard deviation divided by the square root of the sample size.
- Unlike large samples, the division of the sample standard deviation by the square root of the sample size is not an accurate estimator of the standard error.

3.8 Student's *t* distribution

When we are finding the values that contain 95% of small sample means, we need to compensate for the increased variation in the sample standard deviations. As sample size decreases and the variability of the sample standard deviation increases, the probability that the sample standard deviation will be considerably smaller than the population standard deviation will also increase. Consequently, the likelihood that the distribution of sample means will be estimated as less spread about the mean than the true distribution will also increase. Therefore, a reasonable approach would be to draw, as the sample size decreases, the 95% confidence limits increasingly farther from the mean. The difficulty would then reside on where exactly one should draw the confidence limits.

Fortunately, it can be shown that a probability distribution called **Student's** *t* **distribution** will enable us to define where the confidence limits should be correctly placed.

Consider a sample from a normally distributed variable and a standard error estimate calculated from the sample standard deviation. With Student's *t* distribution, similar to what we have seen for the normal distribution, we can relate the relative frequency of the values of a variable with their distance from the mean, measured as the number of estimated standard errors. Unlikely the normal distribution, however, Student's *t* distribution will also take into account the sample size.

For example, take a sample of eight observations from a normally distributed variable and a standard error estimate obtained from the sample data. Student's *t* distribution will tell us that only about 92% of the sample means will be within two times the estimated standard errors on each side of the population mean. If the sample size was 4, for example, only 85% of the sample means will be within two estimated standard errors on each side of the population mean.

This might be getting a little confusing, because with large samples the standard error is also estimated from the data. Why, then, do we use the normal distribution to find confidence limits in large samples, and not Student's *t* distribution? Strictly speaking, we do not, and that can be easily verified in most statistical software packages, which use the *t* distribution to find confidence limits, no matter how large the sample is. But if they used the normal distribution instead, we would see that the difference between the two confidence limits was very small. This is because the *t* distribution approaches the normal distribution as the sample size increases. Technically speaking, we say that the *t* distribution **converges** to the normal distribution with increasing sample size.

Perhaps we should now look at Figure 3.7 for an illustration of Student's *t* distribution. The curve shows the relative frequency of observations as a function of *t* (as is seen on the horizontal axis), where *t* is a quantity that represents the distance to the mean expressed as the number of standard error *estimates*. Overall, Student's *t* distribution is similar in shape to the normal distribution, except that its tails are thicker. There is one curve for each sample size, which is why Student's *t* distribution is said to be a family of distributions. Each curve is referenced according to its **degrees of freedom**. In this particular problem on the construction of confidence limits, the appropriate degrees of freedom are the sample size minus one. Generally, we say the *t* distribution with $n - 1$ degrees of freedom.

For example, with a sample of 5 observations we should use the *t* distribution with 4 degrees of freedom, with 10 observations the *t* distribution with 9 degrees of freedom, and so on. As the number of degrees of freedom (and therefore the sample size) increases, the *t* distribution gradually approaches the normal distribution.

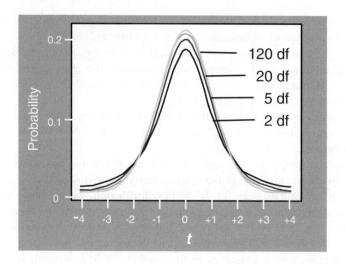

Figure 3.7   Several Student's t *distributions with different degrees of freedom (df).*

Actually, when the number of degrees of freedom is infinite, the two distributions are identical. The convergence of Student's t distribution to the normal distribution is illustrated in Figure 3.8 which shows, for several sample sizes, how many standard errors estimated from the sample standard deviation must be counted on each side of the mean to obtain the 95% confidence limits. For infinite degrees of freedom, one must count 1.96 estimated standard errors, which is the same number we saw previously for the normal distribution.

To sum up, an adequate procedure for finding confidence limits with small samples consists of estimating the standard error from the data, and then using Student's t distribution to find, for that sample size, how many standard error

Degrees of freedom	Number of estimated standard errors (*t*-value)
2	4.30
5	2.57
20	2.09
60	2.00
120	1.98
∞	1.96

Figure 3.8   Number of estimated standard errors on each side of the mean that define the interval of values that contain 95% of the observations, as function of the degrees of freedom.

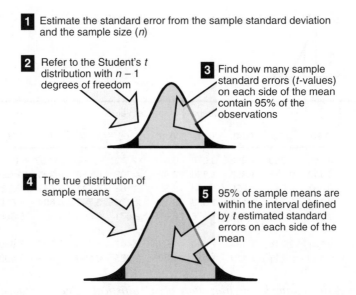

Figure 3.9   Steps in the construction of 95% confidence intervals using Student's t distribution.

estimates need to be counted on each side of the sample mean. We will then obtain the correct confidence limits. The procedure is illustrated in Figure 3.9.

Therefore, the principles and procedures are very similar to those for large samples, except that the normal distribution is replaced by Student's t distribution to find the number of standard errors on each side of the sample mean.

3.9 Statistical tables of the *t* distribution

We can use statistical tables of the t distribution to find the number of standard errors estimated from the data that must be counted on each side of the sample mean to obtain the desired confidence interval, given the sample size. As often happens with statistical tables, not all tables of the t distribution are alike. For example, the table shown in Figure 3.10 displays the values that are exceeded, in both directions, by a certain proportion of the observations. We know that because it is shown by the shaded areas in the figure above the table.

The values in the first column of the table are the number of degrees of freedom of the t distribution. On the first row several proportions are presented. The tabulated values are the distance to the mean of the distribution expressed as the number of estimated standard errors.

For example, we have a sample of six observations, and we want to draw the 95% confidence limits. First we must find the row corresponding

Table A2   Percentage points of the *t* distribution

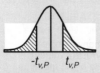

Degrees of freedom	Probability of greater value								
ν	0.90	0.50	0.30	0.20	0.10	0.05	0.02	0.01	0.001
1	0.158	1.000	1.963	3.078	6.314	12.706	31.821	63.657	636.62
2	0.142	0.816	1.386	1.886	2.920	4.303	6.956	9.925	31.598
3	0.137	0.741	1.250	1.638	2.353	3.182	4.541	5.841	12.924
4	0.134	0.741	1.190	1.533	2.132	2.776	3.747	4.604	8.610
5	0.132	0.727	1.156	1.476	2.015	2.571	3.365	4.032	6.869
6	0.131	0.718	1.134	1.440	1.943	2.447	3.143	3.707	5.959
7	0.130	0.711	1.119	1.415	1.895	2.365	2.998	3.499	5.408

Figure 3.10 Example of utilization of a statistical table of Student's t *distribution.*

to 5 degrees of freedom. For 95% confidence limits we find the column corresponding to a proportion of 0.05, that is, 5% of the observations are outside the confidence limits. At the intersection of the row and column, we read off the number 2.571. Therefore, we must count 2.571 standard error estimates on each side of the sample mean to obtain the 95% confidence limits. If we wanted 90% confidence limits, we would look in the column corresponding to a proportion of 0.10, for 99% confidence limits in the column 0.01, and so forth.

Let us return to the example of serum cholesterol in patients with coronary heart disease. Assume that we have obtained the mean and standard deviation, respectively 210 mg/dL and 20 mg/dL, from a sample of seven patients. First we calculate the standard error from the sample data as 20 mg/dL divided by the square root of 7. This value is 7.56 mg/dL and we know there is a considerable chance that this value is smaller than the true value of the standard error. So we find in the statistical table of the *t* distribution how many standard errors estimated from the sample we must count to get the correct 95% confidence limits. At the intersection of 6 degrees of freedom and the probability of 0.05 we read off 2.447. This is the number of estimated standard errors that must be counted on either side of the sample mean to obtain 95% confidence limits. So we multiply 7.56 by 2.447 mg/dL and obtain 18.50 mg/dL. Therefore, the 95% confidence limits are $210 - 18.50$ and $210 + 18.50$, or 191.5 mg/dL and 228.5 mg/dL.

3.10 Estimation with binary variables

So far, the discussion has been restricted to attributes measured in interval scales. However, in clinical research we are often interested in patient attributes that may be characterized only by their presence or absence (e.g., family history of asthma) or that classify subjects into two groups (e.g., males and females, death and survival).

As we saw in Section 2.2, attributes taking only two values are called binary attributes. They represent the most elementary type of measurement and, therefore, convey the smallest amount of information. It is useful to think of binary variables as attributes that may be on or off, because then the above distinction is not necessary. For example, we may think of the 'gender' attribute simply as 'male gender,' and of its values as *yes* and *no*. Similarly, the outcome could be thought as only 'survival,' with values *yes* and *no*. This is the same as for the family history of asthma, which also has the values *yes* and *no*.

We could convey the same information as *yes/no* by using the numerical system. Therefore, we could give the attribute the value 1 to mean that it was present, and 0 to mean that it was absent. This is much more appropriate, because now we can think of binary variables not as categories, but as numerical variables that happen to take only two possible values, 0 and 1.

Furthermore, observations from binary variables are commonly presented as relative frequencies as in, for example, 37% of females or 14% with family history of asthma. If we adopt the 0/1 values for binary variables, those proportions are nothing more than the average of samples from a variable with values 0 and 1. If males have value 0 and females 1, then in a sample of 200 subjects with 74 females the sum of the attribute gender totals 74 which, divided by 200 (the sample size), gives the result 0.37, or 37%.

As a binary variable is a numeric variable, in addition to calculating a mean we can also calculate a variance. The variance of a binary attribute is equal to the product of the proportions with and without the attribute. So, if we denote the population mean of a binary attribute by π, the Greek letter 'p,' then the population variance is equal to $\pi(1 - \pi)$.

Therefore, if we can calculate means and variances for binary variables, then it might be possible to infer the proportion of the population that presents an attribute from the proportion observed in a random sample.

Basically, the rationale for making statistical inferences from samples, as was explained for interval variables, also applies to binary variables. There are, however, some adaptations to be made because binary variables can take just two values, not an infinite number.

The greatest difference from the case of interval variables lies in the assumptions that must be made regarding the distribution of sample means. Unlike interval variables, where we do not know the distribution of sample means unless the samples are large or come from a normal distribution, with binary variables we always know exactly what the distribution is. That distribution is called the binomial distribution.

3.11 The binomial distribution

Imagine samples of four random observations on a binary attribute, such as gender for example, where we know that the distribution in a population is equally divided between males and females. Each sample may have from 0 to 4 females, so the proportion of females in the samples is 0%, 25%, 50%, 75%, or 100%.

It is a simple matter to calculate the frequency with each of these results will appear. We write down all possible combinations of males and females that can be obtained in samples of four, and count in how many cases there are 0, 1, 2, 3, and 4 females. In this example, there are 16 possible outcomes. There is only one way of having 0 females, so the theoretical frequency of this outcome is once out of 16 outcomes, or 6.25%. There are 4 possible ways out of 16 of having 25% of females, which is when the first, or the second, or the third, or the fourth sampled individual is a female. Hence, the relative frequency of this outcome, at least theoretically, is 25%. There are six possible ways of having 50% of females, so the relative frequency of this outcome is 37.5%. There are four possible ways of having 75% of females, so the frequency of this result is 25%. Finally, there is only one possible way of having 100% of females, and the relative frequency of this result is 6.25%.

These results are presented by the graph in Figure 3.11, which displays all the possible proportions of females in samples of four and their relative frequency. Since, as we saw before, the proportion of females in a sample corresponds to the mean of that attribute, the graph is nothing more than the probability distribution of the sample means. This distribution is called the **binomial distribution**. All random

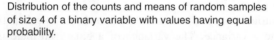

Distribution of the counts and means of random samples of size 4 of a binary variable with values having equal probability.

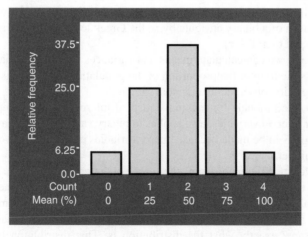

Figure 3.11   Probability distribution of a proportion: the binomial distribution.

binary attributes, like the proportion of patients with asthma in a sample, or the proportion of responses to a treatment, follow the binomial distribution.

The calculation of the frequency of all possible results by the method outlined above can be very tedious for larger sample sizes because there are so many possible results. It is also complicated for attributes whose values, unlike the above example, do not have equal probability. Fortunately, a formula exists that allows us to calculate the frequencies for any sample size and for any probability of the attribute values, but we do not need to know it.

Since the means of binary attributes in random samples follow a probability distribution, we can calculate the mean and the variance of sample proportions in the same way as we did with interval-scaled attributes. If we view a sample proportion as the sum of single observations from binary variables with identical distribution, then the properties of means allow us to conclude that mean of the distribution of sample proportions is equal to the probability of the attribute.

By the same reasoning, we conclude that the variance of sample proportions must be the variance of a binary attribute (the product of the probability of each value), divided by the sample size. If we call π the probability of an attribute having the value 1 (or, if we prefer, the proportion of the population having the attribute) and n the sample size, the variance of sample proportions is, therefore, $\text{var}(p) = \pi(1 - \pi)/n$. The standard deviation, which we call the standard error of sample proportions, is the square root of this quantity.

To sum up, let us review what can be said about the distribution of means of random samples of binary variables:

- The distribution of the sample proportions is known, and is called the binomial distribution.
- The mean of the distribution of sample proportions is equal to the probability of the attribute.
- The standard error of sample proportions is equal to the square root of the product of the probability of each value divided by the sample size.

3.12 Inferences from proportions

We have a sample of random observations on a binary attribute, we know the proportion of individuals in the sample that have the attribute, and, with this, we want to estimate the proportion in the population that has the attribute. We saw in the previous section that, because the attribute is binary, we know that the sample mean is an observation from a variable with a binomial distribution. Therefore, for the calculation of 95% confidence limits we only have to find which values of the proportion in the population would be inconsistent with the proportion observed in the sample. Specifically, we want to know which values of the population proportion would result in a probability of less than 2.5% of obtaining a proportion in the sample as low, or as high, as the one observed in the sample. We do so with the formula for the binomial distribution, which we do not need to know.

If we did know the formula for the binomial distribution we could use it to find **exact confidence limits**. For example, in a sample of size 20 the attribute was present in seven (35%) of the individuals. Using the binomial formula, we could calculate that if the true proportion of the attribute in the population was smaller than 15.4%, the probability was only 2.5% of obtaining seven or more individuals in a random sample. Conversely, if the proportion in the population was greater than 59.2%, the probability was only 2.5% of obtaining seven or fewer individuals in a random sample. We would conclude with 95% confidence that the proportion of the attribute in the population was between 15.4% and 59.2%. In other words, that the 95% confidence interval was 15.4% to 59.2%.

Computing confidence limits with the formula for the binomial distribution is a very tedious procedure. Fortunately, there is a simpler way of finding confidence limits. If we view a sample proportion as the sum of single observations from binary variables with identical distribution, then the central limit theorem applies. Therefore, provided the sample size is large, the distribution of sample proportions will converge to the normal distribution. We can then use the same method as we did for large samples of interval variables. The convergence of the binomial distribution to the normal distribution as the sample size increases can be confirmed visually in Figure 3.12.

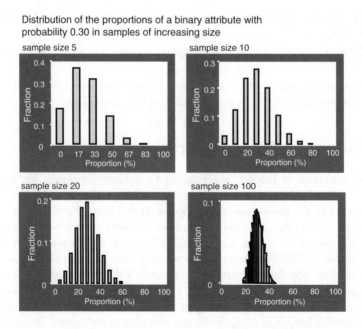

Figure 3.12   The convergence of the binomial distribution to the normal distribution.

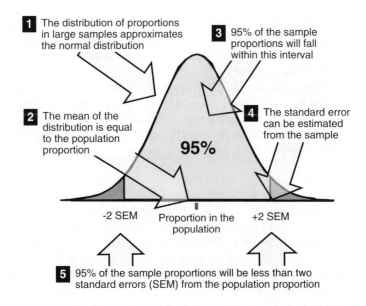

1 The distribution of proportions in large samples approximates the normal distribution

3 95% of the sample proportions will fall within this interval

2 The mean of the distribution is equal to the population proportion

4 The standard error can be estimated from the sample

95%

-2 SEM   Proportion in the population   +2 SEM

5 95% of the sample proportions will be less than two standard errors (SEM) from the population proportion

Figure 3.13   Steps in the construction of the 95% confidence interval of the population proportion.

Let us lay down the method of finding confidence intervals for large samples of binary variables, using the normal approximation of the binomial distribution, by referring to Figure 3.13.

We know that proportions obtained on large samples will closely follow the normal distribution. As was done with interval variables, we estimate the standard error from the data.

As we saw above, the variance of sample proportions is equal to $\pi(1-\pi)/n$. In large samples, this quantity may be estimated from our data as $p(1-p)/n$, where p stands for the sample proportion. Therefore, we may estimate the standard error of sample proportions by the square root of $p(1-p)/n$.

We know that in 95% of the samples the observed proportion will be within 1.96 standard errors on each side of the population proportion. Conversely, with 95% probability, the population proportion will not be more than 1.96 standard errors away, in either direction, from the sample proportion. The 95% confidence limits are, therefore, the observed proportion minus 1.96 standard errors, and the observed proportion plus 1.96 standard errors.

Let us consider an example. In a random sample of size 20 an attribute was observed in seven subjects. We will use the normal approximation to the binomial distribution to obtain 95% confidence limits for the population proportion π. The sample proportion is $7/20 = 0.35$ and we will estimate the standard error of the sample proportions using our data. Therefore, the estimate of the standard error will be the square root of $0.35 \times 0.65/20$, or 0.1067. The lower limit of the 95%

confidence interval will be $0.35 - 1.96 \times 0.1067 = 0.141$ and the upper limit will be $0.35 + 1.96 \times 0.1067 = 0.559$. The 95% confidence interval is, thus, 14.1 to 55.9%. The exact 95% confidence interval obtained with the formula for the binomial distribution is 15.4 to 59.2%.

The remaining question is: How large is a large sample? It is generally accepted that when we have at least five observations on each value of the attribute, it is safe to use the normal approximation of the binomial distribution. Therefore, if the proportion in the sample is 50% we need 10 observations, if it is 1% or 99% we need 500 observations.

If the sample is small, use of the normal approximation results in the wrong confidence intervals, and this will usually be quite obvious, often more so with proportions that are either large or small, because then the sample size requirements for the assumption of normality increase. As standard errors are large and the proportions are close to 0% or 100%, often the confidence limits will be below 0%, that is, negative, or over 100%. In each case, this is an impossibility. Therefore, with small samples we have to use exact confidence limits, which may be obtained either from the formula for the binomial distribution or from tables of the binomial distribution.

3.13   Statistical tables of the binomial distribution

When sample sizes are small, the normal approximation of the binomial distribution cannot be assumed. This occurs when in the sample there are less than five observations on either value of the binary attribute. If one attempts to use the normal approximation, it is almost certain that one will obtain absurd results.

Consider a sample of size 6, where the attribute was present in two (33.3%) individuals. The conditions for the approximation to the normal distribution are not met, but let us ignore this and use the normal approximation to find the 95% confidence limits.

The standard error estimate would be the square root of the product of 2/6 and 4/6 divided by 6, or 0.192. The upper confidence limit, therefore, would be 1.96 times 0.192, added to 2/6 or 0.710; the lower confidence limit would be the same quantity subtracted from 2/6 or -0.043. We would conclude that the 95% confidence interval was -4.3 to 71.0%, which is obviously wrong since proportions cannot be negative.

The only solution to this situation is to find the exact confidence limits. This may be done with a statistical package or with statistical tables of the binomial distribution, an example of which is shown in Figure 3.14.

The tabulated values are the 95% and 99% confidence limits for the specified sample sizes. For example, to find exact confidence limits for the

Table A3   Exact confidence limits for *p*

N = number of trials, x = number of successes, 100p_x = 100x/N

x	100p_x	100 (1-2α) limits 95% 100p_l 100p_h	100 (1-2α) limits 99% 100p_l 100p_h	x	100p_x	100 (1-2α) limits 95% 100p_l 100p_h	100 (1-2α) limits 99% 100p_l 100p_h
		N = 2				*N = 5*	
0	0.00	0.00–84.19	0.00–92.93	0	0.00	0.00–52.18	0.00–65.34
1	50.00	1.26–98.74	0.25–99.75	1	20.00	0.51–71.64	0.10–81.49
2	100.00	15.81–100.00	7.07–100.00	2	40.00	5.27–85.34	2.29–91.72
		N = 3		3	60.00	14.66–94.73	8.28–97.71
				4	80.00	28.36–99.49	18.51–99.90
0	0.00	0.00–70.76	0.00–82.90	5	100.00	47.82–100.00	34.66–100.00
1	33.33	0.84–90.57	0.17–95.86				
2	66.67	9.43–99.16	4.14–99.83			*N = 6*	
3	100.00	29.24–100.00	17.10–100.00				
		N = 4		0	0.00	0.00–45.93	0.00–58.65
				1	16.67	0.42–64.12	0.08–74.60
0	0.00	0.00–60.24	0.00–73.41	2	33.33	4.33–77.72	1.87–85.64
1	25.00	0.63–80.59	0.13–88.91	3	50.00	11.81–88.19	6.63–93.37
2	50.00	6.76–93.24	2.94–97.06	4	66.67	22.28–95.67	14.36–98.13
3	75.00	19.41–99.37	11.09–99.87	5	83.33	35.88–99.58	25.40–99.92
4	100.00	39.76–100.00	26.59–100.00	6	100.00	54.07–100.00	26.59–100.00

Figure 3.14   Example of a statistical table of the binomial distribution.

proportion above, we look for the sample size (*N* = 6) and in the column *x* we find the number of observations with the attribute, which in the example was 2. We then read off the exact 95% confidence limits, 4.33 to 77.72%, or the 99% confidence limits, 1.87 to 85.64%. The result is quite different from the one obtained with the normal approximation.

3.14 Sample size requirements

We have seen in Section 3.6 that as the sample size decreases, the standard error increases and, consequently, the confidence interval widens and the **precision** of the estimate also decreases. The standard error, therefore, can be seen as a measure of the precision of the estimate. This is the reason why authors of scientific papers often present their results as the sample means (the so-called **point estimates**) and their standard errors, thus conveying information on the precision of the study.

Because the sample size is mostly related to the precision of the estimates, when a study is being planned, the decision on the number of observations required must be based on the precision that is desired for the estimates. Put differently, the sample size must be the one necessary to provide confidence intervals of the desired size. For example, one could decide that the precision should be within 2 percentage points of the true population proportion. This means that one wanted the 95%

confidence interval to be smaller than 2 percentage points on each side of the mean. The sample size requirements to meet this specification would be the number of observations necessary to obtain a standard error of about 1 percentage point.

For example, assume that we are planning a study to investigate the proportion of the adult population affected by osteoporosis. Because we are going to estimate a proportion and the standard error depends on the value of the proportion, it would be wise to try to obtain some information on what the proportion of osteoporosis in the population might be. Let us assume that we concluded from a literature review that it must be a value around 15%. The next step would be to decide what precision we wanted for our estimate. Since this will be a study in the community, it is customary in these cases to use a precision of 2 percentage points.

So now we need to calculate a sample size that ensures, with a 95% confidence level, that the sample proportion will not differ by more than 2 percentage points from the population, provided the population proportion is not much different from 15%. This, of course, is the same as saying that the 95% confidence interval should be smaller than 2 percentage points on either side of the sample proportion. All we need to do, therefore, is to calculate the sample size that makes the standard error times 1.96 equal to 0.02, the **error of the estimate**. That is,

$$1.96 \times \sqrt{\frac{\sigma^2}{n}} = error$$

Solving the equation for n, we get

$$n = \frac{1.96^2 \times \sigma^2}{error^2}$$

In this example, 15% is our guess of the population proportion with osteoporosis, 85% the proportion without osteoporosis, and 2% the desired error of the estimate. The calculation of the sample size would then be

$$n = \frac{1.96^2 \times (0.15 \times 0.85)}{0.02^2} = 1225$$

We would need a sample of 1225 individuals to estimate a population proportion of about 15% with an error margin of 2 percentage points.

Sometimes we do not have any information on the expected proportion of the population with an attribute. At other times, we want to estimate the proportions of several attributes with the same study. In either case, the best approach is to assume the worst scenario in terms of sample size requirements. Since the variance of a proportion is largest when the proportion is 50%, decreasing for smaller and larger proportions, we could calculate the sample

Sample size requirements for different errors of an estimate
of a population proportion of 50%.

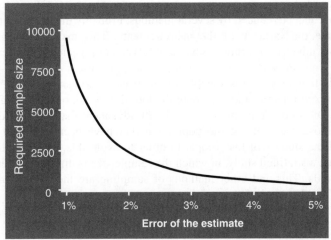

*Figure 3.15   Relationship between sample size requirements and desired error of
the estimate.*

size only for a 50% proportion. This is the **maximum error of the estimates**,
since attributes with smaller or larger frequency will all be estimated with
greater precision.

The procedure is exactly the same if we want to study an interval variable.
Because the standard error depends on the standard deviation, it would be wise to
search the literature for a likely value of the standard deviation of the variable.
Then, we must define the size we want for the 95% confidence interval, and
calculate the sample size in the same way.

Because precision increases only with the square root of the sample size, it is
costly to increase precision above a certain limit. This is illustrated in Figure 3.15,
which relates the sample size requirements with the error of the estimate of a
population proportion of 50%.

It is evident from the graph that sample size requirements increase dramatically
for errors of the estimate below 2 percentage points. This is the reason why the 2
percentage point error is usually the adopted limit of precision when estimating a
population proportion. On the other hand, there are no significant savings in sample
size if the error is greater than 5 percentage points, and this is the usually adopted
lower limit of precision.

It must also be taken into account that the standard error decreases with
decreasing variance. Therefore, another way of increasing precision or, for the same
precision, decreasing sample size, is to reduce variance. This approach is also more
efficient than augmenting the sample size, because the standard error decreases
linearly with the standard deviation. If the standard deviation is reduced to

one-half, the standard error also is reduced to one-half and the precision doubles. To achieve the same result by increasing the sample size, four times more observations would be needed.

Variance can be reduced by several methods. Probably the first one to consider is to reduce the variance of the measurements. This may be done by using reliable instruments, calibration, several observers, or repeated measurements, for example. Another way is to reformulate the population, excluding groups of individuals that may be responsible for increased variance. This is frequently done with experimental studies, where the too ill, too old, or too young are often excluded. Of course, this method has the disadvantage that the results will not apply to those segments of the population that have been excluded from the study, and the study will lose scope. In order to avoid this, an alternative would be to design a stratified study, in which the sample size is different for each group of individuals. This and other methods of sampling are the subject of the next chapter of this book.

4

Descriptive studies

4.1 Classification of descriptive studies

Descriptive studies represent one of the first steps in the formal investigation of a new problem. They consist of the systematic collection of data from a representative sample of a specified population, with the ultimate aim of describing the problem at the finest level of detail. Specifically, the purpose of descriptive studies is to determine the characteristics of a population through the means and proportions of its attributes.

As discussed in Section 1.2, when one is planning an investigation, there are three essential aspects to consider. There must be a clear and unambiguous definition of the population being studied, a sampling plan that will provide an adequate and unbiased sample, and a study design appropriate to the objectives of the investigation. Then, after all the data has been collected, it is necessary to analyze and present the study results with adequate statistical methods.

The basic statistical methods necessary for analyzing the data of a descriptive study have already been presented. They are descriptive statistics, point estimates, and confidence intervals. We have also been through the notions required for the calculation of the sample size required for a descriptive study. In this chapter we will go through a few additional statistical methods that are frequently used in descriptive studies. Because descriptive studies are so important to epidemiology, the science that studies the distribution and determinants of disease in populations, this subject will be presented under the perspective of epidemiological studies, and a few analytical methods that are often used in such studies will be covered. We will also add a bit more to the subject of sample size calculation. The emphasis of this chapter, however, will be on sampling methods and on study designs for descriptive studies and these issues will be addressed in the following sections. For now, only the more general aspects of study design will be presented.

Biostatistics Decoded, First Edition. A. Gouveia Oliveira.
© 2013 John Wiley & Sons, Ltd. Published 2013 by John Wiley & Sons, Ltd.

Descriptive studies can be classified in many ways, since there is considerable lack of uniformity regarding their nomenclature. To keep matters simple, throughout the book the following nomenclature will be adopted.

If the information and the data existed prior to the initiation of the study, the study is called **retrospective**. The opposite of retrospective is **prospective**, that is, the data is collected after the beginning of the study. Some studies have both retrospective and prospective elements. The distinction is of considerable importance, because retrospective studies are more vulnerable to errors, such as missing data and failure to recall past events, and thus they have a tendency to produce data of lower quality. On the other hand, they are less costly than prospective studies and can be executed in a fraction of the time necessary for the latter.

If the data is collected only once from each individual in the sample, the study is called **cross-sectional**. If each individual is observed over a length of time, possibly with repeated measurements of some attributes, the study is a **cohort** study. There are no relative merits of one design over the other – the selection must be based on what is appropriate for the problem under investigation. However, both cross-sectional and cohort studies can be done prospectively or retrospectively, and the considerations above apply unaltered to these studies.

Generally, cross-sectional studies are less costly, easier to plan, manage, and conduct, are less vulnerable to errors, and easier to analyze and interpret than cohort studies. On the other hand, cohort studies provide much more data and allow a more precise characterization of the evolution of a clinical condition than cross-sectional studies. However, the option between the two designs is not based on cost considerations, but rather on the research objectives.

Both types of studies have in common that they aim at obtaining unbiased estimates of population characteristics. This poses stringent requirements on the methodology of sampling, which is much more demanding in descriptive than in analytical and experimental studies. In the next sections we will discuss several sampling methods used in descriptive studies.

4.2 Probability sampling

We have seen how random observations from a population could be used to draw inferences about the characteristics of the population. We have also seen that the estimation of the true mean value of an attribute in the population is made by excluding, with a high degree of confidence, those values that would make very unlikely the set of observations we had made on the population. We have discussed the methods for making inferences from small samples and we saw that, even with small samples, we could obtain valid estimates of population means provided certain conditions were met.

Two conditions required for valid statistical inferences have direct implications on the methodology of sampling. First, we have seen that observations must be random. Second, we have seen that the methodology for the construction of confidence intervals is based on the properties of means and variances that were discussed and which are valid only for independent observations. This means that if the observations are not random the point estimates will be biased, and therefore wrong, and that if the observations are not independent the confidence limits will not be correct, and therefore the precision of the estimates will be less than declared. In the first case, there is no possible correction because we will have no means of knowing the direction and magnitude of the bias. In the second case we may still get valid estimates, but only on condition that the non-independence of the observations has been foreseen and controlled, in which case its impact on the estimates may be accounted for in the analysis and corrected.

In practical terms, this means that we must select the subjects for inclusion in a sample through a random process that prevents the investigator from exercising any type of voluntary or involuntary selection of the individuals. Selecting people at random may turn out not to be random if the investigator avoids, often unawares, people with certain characteristics, like an uncooperative or even an aggressive look. This also means that the investigator must not use an individual who was previously selected for the sample to gain access to further observations in individuals who are, in some way, in close proximity to the first one. If, after interviewing an individual, the investigator collects some more observations from the people who were accompanying the individual, this may bring some gains in time and effort, but the observations may not be independent. Consequently, the study could be severely biased, although still carrying the label of randomized. Therefore, to avoid the introduction of bias in a study, a procedure for the selection of individuals must be defined and strictly followed throughout the study.

In addition to having to select individuals according to a random process, and to define procedures for the inclusion of subjects in the sample, there are two more requirements for a sample to be considered representative. First, every individual in the target population must have some possibility of being selected for the sample. Second, it is possible to know the probability that each individual in the target population has of being selected for the sample and, therefore, the size of the population must be known.

These characteristics define what is called **probability sampling**. The alternative, **non-probability sampling**, selects individuals from the population according to their accessibility or personal characteristics, a process that may leave entire segments of the population with no possibility of being selected for the sample. As it is impossible to know exactly who were the individuals excluded from the sampling process, and how many there were, it is also impossible to assess the direction and magnitude of the bias. A typical example of non-probability sampling is **convenience sampling**, whereby individuals are selected just because they are easily accessible. Convenience samples are used extensively in analytical and experimental clinical studies, where the representativeness of convenience samples, and therefore the validity of the results, rests upon the subjective judgment of eventual sources of bias and their potential impact on the results, rather than on sampling theory. However, as the main goal of descriptive research is to provide unbiased and reliable estimates of the characteristics of a population, every effort should be made to base those estimates on probability samples.

Finally, it is important to mention that probability sampling is not by itself a guarantee of correct population estimates. Probability samples may be severely biased because of **non-sampling errors**, for example, because of high rates of faulty measurements, non-responders, or missing data. Naturally, non-sampling errors have a greater impact on the value of population estimates in the case of small samples. Therefore, large samples have the double benefit over small samples of producing more accurate as well as more robust estimates of population parameters; that is, they are less sensitive to non-sampling errors and to insidious causes of potential bias and they afford greater credibility to the results.

4.3 Simple random sampling

The methodology to obtain a probability sample consists of creating a listing of all elements of the population, assigning a unique number to each element in the list, generating random numbers, as many as the sample size required for the study, and selecting for the sample the elements with a corresponding number. This sampling method is called **simple random sampling** (Figure 4.1).

In survey terminology, the list with the enumeration of all elements of the population is called a **sampling frame**. The elements of the population that are randomized are called **primary sampling units**, or **PSUs**. In the case of simple random sampling, the PSUs are the individuals, but this may not be so in other methods of sampling.

If the population is large, it is tedious to number all elements in the sampling frame and then find each number generated by the computer. In this situation, **systematic random sampling** might be used. Suppose there are 150 000 individuals in the population, of which we want a sample of 500. The **sampling fraction** is, therefore, 1 out of 300. A systematic sampling would be as follows. First, we would take a random number between 1 and 300. Let us assume that number was 156. Therefore, we would select the 156th name in the list, and then every 300th name, that is, the 456th name, the 756th name, and so on. This would be the equivalent to a simple random sample, provided the first individual was selected at random and that there was no cyclic pattern in the list with approximately the same frequency as the interval we defined. **Consecutive random sampling** is a particular form of systematic sampling where the interval between selected elements is one.

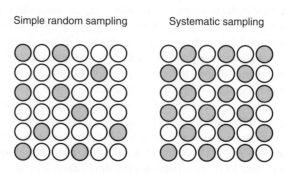

Simple random sampling   Systematic sampling

Figure 4.1   Representation of simple and systematic sampling.

All three methods are valid for obtaining a random sample and if correctly used will produce identical estimates of the population attributes. Sampling frames should be prepared in a way that removes any periodicity in the data and, therefore, should not be sorted by any attribute of the individual, such as given name.

With simple, systematic, and consecutive sampling, estimates of the population attributes are made using the general methods presented in Sections 3.6, 3.8 and 3.12 for means and proportions. However, there are a few issues that need to be addressed in order to completely understand the process of inference from samples. These will be discussed in the next section.

4.4 Replacement in sampling

Throughout our discussions on sampling and estimation we have always assumed that the probability of an attribute remains the same during the entire process of sampling. This will be true only if the sampled elements are returned to the population after being observed, thus keeping the denominator constant. This means that the same individual may be selected more than once for the sample. If we do not allow the same individual to be observed more than once, as is usually the case in clinical research, every time we select an individual for the sample, the population from which we will select the next individual will shrink by one and the probability of the attribute will be different.

For example, consider a population of size 20 and an attribute with probability 0.6. Therefore, in the population there are 12 elements with, and 8 without, the attribute. Suppose that we take a sample of 4 elements, one at a time, returning each sampled element to the population after we have checked whether the attribute was present. Then, the probability of the attribute for the first element is 12/20, as well as for the second, third, and fourth elements,

Now let us repeat the sampling without returning each sampled element to the population. The probability of the attribute for the first element is, of course, 12/20, but because that element was not returned to the population, the population size is now 19. If the first element had the attribute, there would remain only 11 elements in the population with the attribute and thus the probability of the attribute for the second sampled element would be 11/19; if the first element did not have the attribute the probability would be 12/19.

The first method of sampling is called **sampling with replacement** and the second **sampling without replacement**. What are the implications of replacing the sampled elements on the probability distribution of sample proportions?

We have seen before that when the probability of an attribute remains constant throughout the process of sampling, as in sampling with replacement, the sample proportions follow the binomial distribution. Is this also true for sampling without replacement?

Using the same example, let us compare the probability of a sample proportion of 100%, that is, a sample in which all four elements have the attribute, in sampling with and without replacement. The probability of this result is equal to the

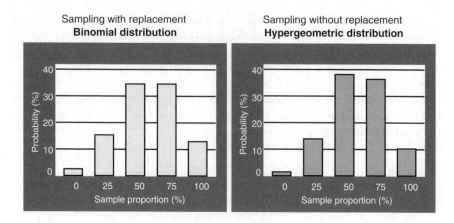

Figure 4.2   Comparison of the binomial and the hypergeometric distributions.

probability of the first element having the attribute multiplied by the probability of the second element having the attribute, by the probability of the third element having the attribute, and by the probability of the fourth element having the attribute. In sampling with replacement the probability of that result is $12/20 \times 12/20 \times 12/20 \times 12/20 = 0.1296$. In sampling without replacement the probability is $12/20 \times 11/19 \times 10/18 \times 9/17 = 0.1022$. Therefore, the distribution of sample proportion is not the same in sampling with and without replacement.

Figure 4.2 shows the probability of sample proportions with (upper graph) and without (lower graph) replacement. It can be seen that in sampling without replacement the probability distribution of sample proportions is not binomial, but a different distribution called the **hypergeometric distribution**.

This difference in probability distributions has implications for the calculation of confidence intervals because the variance of sample proportions, the square of the standard error of the proportion, is smaller in sampling without replacement. This is because, as the sampling process progresses, the sampling fraction remains constant in sampling with replacement but increases in sampling without replacement because the population will be shrinking. Therefore, in the latter case the precision will be greater.

Actually, the two variances differ by a constant factor which is $(N - n)/(N - 1)$, where n is the sample size and N the population size. This factor is called the **finite population correction**, or **fpc**. Its value can be rounded to one minus the sampling fraction, that is, $1 - n/N$. This applies also to attributes measured in interval scales.

Therefore, if sampling was without replacement, the variance of sample means should be multiplied by the fpc in order to obtain the value of standard errors that we will use for estimating population means and proportions. As in clinical research sampling is mostly without replacement, in the calculation of standards errors we would need to account for the fpc. However, if the sampling fraction is 5% or less, which it virtually always is, there is no real gain in including the fpc in the calculation of standard errors because the reduction will be about only 0.7%.

The estimation of population means and proportions from simple random samples is made according to the method described previously. Therefore, point estimates are the sample mean or proportion and the standard errors are obtained by the square root of the sample variance divided by the sample size. If we wish to use the fpc, then the standard error should be multiplied by the square root of the fpc.

Sample size requirements are calculated as explained before, but if the fpc is going to be used then the calculation is slightly different. Let us review how sample size is calculated for binary and interval attributes in sampling with replacement. Remember that the investigator needs to provide an informed guess about the population proportion for binary attributes, and the population variance for interval attributes. The investigator decides upon the error of the estimates and the confidence level. Let us call D the quotient of the error by the z-value corresponding to the desired confidence level (i.e., 1.96 for 95% confidence intervals or, more generally, $z_{\alpha/2}$ for $1 - \alpha$ confidence intervals).

For binary and interval attributes, respectively,

$$n = \frac{\pi(1 - \pi)}{D^2} \quad \text{and} \quad n = \frac{\sigma^2}{D^2}$$

If the fpc is going to be used, the calculations for binary and interval attributes, respectively, are

$$n = \frac{N\pi(1 - \pi)}{(N - 1)D^2 + \pi(1 - \pi)} \quad \text{and} \quad n = \frac{N\sigma^2}{(N - 1)D^2 + \sigma^2}$$

For large population sizes the improvement of estimates in sampling without replacement is negligible. For example, suppose we wish to estimate the proportion of chronic heart failure patients in the population. We start by making an informed guess as to what that number might be and, for this, we use the results from the NHANES survey, which estimated a proportion of 2.62% for the US population over 20 years old. We define an error of the estimate of 1 percentage point, with a 95% confidence level. The calculation of the sample size is

$$n = \frac{0.0262(1 - 0.0262)}{(0.01/1.96)^2} = 980.13$$

As our sample will be without replacement, we calculate the sample size using the fpc. For this, we need to know the population size. Using data from the most recent population census we were able to determine that the population older than 20 years old is 8 249 856. The sample size for sampling without replacement is

$$n = \frac{8\,249\,856 \times 0.0262 \times (1 - 0.0262)}{8\,249\,855 \times (0.01/1.96)^2 + 0.0262(1 - 0.0262)} = 980.01$$

The gain brought about by using the fpc is indeed negligible.

4.5 Stratified sampling

We saw in Section 3.14 that, for a given precision of the estimate and level of confidence, sample size is dictated by the variance of the attribute. Populations are usually heterogeneous regarding an attribute, with some segments of the population having a large, and some a small, mean value. When these groups are merged into a single population, variance is increased (Figure 4.3). Therefore, a more efficient method than simple random sampling might be to sample each group separately. The independent estimates are then averaged, giving each one a weight proportional to its contribution to the overall mean. Then we would obtain independent estimates in each sample. Because the variance in each group is smaller, the number of observations needed to achieve the desired precision may be less than the number required to obtain the same precision ignoring the groups. This method of sampling is called **stratified sampling**. It is called stratified because the technical term for the population segments is **strata**.

For example, imagine that we wish to investigate what the mean value of serum urea is in chronic kidney disease patients on hemodialysis. Suppose we had information that there was a trend toward an increased serum urea level on patients with a longer permanence in the dialysis program. Then, we could divide the population into two strata, for example, one stratum for patients with less than three years' permanence and one stratum for patients with three years or more.

Let us say there was a total of 10 000 patients in the first stratum and 5000 in the second. To perform stratified sampling we would take a simple random sample of, say, 250 patients in each stratum and obtain the mean and standard deviation of serum urea in each sample. The combined mean of the two strata is obtained by a weighted average of the two sample means, and the weight is, naturally, the proportion of the total population represented by each stratum.

For example, let us say that the mean and standard deviation of serum urea was, respectively, 37 mg/dL and 30 mg/dL in the first stratum, and 45 mg/dL and 33 mg/dL in the second. The first stratum represents two-thirds of the population and the second one-third, and these will be the weighting factors. Accordingly, the overall mean would be 37 mg/dL multiplied by 2/3 plus 45 mg/dL multiplied by 1/3, giving

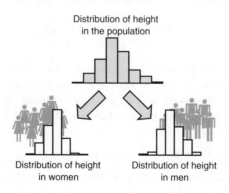

Figure 4.3   Reducing variance by splitting the population in strata.

an estimate of the population mean of 39.7 mg/dL. The standard error of the mean is also a weighted average of the standard error estimates in each stratum. Therefore, we multiply the standard error estimate of the first stratum by 1/3 and the second stratum by 2/3. We cannot add standard errors, but we can sum variances, so we need to square each weighted standard error before we sum them, and then we take the square root to get the overall standard error:

$$\sqrt{\left(\frac{30}{\sqrt{250}}\times\frac{1}{3}\right)^2 + \left(\frac{33}{\sqrt{250}}\times\frac{2}{3}\right)^2} = 1.528$$

It does no harm if we use symbols to explain the calculations. We denote by M_i the mean value of an attribute in each stratum i, by N_i the population size in each stratum i, and by N the total population. The expression

$$M = \sum M_i \frac{N_i}{N}$$

means that, in order to calculate the point estimate M of the population mean μ, we sum across all the i strata the product of the mean M_i of each stratum with the corresponding weight, that is, the fraction of the population that belongs to that stratum, N_i/N.

The symbol $\sum$, which is the Greek letter for 'S,' represents a sum. If there are three strata, called 1, 2 and 3, the above expression translates to

$$M = M_1 \frac{N_1}{N} + M_2 \frac{N_2}{N} + M_3 \frac{N_3}{N}$$

The variance of sample means is calculated with the following expression:

$$\text{var}(M) = \sum \frac{\sigma_i^2}{n_i} \frac{N_i^2}{N^2}$$

which reads, sum across all the i strata the product of the variance of sample means of each stratum with the square of the fraction of the population that belongs to that stratum.

As usual, we substitute the sample variance s^2 for the population variance σ^2. To obtain the standard error we take the square root of $\text{var}(M)$.

If we are estimating a population proportion, we replace σ^2 in the expression by $\pi(1 - \pi)$, the variance of a binary variable. If we wish to use the fpc, then the variance of sample means within the strata should be multiplied also by $(N_i - n_i)/(N_i - 1)$ and the above expression would be

$$\text{var}(M) = \sum \frac{\sigma_i^2}{n_i} \frac{N_i^2}{N^2} \frac{N_i - n_i}{N_i - 1}$$

To sum up, when there is a nominal attribute, or an interval attribute that can be grouped into categories, related with the variable under study, we can split the population in strata defined by the values of that nominal attribute. The variance of the variable in each stratum will be less than the total variance and, with smaller variances, we will get the same precision with a smaller sample size.

Stratified sampling may produce standard errors that are smaller than those obtained with simple random sampling, that is, more efficient estimators (requiring a smaller sample size to achieve the same accuracy), but only if the means are considerably different from one stratum to another. If the means do not differ across strata, the result is evidently the same as with simple random sampling. Because stratified studies are more complex to conduct, this option should be taken only if there is evidence that means do differ across the strata. The strata should be defined in a way that maximizes differences between means, while keeping each stratum as homogeneous as possible.

From the above considerations it is clear that, in stratified sampling, a single study variable will benefit from the increased precision afforded by stratification. Any other variables being studied will not have their accuracy improved unless their means are also significantly different across the same strata.

The efficiency of stratified random sampling can be increased if the sample size is not equal in each stratum, but this is possible only if we have more information about the population. If we know that the standard deviation of an attribute is approximately constant across strata, the sample size in each stratum may be defined in proportion to the size of the stratum. The sampling fraction is constant and, therefore, larger strata will contribute with larger samples. This method is called **proportional stratified sampling** (Figure 4.4) and it will always produce more efficient estimators than simple random sampling, although on occasion the difference may not be substantial.

If we know that the standard deviation of an attribute is different from one stratum to another, the sample size in each stratum may be proportional not only to the size, but also to the standard deviation of the stratum. More specifically, the

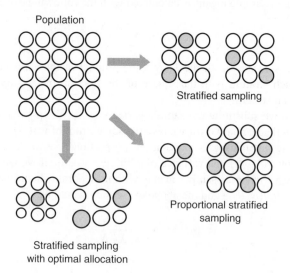

Figure 4.4   The different methods of stratified sampling.

sample size is proportional to the product of the size of the stratum multiplied by its standard deviation. Therefore, the sampling fraction increases in strata with larger standard deviations. This procedure will save observations in strata with small variances and will allocate them to strata with large variances, keeping the precision constant across all strata. For this reason, this method is called **stratified sampling with optimal allocation** (Figure 4.4).

In proportional stratified sampling, the sampling fraction is constant across strata, which means that the probability of each individual in the population of being selected for the sample is the same for all individuals and, consequently, the sampling weights are also the same. Therefore, calculation of the combined mean and the standard error can be made by pooling together all observations and proceeding as if it were a simple random sample. In stratified sampling with optimal allocation, the combined mean and standard error are calculated in the same way as explained for stratified sampling.

The following example will illustrate the gain that may be achieved by each sampling method just described. Recall the example of the survey to estimate the proportion of chronic heart failure patients in the general population. A study conducted elsewhere reported a proportion of 2.62%. We saw previously that the sample size for a survey by simple random sampling with an error of the estimate of 1% with 95% confidence was 980 subjects.

The results of that study also showed that the frequency of chronic heart failure increases with age. Therefore, we will calculate sample sizes for stratified sampling using four age groups as strata. We will need the proportions P_i of individuals with the disease in each stratum, which were reported in the publication, and the number of individuals of the general population in each stratum, N_i.

Figure 4.5 shows the sample sizes needed to achieve the desired accuracy with simple random sampling and with the three modalities of stratified sampling.

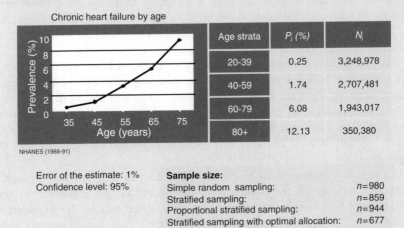

Age strata	P_i (%)	N_i
20-39	0.25	3,248,978
40-59	1.74	2,707,481
60-79	6.08	1,943,017
80+	12.13	350,380

NHANES (1988-91)

Error of the estimate: 1%   **Sample size:**
Confidence level: 95%   Simple random sampling: $n=980$
Stratified sampling: $n=859$
Proportional stratified sampling: $n=944$
Stratified sampling with optimal allocation: $n=677$

Figure 4.5   Sample size requirements for simple and stratified sampling.

For the calculation of sample size in any type of stratified sampling it is convenient to use the formula

$$n = \frac{\sum (N_i^2 \sigma_i^2 / v_i)}{N^2 D^2 + \sum N_i \sigma_i^2}$$

where N is the size of the population, N_i is the size of stratum i, v_i is the proportion of the sample in stratum i, σ_i^2 is the variance of the attribute in stratum i, and D is the error of the estimate divided by $z_{\alpha/2}$.

4.6 Multistage sampling

In the previous section we covered some methods designed to provide extra precision for the estimates or, equivalently, to reduce the sample size requirements while maintaining the desired precision. Those methods, therefore, address both the precision and cost issues. In this section, we will describe methods that are useful only for reducing the costs of a survey but at the expense of decreased precision for the same sample size.

To better explain what **multistage sampling** is, and why it must often be considered, let us resume the example of the hemodialysis study. Assume that there are 15 000 patients in dialysis programs and we need a random sample of 500 patients. In this, as in many other situations, creating a sampling frame with the identification of all 15 000 patients would be exceedingly difficult and costly. On the other hand, those patients are being treated in hospitals, and it would not be so difficult to obtain a listing of all hospitals providing that care. A multistage random sampling of hemodialysis patients could be as follows. First, a simple random sample of size, say, 10 is taken from a sampling frame of the population of hospitals. In each selected hospital, we would create another sampling frame, this time with the identification of all patients undergoing dialysis. From there, we would take a simple random sample of 50 patients.

Thus, we have a first stage consisting of a sample of hospitals, and a second stage consisting of a sample of patients within hospitals. Compared to simple random sampling, there are gains in time, effort, and costs. First, we did not have to list the entire population, only the first stage units and second-stage units within those first-stage units that had been selected. Second, in order to observe the patients we only needed to travel to 10 hospitals, instead of every hospital that happened to have a patient selected in the sample.

If a complete listing of all hospitals with a dialysis program was difficult to obtain, we could divide the country into regions and sample a number of them. Then, hospital lists need be created only for the sampled regions. For example, we could sample five regions, five hospitals within each region, and 20 patients within each hospital. This would be a three-stage random sampling. Alternatively, we could select all patients in each hospital instead of taking a sample of them. This would be called **cluster sampling** (Figure 4.6).

Multistage sampling   Cluster sampling

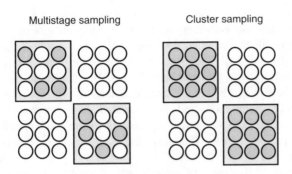

Figure 4.6   Multistage and cluster sampling.

Ideally, in multistage sampling the means and variances should be similar across the last-stage groups and similar to the mean and variance of the population. If the last-stage groups have approximately the same number of elements, cluster sampling is likely to be more effective than multistage sampling.

Of the sampling methods just described, multistage sampling is often the only feasible one because in most situations it is not possible to enumerate all elements in the population, though it is possible to enumerate all the elements of each stage. One problem with multistage sampling, however, is that the analysis is much more complex than with simple or stratified sampling.

This example illustrates how an estimate of the population proportion of a given attribute is calculated from a two-stage sample. Let us return to the hemodialysis study. We said that the total population of hemodialysis patients is 15 000. Suppose there are 50 hospitals and we decide that the first stage will include 3 hospitals selected by simple random sampling, from which we would take a total sample of 500 patients. The first hospital selected had a total of 200 patients in hemodialysis, the second 350, and the third 300. We select a simple random sample of patients from each hospital of, respectively, 100, 230, and 170. The proportion of the attribute (say, serum urea above two times the upper normal level) in each patient sample was, respectively, 30%, 40%, and 10%.

In order to estimate the proportion of the attribute in the total population, we calculate the contribution of each hospital as the product of the observed proportion with the percentage of the total population accounted for by that group. The population of the first hospital accounts for 1.3% ($= 200/15\,000$) of the total population, the second hospital for 2.3%, and the third hospital for 2.0%. So, in the first hospital the attribute is present in 30% of 1.3% of the total population, and it contributes 0.4 ($= 30\% \times 1.3\%$) percentage points to the total proportion in the population. In the second hospital 40% of 2.3% of the population have the attribute, so its contribution is $40\% \times 2.3\% = 0.9$ percentage points, and in the third hospital it is $10\% \times 2.0\% = 0.2$ percentage points.

We then sum the contributions from all sampled groups to obtain the contribution of all the sampled first-stage units. The result is $0.4 + 0.9 + 0.2 = 1.5$ percentage points. The estimate for the entire population is, therefore, this result multiplied by the number of times there are more hospitals in the population than in the first-stage sample. There are 16.7 ($= 50/3$) sets of 3 hospitals in the population, and we estimated that each set of 3 hospitals contributes on average 1.5 percentage points to the total proportion of the attribute in the population. Therefore, the estimate of the proportion in the population is $16.7 \times 1.5 = 25.05\%$.

Calculation of the standard error is more complicated and in multistage sampling is so complex that it should be done only with appropriate statistical software.

Multistage sampling is usually less efficient than any of the sampling methods presented previously; that is, for the same sample size the population estimates have lower accuracy. This is because observations tend to be correlated within each group in the final stage. For example, if hospitals have different admission criteria, then patients within a hospital are more similar to each other than they are to patients in other hospitals. Consequently, the attributes are very homogeneous within hospitals, but very heterogeneous among hospitals. If the admission criteria were the same for all hospitals then the attributes would be heterogeneous within hospitals but very homogeneous among hospitals. The loss of precision in multistage sampling arises because the heterogeneity among hospitals contributes much more to the estimate of the population variance than the heterogeneity within hospitals and, therefore, multistage sampling tends to inflate standard errors in proportion to the degree of correlation of observations in the last-stage groups.

The proportional increase in the estimate of variance, compared to the estimate that would be obtained by simple random sampling with the same sample size, is called the **design effect**. The design effect increases both with the degree of correlation of the observations within each last-stage group and the size of the last-stage group. Therefore, precision of the estimates is gained if the last stages are designed as small as possible.

Despite the loss of precision in multistage sampling, since significant savings are made in the creation of sampling frames and travel, part of the resources can be used to increase the sample size.

In the example above, sampling units were selected by simple random sampling. This does not need to be so. For example, we could improve the precision of the estimate of an attribute by stratifying patients on time of permanence in the dialysis program, and then we would have a **combined multistage and stratified sampling**. The combinations of the methods are manifold, and the decision about which sampling strategy should be employed is frequently a difficult one.

To close our discussion on sampling methods, we will just comment on what is considered the primary sampling unit (PSU) of a multistage or combined sampling.

The PSU is the first stage where its elements are randomized. For example, if we had divided the territory into regions and had selected one hospital at random from every region, the hospital would be the PSU. Had we included every hospital in each region and randomized patients within each hospital, the patients would be the PSU.

4.7 Prevalence studies

Now that we have reviewed the methods for obtaining a probability sample, as well as the notions necessary for determining the sample size required for a descriptive study, we will proceed to cover the main designs of descriptive studies.

Very often, descriptive studies are done to estimate the frequency of a given attribute in the population, such as a particular disease or clinical condition. This is called the **prevalence** of an attribute. The **prevalence rate** is the proportion of the population having that attribute. Prevalence studies are particularly important types of observational studies because the prevalence rate is a **morbidity index** of populations that is central to epidemiology. Knowledge of the frequency of diseases is of major importance for patient management, allowing the clinician to plan efficiently the study of individual patients. For health care managers, prevalence studies are the basis for the definition of health care policies, strategies, and resource allocation.

Prevalence studies are typically cross-sectional. These studies have some particularities regarding terminology and presentation of the results. The population surveyed is usually called the **population at risk** and the subjects having the attribute under study are called **cases**. Prevalence rates are often presented on a 1000 or 10 000 basis instead of percentages, because disease prevalence rates are often rather low. Estimates are often presented only as the point prevalence, with omission of confidence intervals, because studies are usually conducted on very large samples and, consequently, the error of the estimate can be assumed to be very small.

One aspect that needs careful consideration in the design of prevalence studies is the definition of cases, that is, the set of criteria that will allow the classification of each sampled individual as being a case or a non-case. Invaluable help may be obtained from a thorough evaluation of the criteria used in previously published studies, so this is always an obligatory step in the planning of a study. Additional help may be provided by existing guidelines and consensus statements for the diagnosis and management of diseases, which often include a definition of the criteria, methods, and instruments for an epidemiological diagnosis. However, it is necessary to keep in mind that epidemiological surveys are typically conducted on samples of thousands of people and, on most occasions, it is not possible to carry out a full examination of the individuals. Therefore, the criteria, methods, and instruments for diagnosing a disease used in epidemiological surveys are not necessarily the same as the ones used in clinic or hospital-based studies.

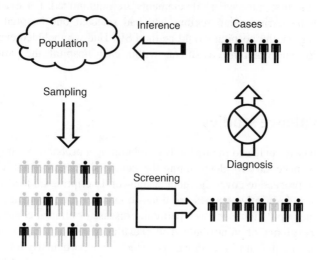

Figure 4.7   Schema of a prevalence study.

Screening instruments are often used in prevalence studies (Figure 4.7). These consist of some kind of measurement that is simple, safe, and inexpensive, and that enables the investigator to exclude the presence of the disease with reasonable confidence. The resources, therefore, can be allocated mainly to investigating the presence of the disease in those subjects that were identified by the instrument. Structured questionnaires are often used in surveys for this purpose. For a screening instrument to be valid, naturally, it must have high sensitivity.

4.8 Incidence studies

Prevalence studies are difficult and expensive endeavors, particularly if the survey is on the general population. There is, however, an easier and less costly approach that will give approximately the same information, provided the disease prevalence is not too high. The idea is to identify only the new cases of a disease occurring in a given time interval, the so-called **incidence** of the disease. Since incident cases become prevalent for the duration of the disease, if there is information on the average duration of the disease, the prevalence rate can be estimated by the product of the incidence and the average duration of the disease. For example, if a disease has an estimated **incidence rate** of 50 new cases a year per 100 000 inhabitants, and if the average duration of the disease is 20 years, then an estimate of the prevalence rate would be 1000 in 100 000 or 1%.

Or course, the importance of incidence rates is not merely to estimate prevalence rates. The incidence rates inform about the dynamics of a disease while prevalence rates are a static portrait of the disease, and often both statistics are necessary for the characterization of the disease. On the other hand, the incidence rate is on many occasions a better measure of morbidity than the prevalence rate. For example, the prevalence of the common cold, that is, the number of people in the population that at a given moment have the common cold, is a rather small value but its incidence is extremely high. In diseases of short duration the prevalence will always be low and, in such cases, disease frequency is better described by its incidence.

In the limit, if the duration is zero, the prevalence will also be zero regardless of the incidence rate. This situation occurs every time we wish to study an event, as often happens in clinical research. Some examples of events that are often the aim of a clinical investigation are death, adverse drug reactions, major cardiovascular events, and tumor response to chemotherapy. Every time we study an event, that is, a change in the patient's condition from one state to another, the frequency measure we should use is the incidence.

One extremely important measure of population health is the **mortality rate**. Mortality is, of course, evaluated only in terms of its incidence, and the methodology of mortality studies is the same as for any other medical event. Mortality studies in patient populations are one of the most common clinical investigations and are often designed as cohort studies. In those studies based on populations with a given disease rather than on the general population, the mortality rate is called the **case-fatality rate**.

Incidence studies require a cohort design. In cohort studies, a sample of the population is selected by one of the sampling methods outlined above and followed over time until the event eventually occurs (Figure 4.8).

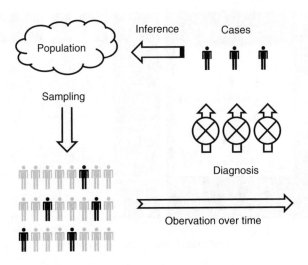

Figure 4.8   Schema of an incidence study.

4.9 The person-years method

The typical design of an incidence study is the cohort study. In these studies, a representative sample of the population is followed over time until the events of interest are observed. These studies are expensive and difficult to manage because they require regular observation of the individuals for a long time.

Cohort studies always have the problem of individuals dropping out of the study. If the study lasts for a long time, measured in years, some people will withdraw from the study, because they moved home, refused to continue, failed to show up for some reason, developed an incapacitating condition, or even died. In addition, the study could be concluded before the event occurred in some individuals.

In these cases, one will have an incomplete follow-up for some individuals. However, it would be unwise to exclude them from the analysis because they may have abandoned the study for some systematic reason that might be related to the condition under study, and excluding them from the analysis would bias the study.

One approach to this problem is the **person-years method**. This consists of pooling together all the follow-up irrespective of the subjects. The result of the sum of all follow-up periods does not convey information on how many subjects were followed, and therefore the designation of person-years. The number of events is also summed, and its division by the number of person-years is the incidence rate. More precisely, the incidence rate estimated with this method is called the **incidence density**.

Figure 4.9 shows an example of a cohort study analyzed with the person-years method. Six subjects were followed over time until the event eventually occurred. This happened in subjects number 2, 3, and 5. The table shows that the sum of the

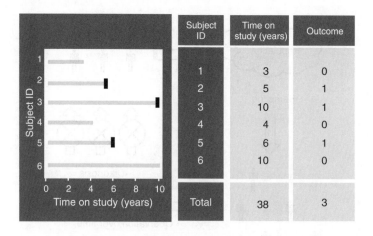

Incidence density: 3/38 person-years

Figure 4.9   Illustration of the person-years method.

follow-up periods of the six subjects was 38 years and that the event occurred on three occasions. The incidence density is, therefore, 3:38 person-years. This number can, of course, be presented in round numbers for easier interpretation, like 79:1000 person-years or 0.079 person-years or 7.9% person-years. The interpretation is straightforward: an incidence density of 7.9% person-years means that, if we observe 100 persons for one year, we expect to observe 7.9 events during that period. If the event can occur several times in the same subject, for example, an adverse drug reaction, that value can also be interpreted as the number of events we expect to observe on one person followed for 100 years.

This method conserves information and avoids excluding incomplete observations from the analysis. However, bias may be introduced if drop-outs are more likely to occur in those that are at greater or lesser risk of developing the event, which is often the case. Therefore, this method should be used only if the number of individuals dropping out from follow-up is relatively low, a commonly accepted level being no more than 15% of the total sampled. We will return to this subject with further details on the analysis of events with cohort studies in the chapter on longitudinal studies.

4.10 Non-probability sampling in descriptive studies

Sometimes we need information on the characteristics of populations, not for scientific purposes or clinical decision making, but simply because we wish to gain some insight on the features, behavior, and patterns in the population that will give us additional clues to the etiopathogeny, pathology, clinical presentation, or clinical course, with the hope that such insights will lead to additional hypotheses about the population that may be verified in further research based on probability samples. Because of their exploratory nature, the cost of conducting such studies in representative samples is often not justifiable and investigators turn to non-probability samples.

For example, a common design of incidence studies consists of sampling physicians or health care centers, estimating the population covered by those units, and recording all new cases of the disease occurring during a given time interval, typically one full year. Suppose 20 primary health centers had been enrolled in the study, each one providing primary care for a defined population, the sum being the total population surveyed in the study, say, 150 000 inhabitants. During a whole year, every new case of the disease was recorded, for example, 75 cases. Therefore, the incidence rate is 1:2000 per year, or 0.05% per year. If the average duration of the disease is eight years, the approximate prevalence rate is 0.4%.

This study design is not as reliable as cohort studies based on probability sampling, because there is no way of knowing whether all new cases were indeed intercepted by the study centers. Inevitably, there will be cases where people will go to other centers or private offices, or they may choose not see a doctor at all. On the other hand, because the subjects are seen at the doctor's office, there are more conditions for doing a thorough evaluation of the disease including follow-up examinations and, therefore, to make a more accurate diagnosis than is usually possible with population-based cohort studies. In addition, this design allows large amounts of clinical data to be collected from each case, which can be used for many different types of analyses, including the development of disease models. Such is the aim of **centralized patient registries**, a methodology widely applied for the study of diseases that is often based on the non-probability samples.

4.11 Standardization

We saw previously that the analysis of observational studies is mainly based on the presentation of summary data, point estimates of prevalence or incidence, and the corresponding standard errors or 95% confidence intervals. Epidemiological descriptive studies of the general population, however, often include additional analyses.

Epidemiological studies usually present estimates of prevalence or incidence rates for the whole population, called the **crude rate**. Because of the dependence of morbidity and mortality indexes on age and gender, it is usual and desirable that separate estimates for males and females, and for each age group, also be presented. These are called **specific rates**. Therefore, a study usually presents, for example, the crude mortality rate and age and gender-specific mortality rates. Of course, specific rates can be obtained and presented for any population variable, not just age and gender.

A frequent analysis in epidemiological studies is adjustment. **Adjustment** is an analytical method used when one wants to compare crude rates between two or more different populations. Because of the dependence of disease on age and gender, a direct comparison of crude rates between two populations can be misleading if they are not similar in age and gender distribution. Adjustment allows us to estimate expected crude rates if the two populations had exactly the same age and gender structure. In epidemiology, the procedure to calculate adjusted rates is called **standardization**.

For example, suppose we wish to compare the crude mortality rate between two regions, called the South and North. The crude annual mortality rate in the South is 300 per 100 000 inhabitants, and in the North 430 per 100 000 inhabitants. The mortality rate is higher in the North, that is a fact, but this may not mean that people in the North have worse health. They could just be older than Southerners. Therefore, we must calculate age-adjusted mortality rates to remove the effect of age on mortality and thus be able to compare mortality rates directly.

In the method of **direct standardization**, we first define the age groups. Then, we use one of the populations as the reference population, for example, the South. In this population we calculate the relative frequency of each age group, as shown in the table in Figure 4.10. In the other population, the North,

	South (reference)				North				
age	pop ×10³	%	deaths	rate (%)	pop ×10³	%	deaths	rate (%)	DAMR
0-4	70	14	210	0.3	80	8	160	0.2	280
5-14	110	22	220	0.2	160	16	160	0.1	220
15-24	100	20	100	0.1	180	18	180	0.1	200
25-34	90	18	180	0.2	150	15	300	0.2	360
35-44	70	14	210	0.3	140	14	420	0.3	420
45-54	30	6	180	0.3	130	13	780	0.6	360
55-64	20	4	200	1.0	90	9	900	1.0	400
>64	10	2	200	2.0	70	7	1400	2.0	400
Total	500		1500		1000		43000		2640
mortality rate (%)			0.30				0.43		0.26

DAMR: direct age-adjusted mortality rate

Figure 4.10   The direct standardization method.

we estimate age-specific mortality rates. We can now estimate what would be the expected mortality rate of the North population if the age distribution were the same as in the South population. It would mean that 0.2% of the age group 0–4 years old, representing 14% of the population, died, plus 0.1% of the age group 5–14 years old, representing 22% of the population, died, and so on until the last age group. In other words, we multiply, in each age group, the age-specific mortality rate of the North population by the relative frequency of that age group in the South population, and sum the result over all the age groups

The final result is the **direct age-adjusted mortality rate**, which we can now compare to the crude mortality rate in the reference population. In this example, the age-adjusted mortality rate in the North population was 264 per 100 000 inhabitants, well below the crude mortality rate of 300 per 100 000 inhabitants in the South. Therefore, it turned out to be the South population that had worse health.

Whenever we wish to compare two rates estimated in two populations it is of major importance to adjust the estimates by population attributes that may influence those rates, because if the distribution of these attributes is not identical in the two populations, direct comparison of crude rates may lead to the wrong conclusions. This is called **confounding** and we will cover this subject in more detail further on. For now, let us just say that, in observational studies, every time we wish to compare estimates from different populations we must consider the need to adjust for one or more population attributes.

This same method is used to make **projections** of population estimates to different populations. For example, if we wish to transpose the results of a prevalence study conducted in a country in order to estimate the prevalence of a disease in a different country, we can use the direct standardization technique to calculate age-adjusted rates, using the age-specific prevalence rates of the country in which the study was conducted and the distribution by age groups in the country to which we want to project the prevalence rates. The example of Figure 4.10 also illustrates, therefore, the projection of mortality rates of the North population to the South population.

A more common problem in epidemiological research is the comparison of morbidity or mortality rates estimated for a segment of the population to the corresponding rates in the general population knowing that the age structure of both populations is different. A situation such as this is typically seen in studies of occupational health, where one might wish to compare disease rates seen among workers exposed to some environment to the rates seen in the general population.

In such problems, the direct standardization method might not be adequate if the estimates of age-specific rates had been obtained from a relatively small sample, because then they would have low precision. In the **indirect standardization** method, the problem with the small number of elements in each age group is avoided by calculating instead what would be the expected number of events if they occurred in each age group with the same frequency as in the general population. The observed and expected number of events can then be directly compared.

For example, consider a study in which we want to compare the annual mortality rate of workers in a certain type of manufacturing activity to the annual mortality rate of the general population. The latter was obtained from official statistics for the entire population and the former from a descriptive study based on a probability sample of workers in that industry. This sample consisted of 6800 workers, and we have the age distribution and the number of deaths occurring during one year in each age group. On the other hand, we have the age distribution of the general population and the age-specific mortality rates. As shown in Figure 4.11, if workers were subject to the same mortality rate as the general population, we would expect that 0.1% of the 950 workers in age group 15–24 years old had died. That is, the expected number of deaths in that age group was 0.95 workers. In the age group 25–34 years old, 0.1% of 2100 workers, or 2.1 workers, were expected to have died. And so forth. In the end, if the mortality among workers were the same as in the general population, we would expect 24 deaths.

The observed number of deaths among workers was 42, that is, 175% of the mortality rate of the population, which suggests an excess mortality among workers.

age	REGION population (a)	deaths (b)	age-specific mortality rate (c=b/a)	FACTORIES n (d)	deaths	expected deaths (e=c×d)
15-24	720,00	720	0.001	950	1	0.95
25-34	600,000	600	0.001	2100	1	2.10
35-44	580,000	1160	0.002	1600	3	3.20
45-54	590,000	4130	0.007	1250	13	8.75
55-64	540,000	5400	0.010	900	24	9.00
Total					42	24.00

Figure 4.11   The indirect standardization method.

The ratio of the observed mortality to the expected mortality, 175% in the example above, is called the **standardized mortality ratio**, a statistic so frequently used in epidemiology that it is usually referred to only as SMR. As the SMR in our study was obtained by sampling, then it is of course subject to sampling variation. Therefore, we must find confidence limits for the SMR. Approximate 95% confidence limits can be obtained by the estimated SMR plus or minus 1.96 multiplied by the quotient of the square root of the observed frequency and the expected frequency, that is, $SMR \pm 1.96 \times \sqrt{(O/E)}$. In the previous example, the result would be 122 to 228%. Given this result, we could say with 95% confidence that mortality was increased among workers.

5

Analytical studies

5.1 Design of analytical studies

We have seen that descriptive studies are an essential instrument to document in detail the various features of the condition being studied. However, at a very early stage in our efforts to describe and understand a given clinical condition we may be interested in exploring associations between variables. The importance of identifying associations is, as was discussed in Section 1.3, because the presence of an association represents the first condition for establishing a cause–effect relationship. Understanding how the various factors present are related and the strength and direction of that relationship allows us, for example, to develop disease models that can be used for several purposes, such as classification, prediction, and identification of opportunities for intervention. The objective of analytical studies is, primarily, to uncover those relationships. Some analytical studies, but not all, also allow us to establish the order factor, which informs us not only of the existence of an association, but also of its direction.

For example, suppose we wish to investigate whether there is an association between body mass index and respiratory development in children. For this purpose, we conduct a cross-sectional survey in a probability sample of children and in each one we measure the body mass index and perform a laboratory test of respiratory function. Further suppose that, while inspecting the data, we found that children with an abnormal low value in the respiratory test had on average a lower body mass index than those with a normal test. This observation may allow us to conclude, under certain conditions that we will discuss later, on an association between respiratory function and body mass in children. However, this study will not allow us to understand the direction of that association because evidence of an association simply means that two attributes go together, not which one, if any, determined the other (unless one is a terminal event, like death). It could be that impaired respiratory development causes children to gain

Biostatistics Decoded, First Edition. A. Gouveia Oliveira.
© 2013 John Wiley & Sons, Ltd. Published 2013 by John Wiley & Sons, Ltd.

less weight, or that poor physical development in children causes a slower rate of pulmonary development, or that some other factor causes both poor physical and respiratory development.

This type of study is called a **cross-sectional analytical study**. It is not unlike the cross-sectional descriptive study regarding design and sampling methodology, but it is very different in three essential aspects: its objective, which is specifically to discover eventual associations between respiratory function and one or more patient attributes; its dataset, which includes measurements of patient attributes suspected of being associated with respiratory development; and its sample size requirements, which may not be the same as those for a descriptive study.

If we wish to understand the sense of an association we need a different study design. We need a design that allows us to evaluate an order factor. A **cohort analytical study** may, in some situations, provide evidence of an association and an order factor. For example, suppose we want to investigate an eventual association between body mass index and stroke in men. We would select a random sample of men over 40 years old who have never had stroke and observe each one of them over a period of time until a stroke occurred. If the body mass index was on average higher in those who suffered stroke, we would be inclined to conclude on an association between body mass and stroke and also, because of the study design, that stroke could not possibly have preceded weight gain. Again, we could not establish causality because there could be another factor that caused both overweight and stroke.

As with descriptive cohort studies, analytical cohort studies are used when one wants to investigate possible associations between events and patient attributes. The study design is similar to that of a cohort descriptive study except for the objectives, data collected, and sample size.

In situations where we are investigating eventual associations between a single characteristic and one or more attributes, and that characteristic can be represented by a binary variable, like the presence of a given disease state or the occurrence of some event, the existence of associations between that variable and each of the attributes is established by splitting the total sample into two groups, with and without the attribute, each group representing a random sample of the population with and without the variable, respectively. The mean value of each attribute can then be compared between the two populations and, if a difference is found, one can conclude on the existence of an association between the attribute and the variable being investigated. We will shortly present the statistical methods that allow one to conclude on the existence of an association between a binary variable and one attribute. For the moment, we will discuss some study designs that are more efficient than the previous ones.

Assume we want to search for possible etiological factors of some disease, for example, chronic hepatitis C. We will conduct a study with the specific aim of investigating the association of several attributes of individuals with the presence of chronic infection by the hepatitis C virus and, therefore, an analytical study is required. Consequently, we design a cross-sectional study on a probability sample of the general population and we decide on a sample size of, say, 6000 subjects. Given that the prevalence of this disease, as is often the case with many diseases, is rather low and on the order of 1.5%, in the end we will have collected data from about 90 cases of chronic hepatitis C. In the analysis we would compare the means of several attributes of the subjects between the small sample of 90 cases and the huge sample of 5910 non-cases.

In a situation like this, a **case–control study** would be much more efficient. In such a study we select, independently, one random sample from the population of cases and one random sample from the population of non-cases. In this fashion we could define the sample size of the two samples, called cases and controls, in a way that would better serve the objectives of the study. Just to give an idea of the savings achieved with a case–control design over a cross-sectional design, we could design a case–control study equivalent to the one in the example above in terms of its ability to show associations, with only 185 cases and 185 controls, that is, with only 6% of the sample size of a cross-sectional study.

Case–control studies are also called **retrospective studies** because, as subjects are observed only once and data about the disease and a set of putative contributing factors are collected at the same moment, then one of them must have existed before the observation took place. In the example of the study on chronic hepatitis C, the presence of the disease is directly observable and verifiable, but information on putative causes, for example, blood transfusions, needle stick injury, dental procedures, etc., can only be obtained by recall, and recall data is prone to inaccuracies.

Stronger evidence of an association between exposure to putative contributing factors and a given disease state may be achieved with a **cohort study**. If the investigation concerns only one factor, if this factor can be represented by a binary variable, and if its prevalence in the population is very low, then we may select approximately equal-sized random samples, one from the population with the factor and the other from the population without the factor, thus saving a large number of observations. The two cohorts of exposed and non-exposed to the factor will then be observed repeatedly for a period of time, allowing for the disease to develop. We can then verify whether the frequency of the disease is different in the two populations and eventually conclude that an association is likely to exist. Both cohorts are selected from the population without the disease, and therefore this study may also provide evidence of an order factor.

This design is also called a **prospective study** because data is collected after the study starts. Therefore, recall is minimized or non-existent and the data is in general of much higher quality than in case–control studies. On the other hand, as pointed out earlier, cohort studies are costly, difficult to manage, subjects have a tendency to drop out from the study, and the investigation often takes a long time. Nevertheless, properly designed and conducted cohort studies are generally considered the most reliable, accurate, robust, and powerful methodology in observational research.

Analytical studies are of major importance to epidemiological and clinical research. The case–control study is prevalent in medical journals, because it is relatively easy to manage and conduct and often provides evidence of associations that may be worth pursuing in experimental studies. Often the greatest difficulty in setting up a case–control study is the selection of appropriate controls.

As a rule, controls should be sampled from the same population as cases, and in some investigations this may require a probability sample of the general population. For example, in a case–control study to investigate the association of patient attributes with hospital survival from an episode of acute pancreatitis, it would be easy to obtain samples of cases and controls because both samples would be selected from the population of patients admitted to the hospital with acute pancreatitis. However, if the aim of the investigation were to identify associations between subject attributes and the occurrence of an episode of acute pancreatitis, then a clinical investigator would have no major difficulty in obtaining a sample of cases from hospital admission for acute pancreatitis, but in order to get the controls a population survey would be required.

To circumvent this difficulty, a partial solution sometimes employed is the **matched case–control study**. In this design, controls are selected at random from the general population according to a set of predefined attributes, for example, age and gender, each case having a corresponding control with the same values in that set of attributes.

A more reliable alternative which is gaining in popularity is the **nested case–control study**. In this design, a cohort is assembled from a random sample of the target population. The subjects are observed regularly and, each time a case emerges, the investigator selects a control at random from the subjects still at risk. For example, assume that we want to investigate patient attributes associated with nosocomial infections. In a nested case–control design we would select a random sample of patients admitted to hospitals and observe this cohort. Whenever a patient developed a nosocomial infection we would select a control at random from the remaining sample. The name of this design stems from the fact that it is a case–control study nested within a cohort.

Compared to the standard cohort study, the nested case–control study has the benefit of requiring the measurement of attributes only in those subjects that are selected as controls rather than in the entire sample, with substantial savings in the cost of the investigation. On the other hand, the population for analysis will be much smaller and the power of the study to identify associations will be significantly reduced. Compared to the case–control study, the nested case–control design has the benefit of a control sample comparable to the cases and of data being collected prospectively, but at a higher cost than in the standard case–control study.

Another approach to selecting controls is the **case–crossover design**, which uses each case as its own control but has the limitation that it can be used only when one wants to investigate the immediate effects of an exposition. For example, assume we want to investigate whether an occasional intake of non-steroidal anti-inflammatory drugs (NSAIDs) increases the risk of acute gastrointestinal bleeding. In a case–crossover study we would select individuals admitted to a hospital with an episode of acute gastrointestinal bleeding, and exclude those with chronic administration of NSAIDs. Each patient would be asked whether a NSAID had been administered in the period of time immediately preceding the episode, say, during the previous 24 hours. This exposure would be compared to the exposure of the same individual in a previous time period, which would be constant for all cases, say, in the day preceding the bleeding episode by two weeks. If the frequency of exposure to NSAIDs on the day of the hemorrhagic episode was significantly higher than on the day serving as the control, then an association between occasional NSAID intake and gastrointestinal bleeding could be established. This design, however, would not allow us to conclude whether chronic administration of NSAIDs would lead to increased risk, in the long run, of upper gastrointestinal bleeding, hence the general designation of **immediate effects model** for this kind of design.

5.2 Non-probability sampling in analytical studies

A major difference between descriptive studies and analytical studies is that the purpose of the former is the estimation of population parameters, while in the latter this is usually not the focus of the research. Rather, the emphasis is on uncovering relationships among variables. On occasion we are interested in estimating the true difference between population means, but in the vast majority of analytical studies this is not the primary research aim. The primary aim is to obtain evidence that an association is very likely to exist.

It is generally accepted that, although the means and proportions of attributes may vary among different segments of a population, the direction and strength of an association between variables does not vary greatly. For example, in a study to investigate the association between glycemic control and retinal changes in diabetics, it is very possible that the proportion of diabetic patients with glycemic control and the proportion with retinal changes vary among the populations observed in different health care centers, but the relationship between the two variables is unlikely to vary much.

This is the reason why non-probability samples are often used, and generally regarded as valid, in analytical studies. Consequently, analytical studies are regularly conducted in health care facilities, with samples recruited from the population of patients attending those centers. Because subjects are selected for the sample because they are easy to access and not through a random process, this sampling method is called **convenience sampling**.

Convenience sampling can be used in descriptive surveys and in analytical studies but, for the reasons explained above, convenience samples are considered much less reliable in descriptive studies than they are in analytical research, provided that some assumptions hold. The basic assumptions are, accepting that relationships between variables do not vary appreciably between health care centers, that the center where the research is being conducted is no different in any particular way from any other health care center, and that patients observed in that center are also no different in any particular way from patients attending other centers.

If all these assumptions are true, then observations of relationships between patient attributes in one health care center could be regarded as random and representative of the relationships in the population, on the condition, of course, that the sample was itself random.

Therefore, patients have to be selected for the sample according to one of the methods described earlier, that is, by simple, systematic, or consecutive sampling.

By far the most common sampling method is consecutive sampling, whereby all patients that belong to the target population that attends the health care center are enrolled into the study. Systematic sampling should in general be avoided because of its sensitivity to periodicities in the sample.

5.3 The investigation of associations

We saw in Section 5.1 an association between two variables was intuitively established when one observed that the values taken by one of the variables were not the same for all values of the other variable. For example, in a case–control study we want to investigate an association between the binary variable that classifies subjects as cases or controls, say, the presence of a given disease, and an attribute of the subjects. We would say that an attribute, for example, age, was associated with the disease if the mean age were different in cases and controls. If the values of the attribute age were unrelated to the values of the attribute disease, then there would be no reason for the mean age to be different between subjects with and without disease.

If we have two binary variables, an association is identified if the proportion of individuals with one attribute is different between the two groups defined by the values of the other attribute. Gender and disease are examples of two binary attributes. An association between gender and disease exists if the proportion of one gender is different between those with and without the disease or, essentially the same, the proportion of individuals with the disease is different in males and females.

Therefore, what we have here is a general rule for establishing an association between attributes. We say there is an association when we find evidence that the mean value taken by one attribute is different for different values taken by the other attribute.

Whenever we look at the results of a case–control study, or for that matter for the results of any analytical study, we must always keep in mind that the observed means and proportions were obtained from samples and that, because of the sampling variation, means and proportions will always be different between the samples of cases and controls. We need, therefore, a method that allows us to decide whether the observed difference between sample means and proportions can be attributed to a true difference between the means and proportions in the two populations, because only then will we have evidence of an association, or that can be explained by random sampling variation. In other words, we need to estimate the true difference between population means and proportions using the results observed in our samples.

In the next sections we will see how statistical methods can help us say how likely it is that an association exists between a binary and an interval variable, and between two binary variables. Afterward, we will see how to investigate associations between two variables with all the combinations of scales of measurement: binary with categorical and ordinal; categorical with categorical; ordinal and interval; ordinal with ordinal and interval; and interval with interval. Finally, we will see how to investigate an association between binary, categorical, and interval variables with a special type of variable, events.

5.4 Comparison of two means

Suppose we want to investigate whether an association exists between diabetes mellitus and total serum cholesterol levels. We design a study and take a sample of 86 individuals with diabetes mellitus and a sample of 90 controls without the disease. In the first sample, the mean total serum cholesterol level was 250 mg/dL with standard deviation 42 mg/dL and, in the second, 230 mg/dL with standard deviation 38 mg/dL. Therefore, as we are studying a binary variable (diabetes) and an interval variable (total serum cholesterol), the appropriate method for showing an association between the two variables is a comparison of cholesterol means between the two populations. If the population means are equal, we cannot conclude on an association, but if they are different, then we will have evidence of the association. In other words, if the difference between the population means is zero, then we will not conclude on an association, but if the difference is not zero, we will say an association is likely. The problem, then, resides in the estimation of the true difference between the two population means.

In this example, the difference between the two sample means is 20 mg/dL. However, if we repeated the study with two other samples, we would certainly obtain another value for the difference between sample means. Because sample means are subjected to sampling variation, so are their differences. Therefore, the difference between sample means is also a random variable. Accordingly, we can use our data to estimate the true value of the difference between population means, as we did when we used our data to estimate a population mean.

For this, the first thing we must do is to ask ourselves what can be said about the distribution of differences between the means of samples obtained from two populations.

If the samples are large, we know from the central limit theorem that both sample means come from normal distributions. From the properties of the normal distribution we know that the difference between variables with a normal distribution also has a normal distribution. Therefore, the differences between means of large samples must have a normal distribution.

We also know that the mean value of sample means is equal to the population mean. From the properties of means we know that when two random variables are subtracted, the mean of the resulting variable is equal to the difference between the means of the two variables.

Returning to our example of total cholesterol and diabetes, from the above considerations we know by now that the difference of 20 mg/dL between our sample means is an observation from a random variable with a normal distribution. We also know that the mean of that normal variable is equal to the true value of the difference between the two population means of total cholesterol, which is the quantity we want to estimate.

Therefore, we already know the distribution of the differences between sample means and the value of one of its parameters, the mean. All that remains for the complete description of that distribution is to know the value of the other parameter, the variance.

As the two samples are independent, we know from the properties of variances that when two variables are subtracted, the variance of the resulting variable is the sum of the variances of the two variables. Remember that the two variables here are sample means, so their standard deviation is called the standard error of the mean. The value of each standard error is equal to $\sigma/\sqrt{n}$ and the variance of sample means is, thus, equal to σ^2/n. Therefore, when we subtract the two variables, the variance of the resulting variable is equal to the sum of their variances, that is, $\sigma_1^2/n_1 + \sigma_2^2/n_2$. The square root of this quantity is the standard error of the difference between sample means.

In conclusion, for the value of 20 mg/dL, the difference between the means of two large samples, we can say that it is an observation on a random variable with the following properties:

- It has a normal distribution.
- Its mean is the difference between the two population means.
- Its standard error is the square root of the sum of the two variances of the sample means.

Because the samples are large, each variance of the sample means can be estimated with good accuracy from our data using the sample variance divided by the sample size.

Therefore, the standard error of the difference between sample means can be written

$$SE = \sqrt{\frac{\sigma_1^2}{n_1} + \frac{\sigma_2^2}{n_2}}$$

and can be estimated from the sample data, with negligible error, by

$$SE^* = \sqrt{\frac{s_1^2}{n_1} + \frac{s_2^2}{n_2}}$$

Now we have everything we need to analyze the problem presented in the beginning of this section (Figure 5.1). To estimate the standard error of the difference between means of large samples, we sum the square of 42 divided by 86 and the square of 38 divided by 90, obtaining 36.56. The square root of this quantity, 6.05 mg/dL, is the estimate of the standard error.

We know, because we have a normal distribution, that 95% of the differences between sample means will be less than 1.96 standard errors away from the difference of population means. We multiply 1.96 by the standard error and obtain 11.85 mg/dL. Therefore, the difference we observed between our samples means

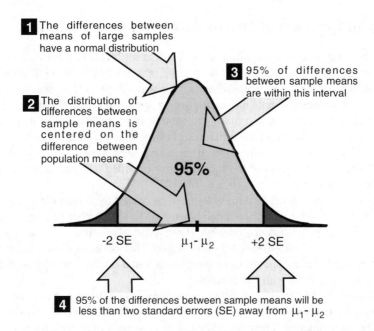

1 The differences between means of large samples have a normal distribution

2 The distribution of differences between sample means is centered on the difference between population means

3 95% of differences between sample means are within this interval

95%

-2 SE μ_1- μ_2 +2 SE

4 95% of the differences between sample means will be less than two standard errors (SE) away from μ_1- μ_2

Figure 5.1   Steps in the construction of 95% confidence intervals for the difference between two population means in the case of large samples.

has a 95% probability of being less than 11.85 mg/dL away from the true difference between population means. Consequently, with 95% probability, the true population difference must be a value no smaller than 8.15 mg/dL (20 mg/dL − 11.85 mg/dL) and no greater than 31.85 mg/dL (20 mg/dL + 11.85 mg/dL).

We conclude that with a probability of 95% the interval 8.15 mg/dL to 31.85 mg/dL contains the true value of the difference between population means. This also tells us that, if the two population means were equal, the probability of obtaining a difference of 20 mg/dL or larger between sample means would be less than 5%. Consequently, our observations are not consistent with the hypothesis of equal population means, and we may exclude that hypothesis with a 95% confidence level.

In summary, the method for investigating an association between a binary and an interval variable is to compare the means of the interval variable between the populations defined by the two levels of the binary variable. Confidence limits for the difference in population means are computed, applying the same concepts used for making inferences from sample means. If the confidence interval does not contain the value 0, we conclude on a difference between population means and on an association between the two variables and, in addition, we obtain an estimate of the true size of the difference between population means. If the confidence interval contains the value 0, this means that any of the two populations may have a higher mean, or they have the same mean. Therefore, in that case we cannot conclude anything.

5.5 Comparison of two means from small samples

Let us now suppose that the mean serum cholesterol levels in the example of the previous section were not obtained from large samples, but from small samples of eight diabetics and nine non-diabetics. As was discussed in the section on inferences from small sample means, we cannot apply the same reasoning as if they were large samples. In the first place, the central limit theorem does not apply and, therefore, the distribution of the differences between sample means can be assumed to be normally distributed only if the attribute also has a normal distribution. In the second place, the estimates of the standard error obtained with the sample standard deviation are not accurate.

We have seen that the latter problem can be solved by substituting Student's t distribution for the normal distribution when finding the number of standard errors that define a given confidence interval. However, this solution is not valid for the present situation of differences between means. This is because the t distribution cannot be used if the standard error estimate is based on the sum of two different variances, as was done with large samples.

The problem has a solution if the variances are equal in the two populations. In that situation, the two sample variances can be seen as two independent estimates of the population variance. Therefore, they can be combined to provide a single, better estimate of the population variance. We do this by calculating a weighted mean of the two sample variances using the number of degrees of freedom of each variance, that is, the sample size minus one, as a weighting factor. So, we add the two sample variances, each multiplied by its weighting factor $(n - 1)$, and we divide the total by the sum of the weighting factors $(n_1 - 1 + n_2 - 1$, or $n_1 + n_2 - 2)$.

We use this estimate of the common population variance of the attribute to calculate each variance of the sample means. Then the variance of the difference between sample means can be estimated by adding the two variances of sample means, obtained in both cases with the same estimate of the variance. The square root of that quantity is the standard error of the difference between sample means.

The number of standard error estimates on each side of the difference between population means that define a given confidence interval is obtained from Student's t distribution, instead of the normal distribution. However, because the variance was estimated by combining two independent variance estimates, the number of degrees of freedom of Student's t distribution is now the total number of observations in the two samples, minus two.

Let us return to our example where we wanted to investigate an association between total serum cholesterol and diabetes using a sample of eight diabetics and nine controls. In order to be able to find the 95% confidence interval for the difference between the means of the two populations we will have to assume that the distribution of total serum cholesterol is normal and that its variance is equal in both populations.

If the variance is equal in both populations we only need one estimate of the population variance. Using both sample variances s_1^2 and s_2^2 we compute an estimate of the population variance σ^2 of cholesterol with a weighted average of the sample variances, as follows:

$$s^2 = \frac{s_1^2(n_1 - 1) + s_2^2(n_2 - 1)}{(n_1 - 1) + (n_2 - 1)}$$

$$s^2 = \frac{42^2 \times 7 + 38^2 \times 8}{8 - 1 + 9 - 1} = 1593.3$$

Now we obtain estimates of the variance of the sample means by dividing the estimate of the variance of cholesterol by the sample size. For the diabetics, the variance of sample means is estimated as 1593.3/9 and for the controls as 1593.3/8. The estimate of the standard error of the difference between sample means is the square root of the sum of the two quantities:

$$SE^* = \sqrt{\frac{s^2}{n_1} + \frac{s^2}{n_2}}$$

The result is 19.4 mg/dL and this is the value of the standard error of the difference between sample means estimated from the data in the sample. In Student's t distribution with $8 + 9 - 2 = 15$ degrees of freedom, 95% of the observations are less than 2.131 standard error estimates from each side of the mean.

In our study we observed a difference between sample means of 20 mg/dL. We now know that this value, in 95% of the cases, does not differ by more than $2.131 \times 19.4 = 41.3$ mg/dL from the difference between the two population means. Accordingly, the difference between the two population means must be a value between -21.3 and 61.3 mg/dL.

In other words, mean serum cholesterol level may be lower in diabetics, down to 21.3 mg/dL, or may be higher, up to 61.3 mg/dL. Since the value 0 is within this interval, serum cholesterol levels may also be equal in diabetics and non-diabetics. Our study, therefore, is inconclusive.

In summary, for the case of small samples an interval estimate of the difference between population means can be obtained if, in both populations, the distribution of the variable is normal and the variances are equal. In that situation, a single estimate of the variance can be obtained by combining the two variances and used for estimating the standard error of the differences between sample means. The t distribution with the appropriate degrees of freedom (total number of observations minus two) will give the number of standard error estimates on each side of the mean that include the desired proportion of observations (Figure 5.2). The confidence limits are found in the usual way.

To conclude, it is convenient to note that, if the attribute has a normal distribution, then it is usually safe to assume that the two population variances are approximately equal. Furthermore, even if the population variances are not equal, the method just presented is still valid as long as the two samples are approximately of equal size. It is also valid if the distribution of the variable is not normal,

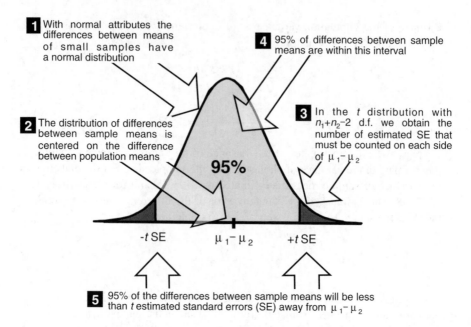

Figure 5.2   Steps in the construction of 95% confidence intervals for the difference between two population means in the case of small samples.

provided that the departure from the normal distribution is not too great. Because this method is still valid for conditions where the assumptions are not completely met, it is said to be **robust**.

5.6 Comparison of two proportions

When an association between two binary attributes is being investigated, the appropriate method is to compare the proportion of individuals to one attribute between the two groups defined by the values of the other attribute. For example, an association between gender and a disease is investigated by comparing the proportion of one gender in those with the disease and without the disease or, equivalently, the proportion of individuals with the disease in males and females.

We know that sample proportions are observations from a binomial variable. From the properties of variances, we can estimate the variance of the differences between sample proportions by the sum of the variances of the two sample proportions. We saw earlier that the variance of the sample proportions is the product of the proportion of subjects with the attribute and the proportion of those without the attribute, divided by the sample size. We also saw that in large samples no significant harm is done if we estimate this quantity from our data. Therefore, we can obtain an estimate of the standard error of differences between sample

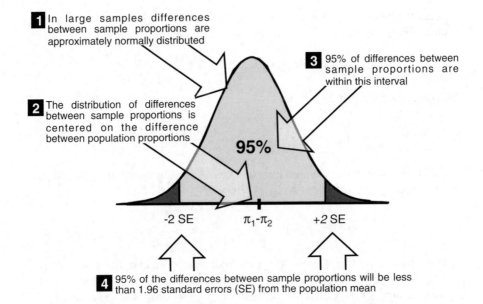

1. In large samples differences between sample proportions are approximately normally distributed

3. 95% of differences between sample proportions are within this interval

2. The distribution of differences between sample proportions is centered on the difference between population proportions

95%

-2 SE π_1-π_2 +2 SE

4. 95% of the differences between sample proportions will be less than 1.96 standard errors (SE) from the population mean

Figure 5.3   Steps in the construction of 95% confidence intervals for the difference between two population proportions.

proportions using the observed sample proportions by adding the two variances of sample proportions and taking the square root.

If the sample sizes are large, we can use the binomial approximation to the normal distribution and obtain from the latter the number of standard errors that must be counted on each side of the mean of the distribution to define a given proportion of observations, typically 1.96 standard errors for 95% confidence intervals. Then the confidence limits for the difference in the population proportions are found by the usual method (Figure 5.3).

For example, consider a sample of 85 individuals with a given disease, and a sample of 70 without the disease. There are 58 males in the first group (67.4%) and 33 in the second (47.1%). We want to find confidence limits for the difference in the proportion of males in the populations with and without the disease, to understand whether the observed difference would be likely if the population proportions of men were equal.

The observed difference in the proportion of males is 20.3 percentage points. The sample sizes are adequate for using the binomial approximation to the normal distribution. We first estimate the standard error of the sample differences with our data. The variance of sample proportions is

$$var(P) = \frac{\pi(1-\pi)}{n}$$

According to the properties of variances, the variance of the difference between sample proportions $P_1 - P_2$ is the sum of the variances of P_1 and P_2:

$$var(P_1 - P_2) = \frac{\pi_1(1 - \pi_1)}{n_1} + \frac{\pi_2(1 - \pi_2)}{n_2}$$

The standard error of $P_1 - P_2$ is

$$SE(P_1 - P_2) = \sqrt{\frac{\pi_1(1 - \pi_1)}{n_1} + \frac{\pi_2(1 - \pi_2)}{n_2}}$$

and an estimate of the standard error can be obtained by substituting p_1 and p_2 for π_1 and π_2. The estimated standard error of $P_1 - P_2$ is

$$SE^*(P_1 - P_2) = \sqrt{\frac{p_1(1 - p_1)}{n_1} + \frac{p_2(1 - p_2)}{n_2}}$$

The variance of the sample proportion of males among the diseased is $0.674 \times 0.326/85$, and among the non-diseased it is $0.471 \times 0.529/70$. The standard error of the difference between sample proportions is the square root of the sum of the two variances. Using the data we collected the result is 0.0784 or 7.84%.

Because we are using the normal approximation, the lower limit of the 95% confidence interval is the observed difference minus 1.96 standard errors, that is, 20.3% minus 15.4% or 4.9%. The upper limit is 20.3% plus 15.4%, or 35.7%. The confidence interval does not include zero, and therefore we reject with 95% confidence the possibility of equal population proportions and conclude on an association between gender and the disease.

5.7 Relative risks and odds ratios

Another way of expressing the differences between two quantities is to take their ratio. If the ratio is 1, then the two quantities are equal; if it is greater than 1, then the first quantity is larger, otherwise the ratio is smaller than 1. Ratios are not commonly used for comparing interval variables, but they are very often used for proportions.

Consider a study for evaluating an association between a binary attribute and a disease. We want to investigate whether the proportion of diseased individuals is different between those having the attribute and those who do not. The ratio of the two proportions tells us how many times the disease is more frequent in those with the attribute than in those without it. For example, if the ratio is 3.5, this means that the disease is 3.5 times more frequent in those with the attribute than in those without it. If the ratio is 0.2, this means that the disease is five times less frequent in those with the attribute. If the ratio is 1, then the two proportions are the same.

Of course, we can also take the ratio of the prevalence of the attribute in the diseased and non-diseased and obtain a measure of association between the

Exposure \ Disease	with COPD	without COPD	Total
ever-smokers	136	4387	4523
never-smokers	72	7980	8052
Total	208	12367	12575

Figure 5.4   Data from a prevalence study of chronic obstructive pulmonary disease (COPD) in the general population.

attribute and the disease. But since we are usually interested in the disease, not in the attributes, it would probably not be very informative to know how much more frequent the attribute is in the diseased. For example, consider the data in Figure 5.4 from a prevalence study of chronic obstructive pulmonary disease (COPD) on a random sample of the general population. The proportion of ever-smokers among the diseased is 65.4% and among the non-diseased 35.5%. The ratio is 1.8, meaning that it is 1.8 times more likely to find a smoker among patients with COPD than in people without the disease. On the other hand, the proportion of patients with COPD among the ever-smokers is 3.0%, while 0.9% of the never-smokers have the disease. The ratio is 3.3, meaning that COPD is 3.3 times more likely in an ever-smoker than in a never-smoker. Both ratios are the same measure of association, but the latter is of greater clinical interest as it is usually the one that we care about.

The ratio of two proportions is called the **risk ratio** or, more commonly, the **relative risk**. It is extensively used in epidemiology as a measure of association. In epidemiological terminology, therefore, the relative risk is said to be the ratio of the prevalence of disease among the exposed to the prevalence among the non-exposed. In our example, 3.3 is the relative risk of COPD among ever-smokers to never-smokers.

An alternative way of presenting proportions is as **odds**. Proportions are the ratio of positive events to the total number of events, while odds are the ratio of positive events to negative events. For example, if in 60 observations we have 20 positive events, the proportion is 0.33 (20 : 60) and the odds are 0.50 (20 : 40).

We can divide the odds of having the disease among those with the attribute by the odds of having the disease among those without the attribute, and obtain a measure that is called the **odds ratio**. The odds ratio, therefore, is the ratio of the odds of disease among the exposed to the odds of disease among the non-exposed. In our example, the odds of COPD among the exposed (ever-smokers) is $136/4387 = 0.031$ and among the non-exposed (never-smokers) $72/7980 = 0.009$. The odds ratio of COPD among smokers is $0.031/0.009 = 3.4$. The odds ratio is used extensively not only in epidemiology, but in clinical research as well.

As we have just seen, the relative risk is the ratio of two prevalences or two incidences. Therefore, this measure can be estimated only from study designs where it is possible to estimate the prevalence or incidence of the cases among the exposed and the non-exposed. Estimates of disease prevalence in the two populations can be obtained in cross-sectional and cohort studies, and relative risks be estimated from these designs, as well as odds ratios. However, in case–control studies we do not have such estimates of disease prevalence. Instead, we have estimates of the proportion of the exposed among those with the disease and among those without the disease. Therefore, we can estimate the odds ratio of the exposure in the diseased to the non-diseased. Fortunately, the odds ratio of exposure in the diseased to the non-diseased is the same as the odds ratio of disease in the exposed to the non-exposed. Furthermore, if the disease has a low prevalence (as is usually the case) the odds ratio is very similar to the relative risk. This is why the odds ratio is also called the **approximate relative risk**.

We can easily verify these properties of the odds ratio. In our example on COPD, the study identified 136 people with the disease among the 4523 exposed and 72 people with the disease among the 8052 non-exposed. The odds of exposure among the diseased are 136:72, among the non-diseased 4387:7980, and the odds ratio of exposure among the diseased to the non-diseased is 3.44. Conversely, the odds of disease among the exposed are 136:4387, among the non-exposed 72:7980, and the odds ratio of disease among the exposed to the non-exposed is again 3.44. Let us now find the relative risk. The prevalence of disease among the exposed is 136:4523 and among the non-exposed 72:8052. The relative risk is, therefore, 3.36, which is very close to the value of 3.44 we obtained for the odds ratio.

These properties of the odds ratio make it a very useful measure of association, in particular because it can be estimated in all types of analytical studies, including case–control studies, which are by far the most common analytical studies in clinical research. Furthermore, the odds ratio has a number of mathematical properties that make it an extremely important quantity in biostatistics.

5.8 Attributable risk

We saw in the previous section that, when one identifies an association between a disease and a factor, relative risks and odds ratios answer the question of how much that factor increases the likelihood of disease. Another question one would normally ask is how much of the disease prevalence (or incidence) is explained by that factor. If the answer was, say, 20%, then we know that we have to search for many other etiological factors. Conversely, if the answer was 90% then we have an important indication that almost all cases could be explained by that factor.

The **attributable risk** was devised precisely to give us that information. It consists of a measure of the proportion of cases that are accounted for by a factor. The attributable risk is also called the **attributable fraction**, and it can be estimated for the exposed (the attributable fraction among the exposed) and for the total population (the attributable fraction in the population).

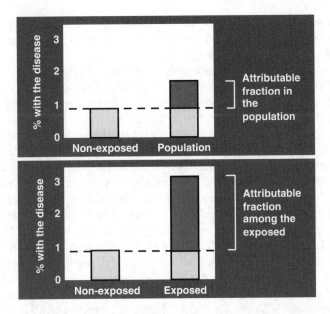

Figure 5.5   Attributable fraction in the population and among the exposed.

The attributable fraction in the population tells us how many cases in the general population are caused by the exposure, assuming of course that there is a causal relationship between the exposure and the disease. This figure is given by the excess prevalence of the disease in the general population, compared to the population that has not been exposed (Figure 5.5, top graph). This gives us a measure of the impact of an exposure on the health of the population.

The attributable fraction among the exposed tells us how many of the COPD cases were caused by the exposure. The reference now is the prevalence of COPD among the non-exposed and this number is compared to the prevalence of COPD among the exposed. The excess prevalence in the exposed must be due to the causative factor (Figure 5.5, bottom graph).

In our example, the prevalence of COPD in the population is $208/12\,575 = 1.65\%$. The prevalence of COPD in the non-exposed is $72/8052 = 0.89\%$. So we have an estimate of the prevalence in the actual population, where some people are exposed and some are not, and an estimate of what the prevalence would be if nobody were exposed. The excess prevalence in the population, $1.65 - 0.89 = 0.76$ percentage points, is therefore attributable to the exposure. This corresponds to a fraction of 45.9% ($= 0.76/1.65$) of the COPD cases in the population. This is the attributable risk in the population.

The attributable risk among the exposed is found by a similar reasoning. The prevalence of COPD in the non-exposed is $72/8052 = 0.89\%$, and the prevalence of COPD in the exposed is $136/4523 = 3.01\%$. Therefore, the exposed have an excess prevalence of $3.01 - 0.09 = 2.12$ percentage points. This corresponds to a fraction of 70.4% ($= 2.12/3.01$) of the exposed, the attributable risk among the exposed.

These results tell us that, assuming a causal relationship between smoking and COPD, about 46% of the cases of COPD in the population are caused by smoking (therefore other factors are responsible for the remaining 54% of COPD cases) and that about 70% of the COPD cases in ever-smokers are caused by smoking (therefore 30% of COPD cases have some other cause).

Shortcut formulas give us a simplified way to find attributable risks from relative risks and odds ratios, which is adequate for mental calculations. The attributable risk among the exposed may be calculated as one minus the reciprocal of the risk ratio, incidence ratio, or odds ratio, whichever is the case. The attributable risk in the population is obtained by multiplying that quantity by the proportion of exposed among the cases. For example, the relative risk estimate is 3.4, meaning that COPD is about 3.4 times more likely in ever-smokers than in never-smokers. The attributable risk among the exposed is one minus the reciprocal of 3.4, or 70.6%, which means that over two-thirds of the cases among ever-smokers are accounted for by smoking. The attributable risk in the population is 70.6% times 65.4% (136/208), which is 46.2%.

5.9 Logits and log odds ratios

We have just seen that we may refer to the likelihood of observing an attribute as a proportion or as odds. Proportions are known to have a binomial distribution which, in turn, converges to the normal distribution when sample size increases. Look at the left graph of Figure 5.6, which shows the frequency distribution of a binary attribute with probability 0.5 in samples of size 40 expressed as proportions.

We saw before how we could make use of those properties of proportions for statistical inference. Odds, however, do not posses such properties. Look at the graph in the middle of Figure 5.6, which shows the frequency distribution of the same attribute, but now expressed as odds. The distribution is highly skewed to the left and will never converge to the normal.

Probability distribution of a binary variable with probability 0.50 in samples of size 40

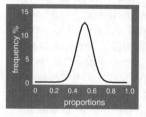 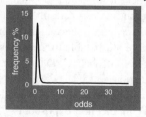 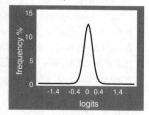

Figure 5.6   Probability distributions of proportions, odds, and logits.

Now look at the graph on the right of Figure 5.6. It shows the distribution of that attribute, but now expressed as the logarithm of the odds. Using the logarithm of the odds, instead of the odds, results in a probability distribution that converges to the normal distribution, exactly as happens with proportions. Therefore, using the logarithm of the odds will allow us to make statistical inferences in the same way as we do with proportions.

The logarithm of the odds is called the **logit**, and we can work with logits in the same way as we do with proportions. For example, because the logits are in a linear scale, we can subtract two logits to find the difference between them. We could not do this operation with odds because they are not in a linear scale.

Logits are important also because the difference between the logarithms of two quantities is equal to the logarithm of the ratio of those quantities. Therefore, the difference between the logits is equal to the logarithm of the odds ratio. This fact allows us, for example, to find confidence limits for odds ratios. We will see further on that it also allows us to obtain a straightforward interpretation of the results of a very important analytical method called logistic regression.

6

Statistical tests

6.1 The null hypothesis

In the previous chapters, we saw how one can use the sample data to create interval estimates of population means and proportions. We do this by finding which values of the population mean or proportion would make the observed sample very unlikely, and exclude those values with known probability.

We also saw how interval estimates allow us to make decisions regarding the plausibility of the existence of a difference between two population means or proportions. We simply verify whether the value 0 is contained in the confidence interval of the difference between population means or proportions, and we conclude on a difference if this value is not contained in that interval.

Therefore, interval estimates allow us to determine the size of the difference that is likely to exist between two populations and also give us a decision rule to decide upon the existence of a difference. However, there are occasions when we may be interested in establishing whether or not there is a difference between populations but the size of the difference is of no particular interest to us.

The statistical tests that we will discuss in the next sections are nothing more than an expeditious method to make a decision about the existence of a difference between population means or proportions. All statistical tests are based on the same principle, which is to evaluate the plausibility of a particular hypothesis about the populations taking into account the results observed in the samples. The tests produce a result expressed in terms of the probability of obtaining a difference between the samples, such as the one we observed if the hypothesis we made was true. If that probability is less than a predefined value we reject the hypothesis.

If we want to operationalize this approach and develop a decision rule, we first need to define the hypothesis that we will evaluate. As a statistical test can only reject hypotheses and may not confirm them, this means that, in order to decide upon the existence of a difference, we must reject the hypothesis that the difference

Biostatistics Decoded, First Edition. A. Gouveia Oliveira.
© 2013 John Wiley & Sons, Ltd. Published 2013 by John Wiley & Sons, Ltd.

between population means is zero. This hypothesis is the starting point of a statistical test and is called the **null hypothesis**. In statistical notation the null hypothesis is denoted by H_0.

When we reject a hypothesis we are bound to accept an alternative. In a statistical test there must be only one alternative, so if we reject the null hypothesis of equality of population means, we will be left with the alternative of accepting that the population means are different. This is called the **alternative hypothesis** and is usually denoted by H_A.

Next we need to define a decision rule for the rejection of the null hypothesis. As a matter of common sense and also by convention, most people will consider it appropriate to reject the null hypothesis when the probability of obtaining a difference between sample means as large as the one observed, when the null hypothesis is true, is less than 5%.

6.2 The z-test

Let us begin with the problem of investigating whether the means of an interval-scaled attribute are different between two populations. To start, we formulate the null hypothesis that the means of the attribute are equal in the two populations, and the alternative hypothesis that the means of the attribute are different in the two populations. Using statistical notation, what we just said was H_0: $\mu_1 - \mu_2 = 0$ and H_A: $\mu_1 - \mu_2 \neq 0$.

Then, we must obtain two random samples, one from each population. In each sample, we compute the mean and the standard deviation. The two sample means, of course, will be different and this will always be so because of sampling variation.

The question is: If the null hypothesis is true, what are the chances of getting a difference between sample means as large as the one we observed? If the chances are very small, then we reject the null hypothesis and conclude on a difference between population means. Conversely, if the chances are reasonable, the result we got through sampling must be accepted as plausible with the null hypothesis and we fail to reject it.

How can we calculate the likelihood of the observed difference assuming that the null hypothesis is true? Let us review what can be said about the distribution of differences between sample means, beginning with the case of large samples.

We know that, according to the central limit theorem, the sample means are normally distributed. We also know that the differences between the two sample means, because of the properties of the normal distribution, are also normally distributed. And we have seen previously that, because of the properties of variances, the variance of the difference between sample means is equal to the sum of the variances of the sample means.

Until now this is exactly the same reasoning we made when we were finding confidence limits for the difference between population means. The difference is that now we are not estimating population parameters, but simply testing a hypothesis.

So the reasoning is as follows: if the null hypothesis is true (people say 'under the null hypothesis') the mean of the differences between sample means must be

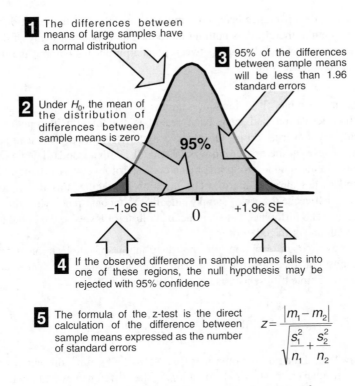

1 The differences between means of large samples have a normal distribution

3 95% of the differences between sample means will be less than 1.96 standard errors

2 Under H_0, the mean of the distribution of differences between sample means is zero

95%

−1.96 SE   0   +1.96 SE

4 If the observed difference in sample means falls into one of these regions, the null hypothesis may be rejected with 95% confidence

5 The formula of the z-test is the direct calculation of the difference between sample means expressed as the number of standard errors

$$z = \frac{|m_1 - m_2|}{\sqrt{\dfrac{s_1^2}{n_1} + \dfrac{s_2^2}{n_2}}}$$

Figure 6.1   Rationale of the z-test for large samples.

zero (Figure 6.1). Accordingly, in 95% of the cases, the sample differences will be less than 1.96 standard errors away from zero. If the observed difference is not, then we reject the null hypothesis with 95% confidence. We say that the observed difference between sample means is within the **rejection region of the null hypothesis**.

People tend to become anxious when looking at mathematical formulas, so now is a good opportunity to show that formulas are really quite harmless. Figure 6.1 shows the formula for the *z*-test. The **z-test** is the statistical test we just described and it is used to test for the differences between two means from large samples. We reject the null hypothesis when the result of the formula (the *z*-value) is greater than 1.96. How does that work?

Look again at the formula. You will recognize that the numerator is the difference between the two sample means. The modulus sign means that we are not concerned with the direction of the difference, only with its absolute value. You will also recognize that the denominator is the standard error of the sample differences, calculated from the sample standard deviation, *s*. What is the meaning of the value 1.96 for *z*, as the value above which we reject the null hypothesis? Well, *z* has that value when the observed difference between sample means is 1.96 greater than its standard error, or larger. When that happens we reject the null hypothesis because this means that the difference between sample means falls into the rejection region. The z-test formula, therefore, is nothing more than an expeditious way of deciding upon the existence of a difference between population means without the need for calculating confidence limits.

Here is an example. Two random samples of 50 and 60 individuals with hypercholesterolemia had each been receiving treatment with one of two different lipid-lowering drugs. On an occasional observation, the mean serum cholesterol level in the first group was 190 mg/dL, with a standard deviation of 38 mg/dL. In the second group, the mean cholesterol level was 170 mg/dL and the standard deviation 42 mg/dL. The question is whether this difference of 20 mg/dL between the two sample means is inconsistent with the null hypothesis of no difference between mean cholesterol levels in the populations treated with one or the other drug.

We have no information on the distribution of cholesterol in those populations, but since the samples are large, we can use the z-test. The standard error of the sample differences estimated from the data is

$$SE = \sqrt{\frac{38^2}{50} + \frac{42^2}{60}} = 7.63$$

We know that, if the null hypothesis is true, 95% of all sample differences must be smaller than 1.96 standard deviations. The difference we got with our samples was 20, which represents $20/7.63 = 2.62$ standard errors and therefore the observed difference in sample means is well inside the rejection region. In other words, our observations are not consistent with the null hypothesis. We therefore reject that hypothesis, and conclude on a difference between population means.

An important point to note is that, when we fail to reject the null hypothesis, we do not accept it either. In other words, if a statistical test does not reject the null hypothesis of equality of means we cannot conclude that the means are equal or even similar. This is a common error made by the uninitiated in the interpretation of the results of statistical tests. Why can we not accept the null hypothesis? Because when we fail to reject the null hypothesis of $\mu_1 - \mu_2 = 0$, this means that the value 0 for the difference between population means is contained within the 95% confidence interval and, therefore, that either population mean may be higher than the other, or they may be equal. This is why we cannot conclude anything when we fail to reject the null hypothesis.

On the other hand, when we reject the null hypothesis we are saying that if the null hypothesis is true the results of our sampling are very unlikely, but by no means impossible. We are left, then, with a small but definite possibility of rejecting the null hypothesis when it is actually true. With the decision rule we have adopted, we know that risk to be less than 5%, but in each situation we can quantify this risk more precisely. The usual terminology is to call the risk the **p-value**.

6.3 The p-value

The p-value can be defined strictly as the probability of obtaining a difference at least as large as the one that was observed in the samples, if the null hypothesis was true.

Therefore, the smaller the p-value, the less plausible the null hypothesis and the greater our belief that a difference between population means truly exists. The p-value, then, also expresses the strength of the evidence in favor of a true difference between means. The smaller the p-value, the stronger the evidence.

As this information about the strength of the evidence is of major importance, it has become current practice to report the exact p-value instead of the simple notation $p < 0.05$, which says only that the difference in sample means is within the rejection region.

We will see here how to find the exact p-value using a statistical table of the normal distribution. Let us continue with the example in the previous section. We saw that the z-value was 2.62, meaning that the difference between sample means is 2.62 standard errors away from the difference between population means, which, under the null hypothesis, is zero. Therefore, we want to find the proportion of observations that, in a normal distribution, are more than 2.62 standard errors away from the mean.

In the table of the normal distribution in Figure 6.2 we find that proportion at the intersection of the row and column that summed together have the value of the z statistic. The tabulated value is 0.0044. Remember that this table only

Table A1   Areas in the tail of the normal distribution

z	0.00	0.01	0.02	0.03	0.04	0.05	0.06	0.07	0.08	0.09
2.2	0.0139	0.0136	0.0132	0.0129	0.0126	0.0122	0.0119	0.0116	0.0113	0.0110
2.3	0.0107	0.0104	0.0102	0.0099	0.0096	0.0094	0.0091	0.0089	0.0087	0.0084
2.4	0.0082	0.0080	0.0078	0.0076	0.0073	0.0071	0.0070	0.0068	0.0066	0.0064
2.5	0.0062	0.0060	0.0059	0.0057	0.0055	0.0054	0.0052	0.0051	0.0049	0.0048
2.6	0.0047	0.0045	0.0044	0.0043	0.0042	0.0040	0.0039	0.0038	0.0037	0.0036
2.7	0.0035	0.0034	0.0033	0.0032	0.0031	0.0030	0.0029	0.0028	0.0027	0.0026
2.8	0.0026	0.0025	0.0024	0.0023	0.0023	0.0022	0.0021	0.0021	0.0020	0.0019

Figure 6.2   Finding the exact p-value using a statistical table of the normal distribution.

shows the proportion of the values that exceed the z-value, but the rejection region is defined in both directions. The exact p-value is, therefore, two times 0.0044, or 0.0088.

The interpretation of a notation such as $p = 0.0088$ is that the probability of two samples having a difference between means, at least as large as the one observed, is only 0.88% if the null hypothesis is true. Alternatively, we may say that, if the null hypothesis is true, then for 99.12% of the time the difference between two sample means will be smaller than the one observed. Consequently, when we conclude that a difference between population means truly exists, there is a probability of 0.88% that we are making the wrong conclusion. This error is called the **alpha error** and it represents the probability of concluding on a difference when actually there is none because the null hypothesis was indeed true. It represents the rate of false positives of the statistical test.

6.4 Student's t-test

When all we have are two small samples, we cannot use the z-test for testing differences between population means. If the population distribution of the attribute is unknown, we cannot assume the normal distribution of sample means, and even if the population distribution is known to be normal, an estimate of the standard error of the difference between sample means based on the sample standard deviations does not have the necessary precision.

The situation is therefore identical to what was said on the construction of confidence limits for the difference between two population means with small samples. Therefore, in the same way as we did for that problem, if we can assume that the two populations have equal variance, then we can test the null hypothesis of no difference between population means with a statistical test known as **Student's t-test**.

Student's t-test works exactly as the z-test, except for the adaptations that were presented in the section on the estimation of differences between population means with small samples. That is, we must assume that the attribute is normally distributed and that the population variances are equal. Then, we use the two variance estimates from the two samples to obtain a common estimate of the population variance. Remember that this is done by calculating a weighted mean of the two sample variances using the number of degrees of freedom as weights. With this common estimate of the variance, we obtain an estimate of the standard error of the difference between sample means. Finally, we establish the cut-off for rejecting the null hypothesis in the same way as we did for large samples. Of course, instead of always using 1.96 standard errors, we must use the number of standard errors given in the table of the t distribution. We calculate the size of the observed difference expressed as standard errors and see if it is within the rejection region.

Figure 6.3 illustrates the procedure. Look at the formula for Student's t-test. The test statistic t represents the difference between sample means expressed in number

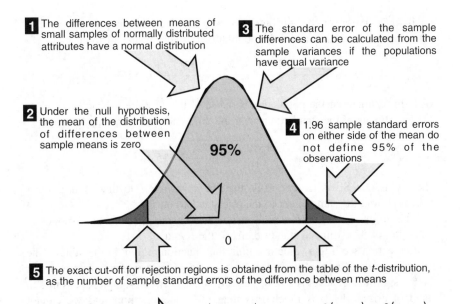

1 The differences between means of small samples of normally distributed attributes have a normal distribution

2 Under the null hypothesis, the mean of the distribution of differences between sample means is zero

3 The standard error of the sample differences can be calculated from the sample variances if the populations have equal variance

4 1.96 sample standard errors on either side of the mean do not define 95% of the observations

95%

0

5 The exact cut-off for rejection regions is obtained from the table of the t-distribution, as the number of sample standard errors of the difference between means

6 The formula for Student's t-test

$$t = \frac{|m_1 - m_2|}{\sqrt{\dfrac{s^2}{n_1} + \dfrac{s^2}{n_2}}} \quad , \quad s^2 = \frac{s_1^2(n_1 - 1) + s_2^2(n_2 - 1)}{n_1 + n_2 - 2}$$

Figure 6.3   The rationale of Student's t-*test in the case of small samples.*

of standard errors. To verify whether the difference is larger than the cut-off for rejection of the null hypothesis, we compare that t-value to the value that, in the table of the t distribution and for the right number of degrees of freedom, is exceeded by 5% of the observations in either direction. If the t-value is larger than the number in the table, we reject the null hypothesis with $p < 0.05$. Otherwise, we cannot conclude on a difference between means.

As we saw in the previous section, we can also obtain the exact p-value from a statistical table of the t distribution.

Here is an example that illustrates the entire procedure of the t-test. Assume that for the comparison of two regimens with lipid-lowering drugs in the previous example we had two samples of 16 and 12 subjects instead of 50 and 60. The mean and standard deviations of the attribute were the same as in the previous example, respectively, 190 ± 38 mg/dL and 170 ± 42 mg/dL. The samples are small but we know that total serum cholesterol has a normal distribution and its variance is equal in the two populations.

For Student's t-test, the first step will be to obtain an estimate of the standard error of the difference between sample means from the observed standard deviation of the attribute. We will use the two sample variances to obtain a combined estimate of the population variance of the attribute

$$s^2 = \frac{38^2 \times (16 - 1) + 42^2 \times (12 - 1)}{16 + 12 - 2} = 1579.4$$

With this estimate of the population variance of the attribute we can now calculate an estimate of the standard error of the difference between sample means

$$SE^* = \sqrt{\frac{1579.4}{16} + \frac{1579.4}{12}} = 15.18$$

The asterisk on SE is to remind us that the value refers to the standard error estimated from the sample, not to the true standard error.

The difference between sample means was 20 mg/dL. This corresponds to $20/15.18 = 1.32$ estimated standard errors, the t-value.

Let us first see if this value is within the rejection region. In the table of the t distribution we see that the rejection boundary for 26 degrees of freedom at the 5% significance level is 2.056. The difference between samples measured in standard errors, 1.32, is much less than the rejection limit and we cannot conclude on a difference between the population means.

We can find the exact p-value using the same table (Figure 6.4). To do so, we look in the row corresponding to 26 degrees of freedom for the value of the test statistic, 1.32. The closest value is 1.315 in the column labeled 0.20. This is the p-value. If the null hypothesis were true, as much as 20% of the sample differences would differ by at least 20 mg/dL, which makes the result of our sampling very plausible with that hypothesis.

Table A2   Percentage points of the t distribution

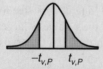

$-t_{v,P}$ $t_{v,P}$

Degrees of freedom	Probability of greater value								
v	0.90	0.50	0.30	0.20	0.10	0.05	0.02	0.01	0.001
23	0.127	0.685	1.060	1.319	1.714	2.069	2.500	2.807	3.767
24	0.127	0.685	1.059	1.318	1.711	2.064	2.492	2.797	3.745
25	0.127	0.684	1.058	1.316	1.708	2.060	2.485	2.787	3.725
26	0.127	0.684	1.058	1.315	1.706	2.056	2.479	2.779	3.707
27	0.127	0.684	1.057	1.314	1.703	2.052	2.473	2.771	3.690
28	0.127	0.683	1.056	1.323	1.701	2.048	2.467	2.736	3.674
29	0.127	0.683	1.055	1.321	1.699	2.045	2.462	2.756	3.659

Figure 6.4   Using the table of the t *distribution to find exact* p-*values.*

Naturally, what was said about the assumption and robustness of the method in the section on interval estimation of the difference between population means with small samples also applies to Student's t-test. In brief, the observations must be independent, the attribute must have a normal distribution, and the two population variances must be equal. Student's t-test is robust and remains valid providing the distribution of the attribute is not too different from the normal and the variances are not too different. Student's t-test is particularly robust if the sample sizes are equal.

6.5 The binomial test

The **binomial test** is a statistical test that compares two proportions obtained from large samples. Therefore, we use this test to investigate associations between binary variables. Again, the foundation of the test is the same as explained in the previous sections, since with large samples we can use the binomial approximation to the normal distribution. However, calculations are made in a slightly different way.

As before, we will test the null hypothesis of no difference between population proportions. Therefore, under the null hypothesis, differences between sample proportions are normally distributed with mean 0. The estimation of the standard error of the differences is unlike the previous tests, however, because here we will pool the observations from two samples. If the null hypothesis is true, then the population proportions are the same and, by pooling all the observations, we are able to get a more accurate estimate of the common proportion and, consequently, of the common variance. Then, we will use it to obtain a better estimate of the standard error. Finally, we will use the formula for the z-test that we saw previously. We want to know how many standard errors the observed difference is away from zero. Hence, we divide the difference between sample proportions by the standard error. The resulting z-value is the distance of the observed difference in sample proportions to zero, expressed as a number of standard errors. If the value of the test statistic is greater than 1.96, this means that the observed difference is larger than the cut-off of the rejection region and we may reject the null hypothesis.

The procedure is as follows. We obtain a common estimate of the proportion in the two populations by adding together all the positive observations, that is, the observations where the attribute was present, and dividing by the total sample size.

Remember that the variance of sample proportions is equal to the variance of the attribute in the population divided by the sample size, that is, $\pi(1-\pi)/n$. Therefore, the variance of the difference between sample proportions will be equal to the sum of the two variances of sample proportions, and the standard error is the square root of that. In short,

$$\text{SE} = \sqrt{\frac{\pi(1-\pi)}{n_1} + \frac{\pi(1-\pi)}{n_2}}$$

Of course, we do not know the value of π, the population proportion of the attribute. We will use the observed proportion of the attribute (p) instead of π to obtain an estimate of the true standard error. As the value of p was obtained from two large samples pooled together, the values n_1 and n_2 that will divide p $(1 - p)$ are large, and the difference between the square root of $\pi(1 - \pi)/n_1 + \pi(1 - \pi)/n_2$ and the square root of $p(1 - p)/n_1 + p(1 - p)/n_2$ will be minimal.

We can now use the formula of the z-test:

$$z = \frac{|p_1 - p_2|}{\sqrt{\dfrac{p(1 - p)}{n_1} + \dfrac{p(1 - p)}{n_2}}}$$

For example, suppose we wish to compare the proportion of a given attribute in two populations. We have a random sample of 90 observations from population 1, and a sample of 110 observations from population 2. In the first sample, the attribute was present in 30 patients (33%). In the second sample it was present in 20 (18%).

The total number of observations is 200, and the number of positive findings is 50. Therefore, under the null hypothesis the common proportion of the attribute in the population is $p = 25\%$. The estimate of the variance of the difference between sample proportions is the sum of the variance of each sample proportion. That is, $0.25 \times 0.75/90$ plus $0.25 \times 0.75/110$, or 0.0038. The square root of this quantity, 0.062, is the estimate of the standard error of the difference between sample proportions.

To obtain the z statistic we divide the difference between sample proportions (0.15) by the standard error (0.062). The result is 2.42, which, being greater than 1.96, allows us to reject the null hypothesis at the 5% error level.

The p-value can be obtained from the table of the normal distribution. At the intersection of row 2.4 with column 0.02 we read off 0.0078. As the table we have been using shows one-sided probabilities, the p-value is twice that value, that is, 0.016.

It is important to note that the procedure for obtaining standard error estimates is different from the one presented when estimating confidence intervals for differences between proportions. This means that it is possible to reject the null hypothesis while the confidence interval includes zero, which is a bit awkward. Regardless, we should still estimate confidence limits and test the null hypothesis, as was explained.

6.6 The chi-square test

We can compare two proportions to a different test. This test uses a different approach to the same question. It is not based on the binomial approximation to the normal distribution. It has an advantage over the binomial test because it can be used to compare several proportions simultaneously. Therefore, we can use the same test to investigate associations between two binary variables, between a binary variable and a categorical variable, and between two categorical variables.

The test is based on the comparison of the frequency distribution of an attribute that was observed in the samples to the frequency distribution we would expect to get if the null hypothesis were true. If the observed distribution is very different from the distribution expected under the null hypothesis, then we may reject the null hypothesis and conclude on a difference between the proportions in the populations.

Let us return to the problem illustrated in the previous section. We had two samples of 90 and 110 subjects, and we observed a binary attribute in 30 and 20 subjects, respectively. The frequency distributions observed in the two samples are shown in the upper table of Figure 6.5. In a two-way table like this one, each frequency shown is called a **cell**, the row and column totals are called the **marginal totals**, and the total number of observations is called the **grand total**.

We now wish to compare this table to the table we would expect to get if the null hypothesis were true. Under the null hypothesis, the two samples are from populations with identical proportion of the attribute. Therefore, we can estimate the common proportion using all the observations, that is, 50/200 or 25%.

Therefore, under the null hypothesis we would expect that 25% of the observations in each sample had the attribute and, thus, we would expect to see the attribute in 25% of the 90 subjects of the first sample, that is, 22.5 subjects, and in 25% of the 110 subjects of the second sample (i.e., 27.5 subjects). We can now construct the table of the expected frequencies under the null hypothesis, which is shown in the lower part of Figure 6.5.

Observed frequencies

	Sample 1	Sample 2	Total
with attribute	30	20	50
without attribute	60	90	150
Total	90	110	200

Expected frequencies under H_0

	Sample 1	Sample 2	Total
with attribute	22.5	27.5	50
without attribute	67.5	82.5	150
Total	90	110	200

Figure 6.5   Observed and expected frequencies under H_0.

The fact that the observed frequency distribution is not identical to the expected distribution under H_0 does not in itself contradict the null hypothesis. Actually, it is quite normal that the two tables are different, because of sample variation. What would not be normal is a large difference between the two tables, because it would mean that our observations were not consistent with the null hypothesis. The key to this problem, therefore, is an assessment of how large the difference is between the table of the observed frequencies and the table of expected frequencies under H_0. In order to evaluate that, we will need to be able to quantify the difference.

One measure of the difference between two tables could be the sum of all the differences between corresponding cells of the tables. This would not work, though, because in each sample the excess frequency in one of the cells is equal to the deficit in the other cell. This means that the sum of the differences between cells will always be zero. Therefore, we need to remove the sign of the differences, and for this we will apply the usual method of squaring all the differences.

Finally, we need to account for the fact that the importance of a difference is related to the number of observations (a difference of 4 in 5 is much greater than a difference of 4 in 50) and thus, in order to be meaningful, we have to express the difference between cells as a proportion of the expected number of observations.

Returning to our example, the difference between the two tables regarding the sample from population 1 is $30 - 22.5 = 7.5$ squared divided by 22.5, plus $60 - 67.5 = -7.5$ squared divided by 67.5. If we calculate the differences between all the cells and then sum all the results we will get the value 6.06. This result, which is a measure of the discrepancy between the two frequency tables, is called **chi-square**, from the Greek letter χ (chi). Of course, the larger the value of chi-square, the greater the difference between the tables.

Now, how can we determine whether this particular value for chi-square represents a small or a large discrepancy between the observed frequencies and the expected frequencies under H_0? If we had a notion of the values that the chi-square takes in a situation where the null hypothesis is true, then we would have something to compare our result to, and we might realize whether it was exceptionally large or not.

Fortunately, we can have that notion because the distribution of the values of the chi-square under the null hypothesis is known, and is called the **chi-square distribution**. Figure 6.6 shows the chi-square distribution for the case of tables with two rows and two columns (2×2 tables).

We may use the chi-square distribution to find out what the proportion is of tables with the same marginal totals that under the null hypothesis have a difference to the expected table smaller than the difference we got. In our example, we would reach the conclusion that 98.6% of the tables we could get by sampling would have a chi-square value lower than 6.06. Or, equivalently, that in only 1.4% of the tables is the value of the chi-square equal to or greater than 6.06. In statistical notation, we would simply say that $p = 0.014$. Therefore, our result would be an exceptional event, were the null hypothesis true, and the logical thing to do would be to reject the null hypothesis at the 1.4% significance level. How did we get this value of

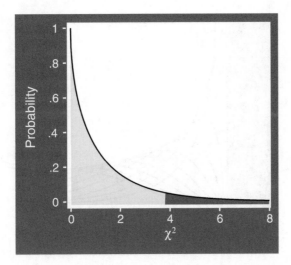

Figure 6.6   The chi-square distribution under the null hypothesis for a 2×2 table. The dark area corresponds to the 5% rejection region.

0.014? By consulting the statistical table of the chi-squared distribution, which we will discuss later on.

This test is known as the **chi-square test**, one of the most common tests seen in the scientific literature. As with other statistical tests, for the rejection of the null hypothesis we may also use the method of setting the rejection region and comparing the chi-square value to the rejection limit. In the case of 2×2 tables, 5% of the values of the chi-square are greater than 3.84 when the null hypothesis is true (the rejection region here is one-sided because the chi-square can never have negative values). The value we got in the example was 6.06, which is clearly within the rejection region (Figure 6.6). In statistical notation we would then say $p < 0.05$.

It is easy to realize that the method we used to measure the discrepancy between the tables of observed and expected frequencies under H_0 can be applied to tables with any number of rows and columns. This is the reason why it was said that this test could be used whenever we needed to analyze nominal variables, whatever the number of categories in each one of those variables. The procedure for the test is always the same; we just need to keep in mind that the distribution of the chi-square cannot be the same for all types of tables.

Similar to Student's t distribution, the chi-square distribution is actually a family of distributions, each one referred to by the number of degrees of freedom (Figure 6.7). In the chi-square test there is a chi-square distribution for each type of table, according to the number of rows and columns. We find out which distribution should be used by multiplying the number of rows minus one by the number of columns minus one. Thus, a chi-square obtained from a 2×2 table follows under H_0 the distribution with 1 degree of freedom, from a 3×3 table the distribution with 4 degrees of freedom, from a 4×6 table the distribution with 15 degrees of freedom, and so on.

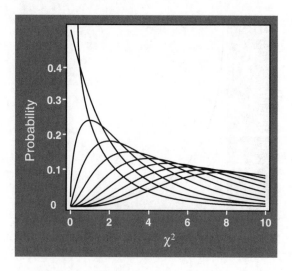

Figure 6.7   The chi-square distribution with 1 to 10 degrees of freedom. The distribution becomes progressively more symmetrical with increasing degrees of freedom.

The procedure for calculating the chi-square statistic can be written as a formula. Recall what was done above: for each cell i, we squared the difference between each observed (O_i) and expected (E_i) value and divided the result by the expected value (E_i); then we summed all those partial results and obtained the chi-square value. If we represent a sum by the symbol $\sum$, what was just said can be represented by

$$\chi^2 = \sum \frac{(O_i - E_i)^2}{E_i}$$

This is the formula of the chi-square test. Using the example in Figure 6.5,

$$\chi^2 = \frac{(30 - 22.5)^2}{22.5} + \frac{(6 - 67.5)}{67.5} + \frac{(20 - 27.5)^2}{27.5} + \frac{(90 - 82.5)^2}{82.5} = 6.06$$

We will discuss the notion of degrees of freedom further, in the next section. Before that, however, let us finish the presentation of the chi-square test by pointing out the restrictions to this test.

The only situation where the chi-square test is not valid is when the samples are too small. It is generally accepted that the samples are too small when one or more cells have zero observations, or when over 25% of the cells have an expected value less than 5. In these situations an alternative test is **Fisher's exact test**, which has no minimum sample size requirements. If one is testing the association between categorical variables, perhaps several categories can be collapsed to increase the number of observations in each cell. Another alternative to the chi-square test for investigating associations with a binary attribute is logistic regression, which we will discuss much later in this book.

6.7 Degrees of freedom

We have already mentioned degrees of freedom on several occasions. For example, we said that the variance was equal to the sum of squares of the differences to the mean divided by the number of degrees of freedom; that in the construction of the confidence intervals of a population mean with small samples we used Student's t distribution with $n-1$ degrees of freedom; that the t statistic of Student's t-test followed Student's t distribution with $n_1 + n_2 - 2$ degrees of freedom; and now that the chi-square test statistic has a chi-square distribution with (rows -1) $\times$ (columns -1) degrees of freedom.

Let us go back to the example in the previous section and look at the table in the upper part of Figure 6.8. If we remove the values of all the cells and keep the marginal totals, then how many cells do we need to know in order to completely determine the table? The answer is: we only need to know one cell. In fact, as illustrated in the table, if we know that, say, the value in the cell of sample 1 with the attribute is 60, this will be sufficient to find the values of all the remaining cells. For that reason, the parameter we estimated from the table, the chi-square, has only 1 degree of freedom.

Attribute	Sample 1	Sample 2	Total
0	30	20	50
1	60	90	150
Total	90	110	200

Attribute	Sample 1	Sample 2	Total
A		10	50
B			70
C	20		80
Total	90	110	200

Figure 6.8   Determination of the degrees of freedom of the chi-square statistic.

Let us now turn to the lower table in Figure 6.8. How many values do we need now? The answer is two. We need to know two values in order to find the remaining values, because only one would not be enough. A chi-square estimated from this table has, thus, 2 degrees of freedom.

When we constructed confidence intervals using Student's t distribution we referred to the t distribution with $n-1$ degrees of freedom. How many values do we need to be able to find all the values we used? Exactly $n-1$, because if we know the total sum of all the observations, we will need to know the individual value of $n-1$ observations to find the value of the remaining one.

When we compared the means of two populations using small samples, where we referred Student's t distribution with $n_1 + n_2 - 2$ degrees of freedom, we used all the data from two samples. To find all the values of the first sample we need to know the values of $n_1 - 1$ observations, and for the second sample we need to know the values of $n_2 - 1$ observations. Therefore, we need to know $n_1 + n_2 - 2$ values.

We can also think of degrees of freedom as the number of values we may change without altering the result. For example, in the case of the variance, we may change all the values except one, because the last value must be chosen so that the total equals the value of the variance. In the case of a 2×2 table, we are at liberty to change only one value: after we enter the value 60 we no longer have the liberty to choose the values of the other cells and keep the same marginal totals.

6.8 The table of the chi-square distribution

The statistical table of the chi-square distribution has a layout similar to the table of Student's t distribution. An example is shown in Figure 6.9. The first column contains the degrees of freedom, and the first row is the probability that a chi-square value exceeds the tabulated value under H_0.

In the example that we have been using to illustrate the chi-square test, we obtained a chi-square value of 6.06. In the table we find the nearest value in the row corresponding to 1 degree of freedom, which is 6.63. The heading of the column containing 6.63 reads 0.01. Therefore, the p-value is about 0.01. The exact value must be a little higher because 6.06 is less than 6.63. Actually, the exact p-value obtained with statistical software is 0.014.

Using the chi-square distribution to evaluate the chi-square statistic introduces a small error, because the latter has a discrete distribution while the chi-square distribution is continuous. Consequently, the p-value of the chi-square test is smaller than it should be. A correction called **Yates' continuity correction** compensates for this and consists of subtracting 0.5 from the difference between each observed value and its expected value. The formula for the chi-square test with Yates' continuity correction is, therefore,

$$\chi^2 = \sum \frac{(|O_i - E_i| - 0.5)^2}{E_i}$$

Table A3 Percentage points of the χ^2 distribution

$\chi^2_{v,P}$

Degrees of freedom	Probability of a greater value, P								
v	0.90	0.75	0.50	0.25	0.10	0.05	0.025	0.01	0.001
1	0.02	0.10	0.45	1.32	2.71	3.84	5.02	6.63	10.83
2	0.21	0.58	1.39	2.77	4.61	5.99	7.38	9.21	13.82
3	0.58	1.21	2.37	4.11	6.25	7.81	9.35	1134	16.27
4	1.06	1.92	3.36	5.39	7.78	9.49	11.14	13.28	18.47
5	1.61	2.67	4.35	6.63	9.24	11.07	12.83	15.09	20.52
6	2.20	3.45	5.35	7.84	10.64	12.59	14.45	16.81	22.46
7	2.83	4.25	6.35	9.04	12.02	14.07	16.01	18.48	24.32

Figure 6.9   Statistical table of the chi-square distribution.

This correction is considered by many as much too conservative and it is debatable whether it is necessary. In medical journals the chi-square test is often presented without the continuity correction.

6.9 Analysis of variance

Analysis of variance, usually referred to as ANOVA or **anova**, is a statistical test used when one wants to compare several means. We may think of it as an extension of Student's t-test to the case of more than two samples. It may be used instead of the t-test and under the same conditions, but anova is usually reserved for the comparison of three or more means.

Currently, anova is not often used as a statistical test for the comparison of means because in many problems anova can be replaced to some advantage by multiple regression. However, anova is used in other situations where there is a null hypothesis to be tested and it certainly is worth knowing the rationale of this method.

The basic idea underlying anova is very simple. Let us begin by looking at Figure 6.10. The upper graph shows the distribution of values obtained with a computer's random number generator of an attribute with a normal distribution and variance 1 on three independent random samples of size 30 from three identical populations A, B, and C with mean 0. The rightmost plot shows the distribution of all values pooled together into a single group. The lower graph shows the distribution of the values of the same attribute in three samples, also random and independent and of size 30, but now from three different populations with means 0, 1 and 2, respectively. Again, the rightmost plot shows all values pooled together.

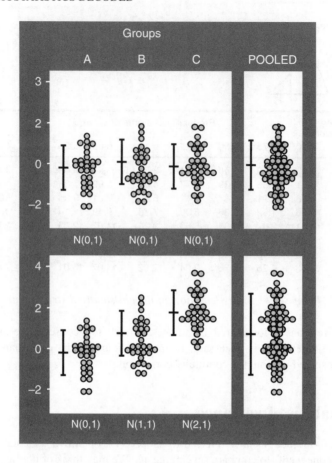

Figure 6.10   Observations in three samples from distinct populations (left panels) of a normal attribute with variance 1 and equal means (top) and unequal means (bottom) and the result of pooling all observations (right panels).

Let us compare the pooled values in the two situations. The difference is that in the first case the variance of the total set of observations is the same as the variance in the individual samples, while in the second case it is greater.

Therefore, if we estimated the population variance of the attribute using either the average sample variance or the variance of the pooled values, in the first situation of equal population means we would obtain similar results. However, in the second situation, where the population means are different, an estimate of the population variance using the variance of the pooled values would be substantially greater than an estimate based on the average of sample variances.

Consequently, one way of detecting a difference among population means could be by comparing an estimate of the population variance of the attribute based on the variance of all observations pooled together (called the **total variance**) with an estimate based on the average sample variance (called the **within-group variance** or **residual variance**). If the total variance was larger than the within-group variance, that would signal a difference between population means. Of course, this reasoning would be valid only under the condition that the population variance of the attribute is equal in the three groups.

Let us now see how we can implement this reasoning. We have seen that there are two ways of estimating the population variance of the attribute, the total variance and the within-group variance.

The total variance can be easily estimated by pooling together all observations, ignoring their distribution by groups and calculating the sample variance in the usual way. We will use the notation s_T^2 for this estimate of the population variance of the attribute.

The within-group variance can be estimated by a weighted average of the sample variances of each group, using the number of degrees of freedom as weights. This is similar to what we did for Student's t-test. Recall that, if it can be assumed that the population variances are all equal, each sample variance is an independent estimate of the common population variance, and we can improve the estimate of the common population variance by calculating a weighted average of the several estimates of the variance, using the number of degrees of freedom as weights. This combined estimate of the population variance, as we saw previously, is obtained by adding all the variances, each one multiplied by its degrees of freedom, and then dividing the result by the total number of degrees of freedom (i.e., the total number of observations minus the number of groups). We will use the notation s_W^2 for the within-group estimate of the variance.

If the null hypothesis that the population means are equal is true, s_T^2 and s_W^2 will estimate the same quantity; if it is false, s_T^2 will estimate a larger quantity than s_W^2. We need a method for comparing the two estimates of the population variance. A natural way is to divide one by the other, that is, s_T^2/s_W^2. This quotient is called the **variance ratio** and is denoted by F. If the null hypothesis is true, F will be close to 1 (close to but not exactly 1, because of the sampling variation of the variances). Values greater than 1 will constitute evidence against the null hypothesis: the greater the value of F, the stronger the evidence.

Actually, the F **distribution**, the probability distribution of the variance ratios when the population variances are equal, is known. Therefore, all we need to do is to use the F distribution to find out which values of F are unlikely when the null hypothesis is true. If the observed F-value is outside the range of values that are expected and explained by sampling variation, we have grounds to question the validity of the null hypothesis.

There is a problem, though. The F distribution can be used only if the distribution of the attribute is normal and if the two estimates of the variance are independent. This is not the case with s_T^2 and s_W^2, which are not independent – the larger the within-group variance, the larger the total variance.

Still, the idea of comparing two estimates of the population variance is appealing, and although this will not work for s_T^2 and s_W^2, perhaps we can think of another way of estimating the population variance.

Let us look again at the top graph in Figure 6.10. This graph shows the situation where the null hypothesis of equal population means is true. We will see how we can obtain an estimate of the population variance of the attribute, other than the total variance and the within-group variance.

If the null hypothesis is true, the means of the attribute in populations A, B, and C are all equal. The sample means, however, will not be equal due to sampling variation. We know that the variance of the means is equal to the population variance of the attribute divided by the sample size, that is, to σ^2/n. Therefore, if we multiply the variance of sample means by the sample size n we will get the value of the population variance of the attribute σ^2.

Of course, we do not know the true value of the variance of sample means, but we can estimate it from our data. We simply compute the variance of the values of the three sample means in the usual way we use to compute variances. For example, suppose the three sample means are 0.87, 0.95, and 1.15. The average of these three values is 0.99. Then, the variance of the sample means is

$$\text{var}(M) = \frac{(0.87 - 0.99)^2 + (0.95 - 0.99)^2 + (1.15 - 0.99)^2}{3 - 1} = 0.0208$$

This is our estimate of the variance of sample means, σ^2/n. Then we multiply the result by the sample size, 30 in this example, and the result is an estimate of the population variance of the attribute, σ^2. We call this estimate the **between-groups variance** and denote it by s_B^2.

Clearly, this estimate of the population variance is independent of the within-group variance – sample means vary randomly, not by virtue of the degree of spread of the observed values. If we return now to the example in Figure 6.10, we can see that when H_0 is true the two estimates of variance s_B^2 and s_W^2 will be approximately equal. However, if the population means are different, s_W^2 will not change while s_B^2 will increase.

As these two estimates of variance are independent, we can now apply the F-test. We calculate the F-value by s_B^2/s_W^2. The F distribution will tell us the range of values taken by the F statistic in 95% of the cases, when two independent estimates of the variance are estimating the same common variance. This is the case when H_0 is true, but not when it is false. If the F-value that we obtain is outside that range, we have evidence that the two variance estimates s_B^2 and s_W^2 do not estimate the same quantity and we will reject the null hypothesis of equality of population means.

In the example all samples had the same number of observations. However, in practice the samples often have different sizes and, thus, calculation of the between-groups variance is more complicated because there is no unique sample size n for multiplying the observed variance of sample means in order to obtain an estimate of σ^2.

There is another way of calculating the between-groups variance. Consider first the total variance. We know that the total variance is the sum of the squared differences between each value and the mean of the pooled observations, divided by the number of degrees of freedom. We will call this mean the grand mean. Now, if the observations are aggregated in different groups, as is the case here, the difference of each observation to the grand mean is equal to the difference of that observation to its group mean plus the difference of its group mean to the grand mean. For example, if the grand mean is 20 and an observation is 8 and its group mean is 13, the deviation of the observation to the grand mean is 12. This is the same value as the sum of the deviation of the observation to its group mean $(13 - 8 = 5)$ and the deviation of its group mean to the grand mean $(20 - 13 = 7)$.

It so happens that the sum of squares of deviations of individual observations to the grand mean is also equal to the sum of squares of deviations from each observation to its group mean plus the sum of squares of the deviation of its group mean to the grand mean. In other words, the total sum of squares can be partitioned into two components, the within-group sum of squares and the between-groups sum of squares.

With this knowledge, we can compute the between-groups variance when samples are of different sizes (as well as when they are the same size) simply by subtracting the within-group sum of squares from the total sum of squares, which will give us the between-groups sum of squares, and then dividing this sum of squares by the number of degrees of freedom. As this estimate of the population variance of the attribute is obtained from the means of the groups, the number of degrees of freedom must be, of course, the number of groups minus one.

Let us look at an example. Figure 6.11 shows the results observed on samples from three populations. Sample A is from patients with a hereditary cardiomyopathy, sample B is from asymptomatic mutation carriers, and sample C is from healthy controls. The attribute that was measured was the left ventricular ejection fraction (LVEF). We want to verify whether the LVEF is associated with the disease status and, for that purpose, we will use anova to test whether the three samples are from populations with equal means.

We will need to compute the total sum of squares (SSq_T) in order to calculate the between-groups sum of squares (SSq_B). We obtain the SSq_T from

Group	n	mean	SD
A	50	45.01	6.65
B	24	41.06	8.00
C	19	41.41	4.91
Total	93	43.26	6.92

Figure 6.11 Left ventricular ejection fraction in three groups classified according to mutation status in hereditary cardiomyopathy. SD: Standard Deviation.

the total sample variance by multiplication with its degrees of freedom $(n-1)$:

$$\text{Total variance} = 6.92^2 = 47.89$$
$$\text{SSq}_T = 47.89 \times 92 = 4405.88$$

To calculate the between-groups sum of squares we subtract from the total sum of squares the within-group sum of squares. We can compute the latter from the three variances

$$\text{SSq}_W = 6.65^2 \times 49 + 8.00^2 \times 23 + 4.91^2 \times 18 = 4072.85$$

Now we can calculate the between-groups sum of squares by $\text{SSq}_T - \text{SSq}_W$:

$$\text{SSq}_B = 4405.88 - 4072.85 = 333.03$$

We obtain the two independent estimates of the population variance of LVEF, s_W^2 and s_B^2, by dividing each sum of squares by its number of degrees of freedom. The number of degrees of freedom for s_W^2 is $n_1 - 1 + n_2 - 1 + n_3 - 1$ and for s_B^2 it is the number of groups minus 1:

$$s_W^2 = 4072.85/90 = 45.25$$
$$s_B^2 = 333.03/2 = 166.52$$

These estimates are called **mean squares** and they represent the two independent estimates of the population variance of the attribute. If H_0 is true we expect the two estimates to yield similar values. It is apparent from the results that the variance estimate computed from the sample means is greater than the estimate computed from the average of sample variances. The variance ratio is

$$F = 166.52/45.25 = 3.68$$

Like the t and the chi-square distributions, the F distribution is also a family of distributions, but here we have two kinds of degrees of freedom. Which one will we use? We will use both. Actually, there is one F distribution for each combination of degrees of freedom of the between-groups and within-group estimates of variance.

The limit of the rejection region of the F-test can be found in a statistical table of the F distribution by looking for the tabulated value corresponding to the degrees of freedom of the numerator (the degrees of freedom of s_B^2) and of the denominator (the degrees of freedom of s_W^2) of the variance ratio. For example, the table of the F distribution tells us that for 2 and 90 degrees of freedom the limit of the 5% rejection region is 3.10. If the F-value we obtained was greater than that value we would reject the null hypothesis.

In the F distribution, the rejection region is one-sided, that is, defined only in the direction of an F-value greater than 1, because those are the only F-values that we are interested in. Figure 6.12 shows the F distribution with 2 and 90 degrees of freedom with the one-sided 5% rejection region.

Thus, anova allows us to test if several sample means are all from populations with identical distributions of the attribute. As the classification into groups can be

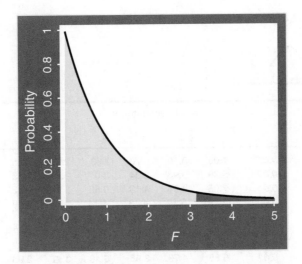

Figure 6.12   The F *distribution with 2 and 90 degrees of freedom. The dark area under the curve corresponds to the 5% rejection region.*

regarded as the values of a nominal variable, anova is actually a test of association between an interval and a nominal variable, in the same way that Student's *t*-test tests the association between an interval and a binary variable.

On several occasions throughout the above discussion of the anova test we referred to the assumptions we had to make about the data. That is, we mentioned that the attribute must have a normal distribution in each group and that the variance of the attribute must be the same in all groups. We call this condition of equality of variances **homoscedasticity**, the opposite being **heteroscedasticity**. Similar to Student's *t*-test, which has some affinities, anova is rather more sensitive to deviations from the condition of homoscedasticity than from the condition of a normal distribution of the attribute.

In the next section we will see how to use a statistical table of the *F* distribution. There is one last issue, though. This test allows us to conclude that the means are not equal across groups, but does not inform us which groups have different means. Intuitively we might think it would be appropriate to test pairwise comparisons of all groups, but we will see how this would lead to the problem of multiple comparisons.

6.10 Statistical tables of the *F* distribution

The *F* distribution is tabulated, and one example is shown in Figure 6.13. Tables of the *F* distribution often are, as in the example, three-way tables. One entry is for the degrees of freedom for the numerator, one is for the degrees of freedom for the denominator, and the third entry is for the *p*-values. Tabulated

Table A4 Percentage points of the *F* distribution

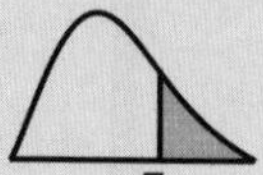

$F_{P,v1,v2}$

df for denominator, v_2	P	df for numerator, v_1: 1	2	3	4	5	6	7
60	*0.05*	4.00	3.15	2.76	2.53	2.37	2.25	2.17
	0.025	5.29	3.93	3.34	3.01	2.79	2.63	2.51
	0.01	7.08	4.98	4.13	3.65	3.34	3.12	2.95
	0.005	8.49	5.79	4.73	4.14	3.76	3.49	3.29
120	*0.05*	3.92	3.07	2.68	2.45	2.29	2.17	2.09
	0.025	5.15	3.80	3.23	2.89	2.67	2.52	2.39
	0.01	6.85	4.79	3.95	3.48	3.17	2.96	2.79
	0.005	8.18	5.54	4.50	3.92	3.55	3.28	3.09

Figure 6.13 A section of a statistical table of the F *distribution.*

values are the values exceeded with the probability indicated in the column P. As usual, we can use this table to find the limits of the rejection region or to find the approximate p-value of a given F-value.

To see how to use the table, let us resume the previous example. We had obtained an F-value of 3.68 with 2 and 90 degrees of freedom. Therefore, we search the table for the intersection of 2 degrees of freedom for the numerator and 90 for the denominator. Such a combination is not tabulated, the nearest entries being 60 and 120 degrees of freedom for the denominator, so the 5% rejection limit for 90 degrees of freedom is somewhere between 3.07 and 3.15. Therefore, the F-value of 3.68 exceeds any of those values and, thus, is within the rejection region. We may conclude on a difference in population means.

Let us look at whether we can reject the null hypothesis at the 2.5% level. The 2.5% rejection limit is somewhere between 3.80 and 3.93, so we cannot reject the null hypothesis at the 2.5% significance level. The p-value must be between 0.05 and 0.025. Actually, its exact value obtained with statistical software is 0.029.

7

Issues with statistical tests

7.1 One-sided tests

In the previous chapter we mentioned one-sided rejection regions when we discussed the chi-square test and the F-test. In both cases the rejection region was defined only in the direction of positive values, but for different reasons. In the chi-square test, the reason was because it is just not possible for the chi-square statistic to have negative values; in the second case, it was because we were only interested in F values greater than unity, since then there would be evidence of a difference between population means.

All statistical tests may have one-sided rejection regions. One-sided statistical tests are exactly the same as two-sided tests, except for the way the null hypothesis is formulated. Until now we have always assumed that a test had two-sided rejection regions, and every time we rejected the null hypothesis (H_0) we had to accept the alternative hypothesis (H_A) of a difference between population means. In statistical notation, if we denote the population means by μ_1 and μ_2, we would write H_0: $\mu_1 = \mu_2$; H_A: $\mu_1 \neq \mu_2$.

However, we can define H_0 and H_A in other ways, as long as the two hypotheses cover all possibilities. For example, we can formulate H_0: $\mu_1 \leq \mu_2$ and, in that case, H_A: $\mu_1 > \mu_2$. We can also formulate H_0: $\mu_1 \geq \mu_2$ and H_A: $\mu_1 < \mu_2$. In words, this means for the first case that if we rejected H_0 we would conclude that μ_1 is greater than μ_2; and for the second case that μ_1 is less than μ_2.

We can also test the null hypothesis that the population mean is equal to a given value x. This would be a two-sided test, and the null and alternative hypotheses would be H_0: $\mu = x$ and H_A: $\mu \neq x$. If we want to demonstrate that the population mean is greater, or less, than a given value x then the tests are one-sided and the hypotheses would be, respectively, H_0: $\mu \leq x$; H_A: $\mu > x$ and H_0: $\mu \geq x$; H_A: $\mu < x$.

Differences between population means can also be tested against a given value. In this case, it is rare to test the two-sided hypothesis, but it is very common to test

Biostatistics Decoded, First Edition. A. Gouveia Oliveira.
© 2013 John Wiley & Sons, Ltd. Published 2013 by John Wiley & Sons, Ltd.

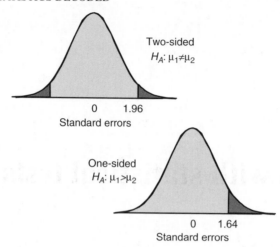

Figure 7.1   Difference between two-sided and one-sided tests.

the one-sided hypothesis. For example, we may be interested in showing that the mean difference in outcomes between a new drug and the standard treatment is greater than a given value x, and we declare the new drug to be ineffective if the difference in outcome is less than x or the new drug is worse than the standard treatment. In this situation, it makes sense to formulate a one-sided hypothesis in the form H_0: $\mu_1 - \mu_2 \leq x$ and H_A: $\mu_1 - \mu_2. > x$.

Therefore, a **two-sided statistical test** is a test where the alternative hypothesis is that a difference exists in either direction. When the alternative hypothesis is that a difference exists in one of the directions, then it is called a **one-sided statistical test**.

As we said earlier, the procedure for a one-sided test is exactly the same as for a two-sided test, but with a single difference: since we are looking for a difference in only one direction, the rejection region is not defined in both directions; it is defined only in the direction where the difference interests us. Figure 7.1 shows the difference between the rejection regions of a two-sided test and a one-sided test for the comparison of two means, for the same significance level of 5%.

As the figure shows, in the case of a one-sided test the rejection region is defined as the values of the differences between sample means that exceed 1.65 standard errors, the value above which are 5% of the differences between two sample means. If the difference is less than −1.65 the null hypothesis should not be rejected.

One-sided tests are used in two distinct situations and it is important to separate them because of different implications in the definition of the rejection region. This distinction is important because failure to set the rejection region appropriately may lead to incorrect conclusions.

In situations where there is no way that the difference could be in one of the directions, a one-sided test would be appropriate. In these cases, the rejection region should be set at 5% on a single direction, as in the lower graph of Figure 7.1. This is a very strong assumption and it is seldom possible to be absolutely sure that it is true. For this reason this type of one-sided hypothesis is rarely found.

More often one will want to demonstrate that a population parameter (the mean or the difference between two means) is greater than a given value x, and it is understood that the population parameter may be less than x but it happens that we are interested in departures from x only in the direction of a greater value. In this case the rejection region should be set at 2.5% in the direction that interests us, therefore keeping the alpha error at the 5% level. The same applies if one wants to show that a population parameter is less than a given value.

The following example illustrates how we can test a one-sided hypothesis. Let us again use the example of the observational study where we want to compare occasional measurements of blood lipids between two groups of subjects that had been receiving either drug A or drug B. Suppose we wanted to show that subjects treated with drug B have HDL cholesterol levels that are more than, say, 5 mg/dL higher than those treated with drug A. Assume that we had a sample of 120 subjects treated with drug A and 140 treated with drug B, and that the mean serum HDL cholesterol level in those in group A was 52 mg/dL with standard deviation 14 mg/dL, and in group B the mean was 61 mg/dL with standard deviation 16 mg/dL. We will use Student's t-test for this analysis.

We want to be able to reject the null hypothesis that the difference between the population means of group B and group A is 5 mg/dL. It is quite possible that drug A achieves higher HDL cholesterol levels than drug B, but this is of no interest to us – we only wish to check whether drug B achieves an increase of 5 mg/dL or more than drug A. So the null hypothesis will be formulated one-sided, but because it is possible that a difference in HDL cholesterol exists in either direction, we will set the one-sided rejection region at $\alpha/2$, that is, at 2.5%.

The null and alternative hypotheses are H_0: $\mu_B - \mu_A \leq 5$ mg/dL and H_A: $\mu_B - \mu_A > 5$ mg/dL. Recall the reasoning we made when we discussed the t-test for evaluating the null hypothesis of equality of population means. Then, the hypothesis was that the distribution of the differences between sample means had zero mean, and departure of the observed difference between the two sample means from zero constituted evidence against the null hypothesis. Now the hypothesis is that the distribution of differences between sample means has mean 5 mg/dL and departure of the observed difference from that value in the direction of a higher value will constitute evidence against the null hypothesis of a lesser difference.

In order to test this one-sided hypothesis we will need to write the formula for the t-test in a slightly different way. Remember that the formula for the comparison of two means is

$$t_{\alpha, n_1 + n_2 - 2} = \frac{|m_1 - m_2|}{\sqrt{\dfrac{s^2}{n_1} + \dfrac{s^2}{n_2}}}$$

where s^2 is a weighted average of the two sample variances and t is the difference of the absolute observed difference between sample means expressed as the number of estimated standard errors away from zero. So we could write this formula, without changing its meaning, as

$$t_{\alpha,n_1+n_2-2} = \frac{|m_1 - m_2| - 0}{\sqrt{\dfrac{s^2}{n_1} + \dfrac{s^2}{n_2}}}$$

and in the case of our example we could write it as

$$t_{\alpha,n_1+n_2-2} = \frac{m_B - m_A - 5}{\sqrt{\dfrac{s^2}{n_1} + \dfrac{s^2}{n_2}}}$$

where now the t-value represents the difference between sample means to 5 mg/dL in the direction of greater values, expressed as the number of estimated standard errors.

Let us carry out the calculations using the data in our example. We start by calculating the weighted average of the two sample variances

$$s^2 = \frac{14^2 \times 119 + 16^2 \times 139}{119 + 139} = 228.33$$

Then we apply the formula of Student's t-test

$$t_{0.025,258} = \frac{61 - 52 - 5}{\sqrt{\dfrac{228.33}{120} + \dfrac{228.33}{140}}} = \frac{4}{1.88} = 2.13$$

The table of Student's t distribution tells us that, for 258 degrees of freedom, 2.5% of the observations are more than 1.97 estimated standard errors away from one side of the mean.

The t-value 2.13 exceeds that value and is therefore within the rejection region. Accordingly, we reject the null hypothesis that the mean HDL cholesterol in subjects treated with drug B is 5 mg/dL or less than in subjects treated with drug A, and we accept the alternative hypothesis that the mean HDL cholesterol is more than 5 mg/dL higher with drug B than with drug A.

We can find the exact p-value from the table of the t distribution, which is 0.034. Therefore, we conclude at the 3.4% significance level that in subjects treated with drug B the mean HDL cholesterol is more than 5 mg/dL higher than in subjects treated with drug A. Figure 7.2 shows the distribution of the differences between sample means under the null hypothesis. H_0 is rejected if the observed difference is greater than 8.7 mg/dL ($= 1.97 \times 1.88$ mg/dL).

We can reach the same conclusion by looking at the 95% confidence interval of the difference between population means. The interval is 5.3 to 12.7 mg/dL and, as the lower limit is greater than 5 mg/dL, we conclude with 95% confidence that mean HDL cholesterol is at least 5.3 mg/dL higher with

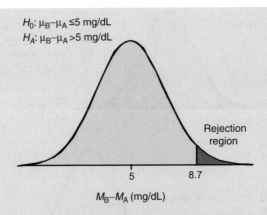

Figure 7.2   Distribution of differences between sample means under a one-sided null hypothesis.

drug B than with drug A. The *t*-test gives us additional information on the strength of the evidence as reflected in the *p*-value.

7.2 Power of a statistical test

As in any other imperfect method of decision making, with statistical tests there are also cases of wrong conclusions. The methodology of statistical tests is based on the definition of a decision rule that assures, on the condition that any assumptions made are valid, that one will wrongly conclude for a difference in only 5% of the times that the null hypothesis is actually true. This error is called, in statistical terminology, the **type I error** or **alpha error**.

In other words, the method assures a rate of false positives for the test of only 5%. In 95% of the cases when the null hypothesis is true the test will not reject the hypothesis. The rate of true negatives of a statistical test when the null hypothesis is true, the **specificity** of the test, is thus set at 95%.

Just as there are false positives, there are also false negatives, that is, situations where the null hypothesis is not rejected although it is actually false. This error is called the **type II error** or **beta error**. Unlike the alpha error, which is set by us, the beta error depends on several factors. Figure 7.3 illustrates the alpha and beta errors corresponding to the dark shaded areas. The proportion of well-classified cases is represented by the light shaded areas.

The chart at the bottom of the figure illustrates what happens when the null hypothesis is false. The distribution of the differences of the sample means is centered on the true value of the difference between the population means. In all cases in which the differences between sample means are less than 1.96

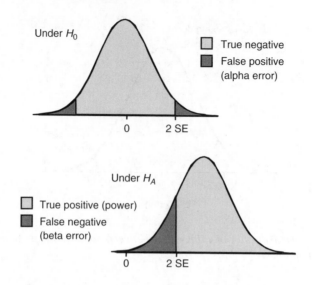

Figure 7.3   The alpha and beta errors of a statistical test.

standard errors away from zero the null hypothesis will not be rejected, although it is false. In all other cases, the null hypothesis will be rejected and the proportion of these cases corresponds to the rate of true positives, better known as test **sensitivity** and, in the context of statistical tests, called the **power** of the test.

If we wish to increase the power of a test we will have to reduce the dark shaded area. In theory, this could be done by having the distribution further away from zero (i.e., increasing the difference between the population means) or by decreasing the variance of the distribution (i.e., decreasing the standard error). Since we cannot act on the difference between the means, or on the variance of the attribute, we are left with the possibility of increasing the sample size, which will decrease the standard error and, consequently, increase the power of the test

7.3 Sample size estimation

When planning a study that aims to make comparisons between the means of two populations, it is important to plan the size that the samples must have in order to be able to detect a difference, if a difference truly exists. We are now able to understand that for this calculation we first need to define the magnitude of the difference between population means that we do not want to miss. Then we need to define how much power we want our study to have. In general, a power between 70 and 90% is desired. Then we need to calculate the value that the standard error must have for the test to have the desired power, if that difference really exists. Knowing the approximate value of the variance of the attribute in the population, we can then determine the sample size. We may know the approximate value of the variance using the data from published studies or, in their absence, through a small number of observations on a population identical to the one that will be studied.

We will now see how we can calculate the sample size required for detecting a specified difference δ between population means, if such difference truly exists. The upper graph in Figure 7.4 displays how the differences between sample means are distributed if the null hypothesis is true, with the rejection regions shown as dark- shaded areas. We will reject the null hypothesis if the observed difference in sample means is greater than R. What is the value of R? If we set the alpha error at 5%, the value will be 1.96 times the standard error; if we set it at 1%, the value will be 2.57 times the standard error. In general we may write that R is at $z_{\alpha/2} \times SE$, where $z_{\alpha/2}$ is the value of the standardized normal deviate exceeded with probability α in both directions.

When we want to calculate the sample size we also need to specify the alternative hypothesis and we assume that the true value of the difference between population means is δ. The lower graph of Figure 7.4 displays how the differences between sample means are distributed if that hypothesis is true. The alternative hypothesis will be accepted if the observed difference between sample means is greater than R and we want that to happen with specified probability $1 - \beta$, represented by the light-shaded area. So what is the value that R must have so that the light-shaded area has the specified probability? It will be $\delta - z_{\beta} \times SE$, where z_{β} is the value of a standardized normal deviate exceeded with one-sided probability β.

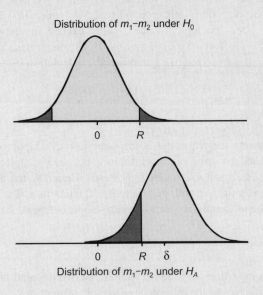

Distribution of m_1-m_2 under H_0

Distribution of m_1-m_2 under H_A

Figure 7.4   Distribution of the differences between sample means under the null hypothesis and an alternative hypothesis with specified difference δ. The null hypothesis is rejected when the observed difference is greater than R.

Therefore, the value R is such that $R = z_{\alpha/2} \times SE$ and $R = \delta - z_\beta \times SE$ and we can write

$$z_{\alpha/2} \times SE = \delta - z_\beta \times SE$$

or

$$\delta = (z_{\alpha/2} + z_\beta) \times SE$$

If the two samples are to be of the same size, then

$$SE = \sqrt{\frac{\sigma^2}{n} + \frac{\sigma^2}{n}} = \sqrt{\frac{2\sigma^2}{n}} = \sigma\sqrt{2/n}$$

Solving the equation for n we get

$$n = \frac{2(z_{\alpha/2} + z_\beta)^2 \sigma^2}{\delta^2}$$

Let us now calculate the sample size using the previous example. We estimate that in subjects treated with drug A an occasional reading of HDL cholesterol has standard deviation 14 mg/dL, and we want to calculate the sample size required to show with a two-sided significance level of 5% (or perhaps preferably a one-sided significance level of 2.5%) and 80% power that subjects treated with drug B have a mean HDL cholesterol at least 5 mg/dL higher.

In the statistical table of the normal distribution we look for the z-value exceeded in one direction with probability 2.5%, which is 1.96. This is the value of $z_{\alpha/2}$. Then we look for the z-value exceed in one direction with probability 20% ($= 100 - 80\%$), which is 0.84. This is the value of z_β. The specified difference δ is 5 mg/dL. Therefore, the calculation is

$$n = \frac{2 \times (1.96 + 0.84)^2 \times 14^2}{5^2} = 122.9$$

We will need 123 subjects per group.

For proportions the reasoning is the same but an adaptation is necessary because the standard error of the difference between sample proportions is different for the null and alternative hypotheses. Under H_0 the proportion with the attribute is identical in both populations and equal to π. The standard error of the difference of sample proportions with samples of equal size n is

$$\sqrt{\frac{2\pi(1 - \pi)}{n}}$$

Under H_A the proportion with the attribute in the two populations is π_1 and π_2, and the standard error of the difference of sample proportions is

$$\sqrt{\frac{\pi_1(1 - \pi_1) + \pi_2(1 - \pi_2)}{n}}$$

Therefore, the sample size calculation becomes

$$n = \frac{\left[z_\alpha \sqrt{2\pi(1 - \pi)} + z_\beta \sqrt{\pi_1(1 - \pi_1) + \pi_2(1 - \pi_2)}\right]^2}{\delta^2}$$

For example, suppose the proportion of subjects treated with drug A with HDL cholesterol above 60 mg/dL is 30% and we want to know how many subjects we would need in a study to show, with 70% power at the two-sided 5% significance level, that with drug B the proportion will be increased to 45%.

Using the above formula with $\pi_1 = 0.30$, $\pi_2 = 0.45$, $\pi = 0.375$, $\delta = 0.15$, $z_{\alpha/2} = 1.96$, and $z_\beta = 0.52$ we estimate that 128 subjects per group will be necessary:

$$n = \frac{\left[1.96 \times \sqrt{2 \times 0.2344} + 0.52 \times \sqrt{0.21 + 0.2475}\right]^2}{0.15^2} = 127.5$$

There are occasions when we may consider forming samples of unequal size. For example, we may find it difficult to enroll controls in a case–control study or exposed in a cohort study. In such cases we may increase sample size by recruiting more subjects from the population that is more accessible, for example, using an unbalanced rate of 2 : 1 or 3 : 1.

It must be noted, however, that for a given total sample size, maximum power is obtained with equally sized groups and that power decreases almost linearly with increasing unbalanced rates. Conversely, for a given power, minimum total sample size is achieved with equally sized groups and total sample size increases almost linearly with increasing unbalanced rates. In addition, some statistical tests such as Student's t-test are most robust when sample sizes are equal.

Resorting to unbalanced samples in order to afford more power to a study may be cost effective up to a certain degree of unbalance, but after a certain extent the gains in power are not significant. Figure 7.5 shows the increase in power obtained by increasing the size of one sample without changing the size of the other. It is apparent that little is gained by an unbalance greater than 3 : 1.

7.4 Multiple comparisons

We have seen in the section on anova that this test can detect a difference between several population means but does not inform about which groups are different from the others. At first sight we might clarify the question by comparing all groups in pairs with t-tests or with anova. However, that procedure would quickly lead us to the wrong conclusions. This is a problem known as multiple comparisons.

Why is this procedure inappropriate? Quite simply, because we would be wrongly testing the null hypothesis.

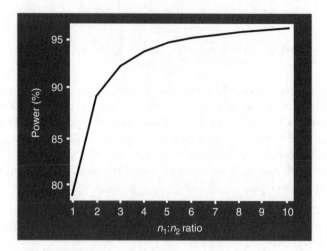

Figure 7.5   Effect of increasing sample size ratios on the statistical power to detect at the 5% significance level a difference of 5 on an attribute with standard deviation 14 and a sample size n_2 of 120.

We saw in the discussion on the *t*-test that for the comparison of two population means we formulate the null hypothesis of equality of means and take a large difference between two samples taken at random for those populations as evidence against the null hypothesis. This is not what we do when we make a series of pairwise comparisons between a number of sample means. What we actually do is first to select the largest difference between any two means and then conclude that such a large difference is unlikely if the null hypothesis is true. The more pairwise comparisons we do, the more likely we are to obtain an uncommonly large difference between any two means and the more likely we are to conclude on a significant difference between population means when there is none.

A computer simulation helps to understand the consequences of multiple comparisons. The results of the simulation are shown in Figure 7.6. The means of two series of 10 000 computer-generated random samples of a standard normal variable (a random variable with a normal distribution with mean 0 and variance 1) were calculated. The curve on the left shows the distribution of the differences between two sample means, one from each series. As H_0 is true, the mean of the distribution is zero.

Then four series of 10 000 random samples of the same standard normal variable were generated. All the pairwise differences of the six possible combinations were obtained and the largest difference was retained. The curve on the right shows the distribution of the largest difference between four samples taken two by two and, as the four series are from identical variables, this curve corresponds to the actual null hypothesis.

The limits of the rejection regions are set in the usual manner, as if we had observed only one difference between two samples, disregarding the fact that we had

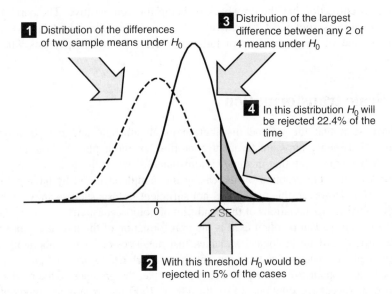

Figure 7.6   Comparison of the distribution of the differences between means under the null hypothesis. On the left is the probability distribution of an observed difference between two sample means. On the right is the probability distribution of the largest difference between 2 of 4 means.

indeed selected the largest difference out of six comparisons. However, as can be seen in the figure, in as many as 22.4% of the times the largest difference between any two of four samples will be greater than the rejection limit. Still disregarding the fact that six comparisons were made, if a difference happened to be greater than the rejection limit one would conclude on a difference between the two population means at the 5% error level. However, the error level was actually 22.4%.

This is the reason why, after an anova concluding on a difference between population means, we cannot just compare all possible pairs of means with t-tests. In fact, this applies to all situations where we want to make multiple tests. Several statistical tests are adequate for these situations because they take into account the number of planned comparisons. Some of the more commonly used ones are **Tukey's test** and **Duncan's test**. An alternative that is more expeditious is the **Bonferroni correction**. This method compensates the likelihood of a false positive test in multiple comparisons simply by dividing the desired significance level by the number of comparisons made. In the above example, in which six comparisons were made, a significance level of 0.0083 should be adopted to ensure that the null hypothesis would be rejected with an error of 5%. In the statistical table of the normal distribution, we can see that this means the rejection threshold will be defined by 2.64 standard errors on each side of the mean instead of 1.96. Alternatively, one can multiply the p-value obtained at each comparison by the number of planned comparisons, thereby obtaining the actual value of p. The

Bonferroni correction has the drawback of being too conservative. The correction of the *p*-value is somewhat excessive and, as a result, the false negative rate increases, that is, the probability of failing to reject the null hypothesis when it is false.

7.5 Scale transformation

We have seen that the normal distribution of an attribute and the equality of variances between groups are frequent conditions of the application of statistical tests. We have also seen that in some circumstances, such as in Student's *t*-test, we can work around the problem by increasing the sample size and by having equal sample sizes. However, this measure is not sufficient in tests such as anova, which are very sensitive to deviations of the condition of homoscedasticity.

A solution that can be often used is the transformation of the measurement scale of the attribute. Many biological variables that are assessed in the laboratory use dilution methods to determine their concentration, which is a logarithmic process, but the results are reported on a linear scale. For example, the presence of the product is screened in successive dilutions 1 : 10, 1 : 100, 1 : 1000 and results are reported as 1, 2, 3. For this reason, many biological attributes have a logarithmic distribution, which is characterized by a pronounced asymmetry of their frequency distribution with a sharp deviation toward lower values. If we apply a **logarithmic transformation** to their values, we will obtain a new variable with a normal distribution.

The logarithmic transformation is surely the transformation toward normality more often used in clinical research. Other transformations occasionally use the square root and reciprocal transformations. Figure 7.7 shows the appearance of some distributions that can be transformed into the normal distribution using the indicated operation.

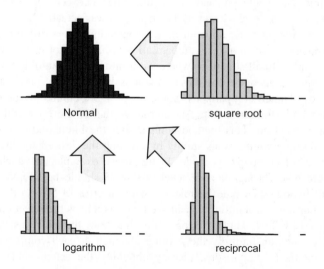

Figure 7.7   Commonly used transformations to the normal distribution.

It is quite common for several transformations to be tested before a good approximation to the normal distribution is achieved. It is also possible to use two or more transformations on the same variable, for example, the logarithm of the reciprocal. Still, there are occasions when it is not possible to find a transformation that gives the attribute the required properties for analysis by one of the statistical tests that we have discussed. For these situations there is an approach to the theory of statistical tests that is different from the statistics of the normal distribution that we have been discussing so far in this book. It is known as **non-parametric statistics**.

7.6 Non-parametric tests

Non-parametric statistics approach hypothesis testing from a different perspective. Generally, non-parametric tests are based on the determination of the probability, under the null hypothesis, of the particular set of observed values. As such, no assumptions about the probability distribution of variables are made and therefore there are no distribution parameters to estimate. Hence the designation of non-parametric.

Let us consider an example to illustrate the basis of non-parametric methodology. Suppose we want to determine whether there are differences in body weight between men and women. For this, we get two random samples from each population and measure the weight of each subject. Now if we rank all weights in ascending order we would expect, if the null hypothesis of no difference between the average weight is true, that men and women would be completely mixed, almost in alternate order. However, if men were on average heavier than women, then we would expect to find more women than men in the lower values of weight and, conversely, more men than women in higher values. This ranking of the elements of the two samples, in turn, would not be likely if the null hypothesis were true. The ranking of the values of the two groups can therefore inform us about the plausibility of the null hypothesis.

Let us create a statistic, call it U, which captures the rank order of the weights. One way to do this is by counting, for each man in turn, how many women have a lower weight and then adding all these values.

We thus obtain a single value. If males are lighter, there will be few women weighing less than every man and U will have a low value. If heavier, for every man there are several women who weigh less and U will have a high value. Intermediate values of U mean that the weights are distributed regardless of gender.

The decision to reject the null hypothesis will depend, therefore, on the value of U. Under the null hypothesis, intermediate values of U should be the most frequent, while very low or very high values of U would be unlikely. We therefore need to determine the probability of occurrence of each value of U under the null hypothesis.

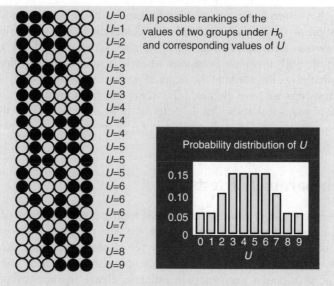

Figure 7.8   Rationale of a non-parametric test for comparing two independent samples.

To do this we need to identify all possible orderings of men and women under the null hypothesis and then calculate the value of U for each one. Figure 7.8 exemplifies the procedure for a case of two samples of three individuals (males in dark circles and females in light circles), ranked in ascending order of weight from left to right. Suppose that after sorting the weights our sample was ordered as FFMMFM. We count two females lighter than the first man from the left, plus two females lighter than the second man from the left, plus three females lighter than the third man from the left. Thus $U = 7$. We could have defined U the other way around, as the number of men with lower weight than each female in the sample, and then $U = 2$ because there are two men lighter than the third female from the left.

Now we find the probability of occurrence of each value of U under H_0 by dividing the number of times a given value occurs by the total possible orderings. In this example, there are 20 possible orderings and the probability of the value of U being, say, 4 is given by $3/20 = 0.15$ and of being 7 is $2/20 = 0.10$. Calculating the probability of each value of U gives us the probability distribution of U under H_0, which is shown on the chart in Figure 7.8. From there, we can find the probability of a value of U at least as extreme as the observed value.

In our example we obtained a value of $U = 7$ or $U = 2$. From the probability distribution of U we calculate the probability of values of U equal to or greater than 7 as $0.10 + 0.05 + 0.05$ or 20%. In the other direction, the probability of a value of U equal to or less than 2 is also $0.10 + 0.05 + 0.05$ or 20%. Therefore, the two-sided p-value is 0.40.

There are non-parametric tests for many of the problems that we have addressed so far and still others for problems that we will address further on. We previously mentioned **Fisher's exact test**, which is used to test the association of two nominal variables. This test replaces the binomial test and the chi-square, but unlike these it does not have restrictions on its application, except for the independence of observations.

The test we just discussed is called the **Mann–Whitney U test** (also called the Mann–Whitney–Wilcoxon U test and the Wilcoxon rank-sum test), which is used to test the association between a binary variable and an interval variable and therefore can be used in place of the Student's t test, but without restrictions to its application other than the independence of observations. This test can also be used to test for an association between a binary and an ordinal variable, and in this context can be interpreted as a test for the comparison of two population medians.

A non-parametric alternative to the analysis of variance is the **Kruskal–Wallis test** which in practice has an identical result to an anova wherein the variable values have been replaced by their rank in an ascending sorting of all data, disregarding the groups. This test does not require an attribute to have a normal distribution, but assumes that its distribution is identical in all groups. In all the tests mentioned, it is also necessary that the samples are independent.

We have just seen that the notion that non-parametric tests have no restrictions on their application is not correct. Several of these tests also make assumptions about the data and in the remainder of this book we will mention the non-parametric equivalents to the statistical tests of the normal distribution, whenever there are any, always making reference to their conditions of application.

Another widespread and also incorrect notion about non-parametric tests is that they must be used in the case of small samples. We already know that the decision about which test to use does not have to concern only the sample size. Moreover, non-parametric tests are almost always less powerful than their analogs from the statistics of the normal distribution when their assumptions are met, and on some occasions it is even mathematically impossible for a non-parametric test to detect a difference below a certain sample size. For example, with the Mann–Whitney–Wilcoxon test, it is impossible to show a statistically significant difference with less than four observations per group.

Therefore, the statistics of the normal distribution have the advantage of greater power of the tests and also of providing estimates of differences between population means and proportions. However, these tests cannot be used with data measured in ordinal scales because, as we have seen, with these scales the differences between values have no meaning. Non-parametric tests have the advantage, in general, of making no assumptions about the distribution of the attributes, or of their scale of measurement. However, they are not always free of assumptions and it is necessary to take into account the conditions of application of each test.

8

Longitudinal studies

8.1 Repeated measurements

Longitudinal studies quite often offer the opportunity to perform repeated measurements in the same subjects. These studies are very common in clinical research. A cohort of individuals is followed up over a period of time and data from several attributes is recorded at more or less regular time intervals during the observation period. Data from a set of attributes that are recorded repeatedly in the same subjects is called **panel data**.

These studies allow us to investigate the change in an attribute over a period of time in a sample of individuals. That is, we can evaluate whether the values of certain attributes have changed between an initial and a later observation. This information enables us to assess, for example, if a clinical parameter changes during the course of a clinical condition or to understand whether a particular therapeutic intervention is associated with a modification of the course of disease.

We are, therefore, faced with a problem of comparing the average values of an attribute at two distinct moments in time. This problem, however, cannot be solved using the methods we have discussed so far, because all of them apply only to the case of independent samples. This is not the present situation. Here we have a single sample and obviously we cannot assume independence of the two measurements in each individual.

In the following sections we will see how we can analyze data from repeated measurements with statistical methods appropriate for non-independent observations.

8.2 The paired Student's *t*-test

When we conduct a study in a sample of individuals and perform two observations of the same attribute, measured on an interval scale, at two different moments in time or under two different conditions, the question that we are trying to answer is whether

Biostatistics Decoded, First Edition. A. Gouveia Oliveira.
© 2013 John Wiley & Sons, Ltd. Published 2013 by John Wiley & Sons, Ltd.

the population mean value of the attribute increases or decreases between the first and second observation. Rephrasing this, we want to assess whether the population mean difference between the first and second observation in each subject is zero.

Therefore, this problem comes down to the estimation of the population mean difference between the first and second observation in the same subjects and to the construction of a statistical test that will tell us whether the average difference between the two measurements in the same subjects as observed in the sample can be considered an extreme value under the hypothesis of no difference in the population.

In practice, what we do is to create a new variable whose value is the difference between the first and second observation in each subject. We will denote this new variable by D. If x_1 and x_2 are the first and second measurements in a given individual, then $D = x_1 - x_2$. From this moment on, the general procedure for estimating the value of a population mean can be applied. From the central limit theorem we know that in large samples the sample mean of D comes from a normal distribution, and from the properties of means we know that the mean of that distribution is the population mean of D, the value we want to estimate. We can estimate the value of the standard error of the sample means of D from our data by dividing the standard deviation of D by the square root of the number of observations. Then we can set the 95% confidence limits at 1.96 standard errors on each side of the sample mean of D.

In the case of small samples, the sample means of D are not normally distributed unless the attribute has a normal distribution, because then D is the difference of two normal variables and, by the properties of the normal distribution, D will also have a normal distribution as well as the sample means of D. As in a small sample the standard error of the means of D will be estimated from the observed data, we must find in the table of Student's t distribution how many standard errors estimated from the data we have to count on each side of the sample mean of D to set the correct confidence limits. Because we only estimate a single parameter, the standard deviation of D, we will refer in the statistical table to Student's t distribution with $n - 1$ degrees of freedom.

To construct the corresponding statistical test (Figure 8.1), called the **paired Student's t-test**, we formulate the null hypothesis that the population mean of D is zero. Then we divide the absolute value of the sample mean of D by its standard error to find how many times the sample mean of D is larger than its standard error. The resulting value, t, follows Student's t distribution with $n - 1$ degrees of freedom. We then read in the statistical table what the 5% rejection limit for that distribution is and, if the value of t exceeds that limit, we reject the null hypothesis.

Suppose we observed a sample of six individuals and in each of them we measured some attribute with a normal distribution under two different conditions, for example, before and after they have been exposed to a certain drug. To test the null hypothesis of no difference between the first and second measurement, we begin by calculating for each individual the difference between the first and second measurement. Then we calculate the mean and

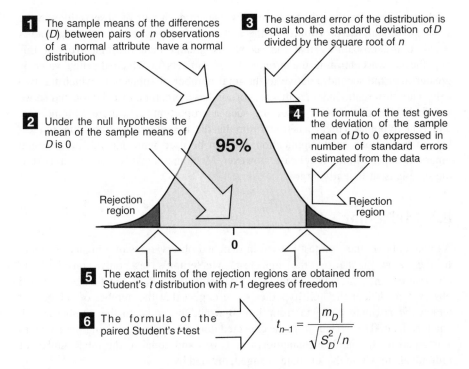

1 The sample means of the differences (*D*) between pairs of *n* observations of a normal attribute have a normal distribution

3 The standard error of the distribution is equal to the standard deviation of *D* divided by the square root of *n*

2 Under the null hypothesis the mean of the sample means of *D* is 0

95%

4 The formula of the test gives the deviation of the sample mean of *D* to 0 expressed in number of standard errors estimated from the data

Rejection region

Rejection region

0

5 The exact limits of the rejection regions are obtained from Student's *t* distribution with *n*-1 degrees of freedom

6 The formula of the paired Student's *t*-test

$$t_{n-1} = \frac{|m_D|}{\sqrt{s_D^2/n}}$$

Figure 8.1   Steps in the paired Student's t-*test.*

standard deviation of those differences. Let us say that the result was 5 with standard deviation 4.44. We estimate the standard error by dividing the standard deviation by the square root of the sample size, that is, $4.44/2.45 = 1.81$. The *t*-value of the paired *t*-test is $5/1.81 = 2.76$. In the table of the *t* distribution with $6 - 1$ degrees of freedom the two-sided 5% rejection limit is 2.57 and, therefore, the sample mean of the differences between the two observations is within the rejection region and we reject the null hypothesis. The *p*-value corresponding to 2.76 standard errors away from the mean is 0.04.

The lower limit of the 95% confidence interval, therefore, is $5 - 2.57 \times 1.81 = 0.35$ and the upper limit is $5 + 2.57 \times 1.81 = 9.65$.

The non-parametric equivalent of the paired *t*-test is the **Wilcoxon signed-rank test**. This test can be used for large samples and for small samples when the attribute does not have a normal distribution. It is important to note, however, that the Wilcoxon test is based on the ranking of the differences between the two observations in each individual. As it is based on the differences between values of an attribute, this test cannot be used with attributes measured in ordinal scales. In this situation, the appropriate non-parametric test is the **sign test**.

In the sign test, the null hypothesis is that, in each pair of observations in the same individual, the probability of the first value being greater than the second is equal to the probability that the second is greater than the first. To test this hypothesis, we determine the proportion of subjects in which the first value is greater than the second, relative to the total number of subjects in which the two values are different. Under H_0 this proportion has a binomial distribution and so we can find the probability of observing such a proportion, or a more extreme one, under the hypothesis that H_0 is true. From this discussion it follows that the sign test has no restrictions on its application and can be used with interval, ordinal, and binary attributes. In the latter case, however, McNemar's test is more often used, if the sample is sufficiently large.

8.3 McNemar's test

As noted above, the sign test is used in matched observations of a binary attribute. In the same manner as the sign test, **McNemar's test** only considers the observations in which the attribute value changed between the first and second observation. Under the null hypothesis, we expect that the number of subjects in which the attribute changed from 1 to 0 would be equal to the number which changed from 0 to 1. Therefore, the expected number of subjects changing from 0 to 1 is equal to the number changing from 1 to 0 and equal to the total number of individuals in whom the attribute changed, divided by 2.

We can create a measure of the discrepancy between the observed and the expected values in the same way as we did for the chi-square test. That is, we calculate, in those who changed from 0 to 1, and in those who changed from 1 to 0, the square of the difference between the observed and expected value divided by the expected value. Adding the two results, we obtain a quantity which follows a chi-square distribution with 1 degree of freedom.

For example, if the attribute changed from 0 to 1 in 52 individuals and from 1 to 0 in 32, there would be a change in the attribute value between two observations on 84 individuals. Under H_0 we would expect that 42 had changed from 0 to 1 and 42 from 1 to 0. McNemar's test is

$$\chi_{(1)}^2 = \frac{(52 - 42)^2}{42} + \frac{(32 - 42)^2}{42} = 4.76$$

We can obtain the test statistic faster by squaring the difference between those who changed from 0 to 1 and from 1 to 0, and dividing the result by their sum. In this example, it would be $52 - 32 = 20$ squared, divided by $52 + 32 = 84$. The result, 4.76, is greater than 3.84, the 5% rejection limit for a chi-square distribution with 1 degree of freedom. Thus, we reject the null hypothesis at the 5% significance level and conclude that it is more frequent that individuals change from 0 to 1.

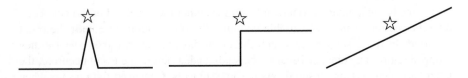

Figure 8.2   Types of events.

8.4   Analysis of events

One of the most important investigations in clinical research is the **analysis of events**. 'Event' refers to an instant in time when a clinical manifestation of short duration arises (e.g., seizure, hypoglycemic coma, myocardial infarction, stroke), or there is a well-defined transition of the condition of the patient to a different state (e.g., death, tumor response to treatment, clinical cure), or when the patient's clinical condition reaches a certain stage of evolution (e.g., normalization of blood pressure, reaching a predefined CD4 count). Figure 8.2 illustrates these three types of events.

Studies aimed at investigating events are always based on longitudinal designs. As we saw in the discussion on study designs, such studies may be one-sample cohort studies on a probability sample of the target population, or cohort studies in the populations exposed and not exposed.

Events may occur only once (e.g., death) or multiple times (e.g., seizures). Events are special attributes, because they must always be represented by two variables. In the case of events that can occur only once, or when investigating only the first occurrence of an event that may occur several times, the event is represented by a binary variable with value 1 if the event occurred and 0 otherwise, associated with an interval variable recording the time elapsed since the moment the subject was first observed until the moment the event occurred. In the case of events that may occur several times, one variable is a count of the number of events observed in each subject and the other is the total time of observation of each subject.

Because of this dependence on time, we need special statistical methods for the analysis of events. In the following sections we will first discuss analytical methods for the case of events that occur only once, and then the methods used in the case of multiple events.

8.5   The actuarial method

We begin by discussing the methods used in the analysis of events that occur only once. We have seen that the data comes from longitudinal studies and, because individuals need to be observed for a considerable length of time, quite often data on the time to the event cannot be obtained for all individuals. We have also seen that the information regarding events has to be encoded in two variables for each event, one being a binary variable indicating whether the event did or did not occur, the other being an interval variable counting the observation time until the event occurred. In those subjects in which the event did not occur, either because they dropped out of the study, or because the study ended before the event occurred, the latter variable contains censored data in some subjects. **Censored data** occurs when we know that the true value exceeds the measured value, but we do not know by how much.

We have seen in Section 4.9 that it is possible to analyze the data by the person-years method, but in this section we will discuss a different approach, generally referred to as **survival-time analysis**.

Survival-time analysis estimates over time the proportion of individuals in whom the event has not occurred. If their data was not censored, that is, if all individuals in the study had been observed until the event occurred, the proportion of individuals in whom the event had not occurred up to a given time point would be easily obtained by dividing, at that time point, the total number of individuals in whom the event has not yet occurred by the total sample size. However, as there are virtually always censored data, it is necessary to discount from the total at risk those individuals who have been censored before that time point. This is what survival-time analysis does.

The **actuarial method** is illustrated in Figure 8.3. The method estimates first the proportion of individuals free of the event in consecutive periods of time of equal duration, by dividing the number of individuals free of the event at the end of each period by the total number of individuals still under observation at the beginning of that period. For example, if at the beginning of month 6 there were still 82 individuals under observation (we call them the **number at risk**), and during that month the event occurred in 5, the estimated proportion of individuals free of the event at the end of month 6 is $(82 - 5)/82$.

As some subjects may have been censored during that month, we halve each one of those because, on average, censored subjects were observed for half of the length of the period. If there were, say, 7 subjects censored during month 6, we would count them only as 3.5 subjects. The estimated proportion of subjects free of the event at the end of month 6 would be, therefore, $(82 - 5)/(82 - 7/2)$ or $77/78.5 = 98.09\%$.

Now suppose that in only 91% of the initial sample the event did not occur up to month 6. What will be the proportion of individuals in the initial sample that will reach month 7 without having the event? It will be 98.09% of the 91% who reached month 6 without the event, that is, $0.9809 \times 0.91 = 0.893$ or 89.3%. This is called the **cumulative probability** of the event not occurring.

In summary, the actuarial method estimates for each equally sized period of time the probability of the event not occurring, taking into account the individuals actually observed during each time period. Then the method estimates the probability of the event not occurring in a subject at any given time by calculating the cumulative probability of the event not occurring since the beginning of the observation period. This procedure is illustrated in Figure 8.3.

1 The longest observation time is divided into periods of equal length.

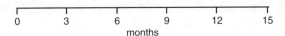

2 The proportions of survivors in each time period are calculated from the number still at risk at each period.

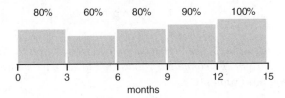

3 For the individual survival probabilities in each time period we estimate the cumulative probability of survival up to that time period.

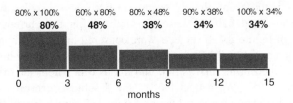

Figure 8.3   The actuarial method.

With the actuarial method, we first look for the subject with the longest observation time and we divide that time into equal intervals. In the example illustrated in Figure 8.4, the initial sample consisted of 74 subjects and the longest observation period was 15 months.

We may divide the 15 months into intervals of, say, 3 months. Then we estimate the probability of a subject reaching the end of the first time interval, that is, the first 3 months without the event having occurred. In the first interval, 6 subjects dropped out of the study and in 5 the study ended before they had completed 3 months of observation (because they had entered the study less than 3 months before the study was terminated).

As these 11 subjects who were censored before 3 months count as half, the total number at risk in the first interval was $74 - 5.5$, or 68.5 subjects. From this number we subtract the 6 subjects who reached the event during this period to obtain the total number who reached the end of the interval without the event occurring, that is, 62.5. Now we estimate the probability that a subject reaches

Interval (months)	Begin total	Censored in the interval	Events in the interval	Number at risk	Probability of the event not occurring	Cumulative probability of the event not occurring
0-3	74	11	6	68.5	0.912	0.912
3-6	57	8	5	53	0.906	0.826
6-9	44	4	3	42	0.929	0.767
9-12	37	7	4	33.5	0.881	0.676
12-15	26	5	2	23.5	0.915	0.618
	a	b	c	d $=a-\frac{1}{2}b$	e $=(d-c)/d$	f $=e_i(f_{i-1})$

Figure 8.4   Illustration of the actuarial method showing on the left the study data and on the right the computed values; the lower part shows the method for calculating those values.

the end of the interval without the event occurring, by dividing one value by the other, 62.5/68.5. The result is 91.2%.

We do the same for the next interval, discounting from the total sample those in whom the event already occurred, those who were censored in the previous period, and half of those censored during the current period. In the example of Figure 8.4, 8 subjects were censored in the 3–6 months period. The population at risk in the beginning of this interval is 74 of the initial sample minus 11 censored in the first period and minus 6 in whom the event already occurred, that is, 57. We deduct half of the 8 subjects censored during this period to obtain the number at risk in this interval (53 individuals). In 5 of these the event occurred during the 3–6 months period. Thus, the probability that an individual who has entered this period reaches the end of the period without the event occurring is estimated to be $53 - 5 = 48$ divided by 53, or 90.6%. But as only 91.2% of the initial sample have entered this period, by 6 months only 90.6% of the 91.2% of the initial sample, or 82.6%, have not reached the event. This value was obtained, naturally, by multiplying the two proportions. We repeat the procedure for each time interval, obtaining for each one the probability of an individual being free of the event since entry into the study.

We have discussed here how to estimate the cumulative probability of an event not occurring. If we wanted to estimate the probability of the event occurring, the procedure would be identical. The actuarial method provides estimates for any of the periods of time. In the example, the estimate of the probability that the event did not occur up to 12 months is 67.6%. Additionally, this method allows us to evaluate the time course of the probability of an event not occurring, known as the **survivor function**, which in many situations is also informative. The most common way of showing the survivor function is with an actuarial curve. Figure 8.5 illustrates the actuarial curve relating to the data of Figure 8.4.

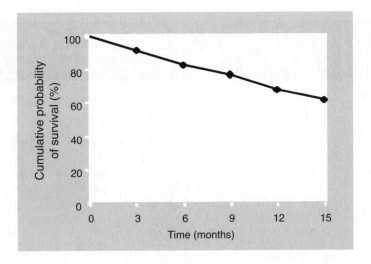

Figure 8.5   Actuarial curve.

8.6 The Kaplan–Meier method

The Kaplan–Meier method is identical to the actuarial method, the only difference being that time intervals are defined by the smallest unit of time used to count observation time. This method is more accurate than the previous one because the probabilities are computed at each unit of time instead of in intervals of arbitrary length.

Figure 8.6 illustrates the procedure for calculating the cumulative probabilities of survival in a cohort of 21 subjects with the Kaplan–Meier method. The time unit used is the day. The first step is to arrange the study data on a table ordered by increasing length of observation time. Each entry in the table corresponds to a study day when something happened, that is, either one or more subjects reached the event (died, in this example) or were censored.

Then we calculate for each of those days the number of individuals still at risk. In the example, 21 subjects initiated the study. On day 30 there were still 21 subjects at risk but then 1 subject died and, therefore, after that day only 20 patients remained under observation. On day 50 there were 20 subjects at risk, 1 subject died and 1 was censored, leaving 18 under observation. On day 51 a subject died, leaving 17 still at risk. On day 66 a subject was censored, leaving 16 in the study. We continue until the last individual, the one who was observed for the longest time in the study.

We can now estimate the probability of survival in each day. On the first day of observation no one died, so the probability of survival on day 1 is 100%. On the second day there were no deaths and, thus, the probability of survival on

Subject number	Days on study	Censored	Died	Number at risk	Probability of survival	Cumulative probability of survival
3	30		1	21	0.952	0.952
5	50		1	20	0.950	0.905
7	50	1		19		
1	51		1	18	0.944	0.855
4	66	1		17		
2	82		1	16	0.938	0.801
6	92		1	15	0.933	0.748
8	120	1		14		
	a	b	c	d	e $=(d-c)/d$	f $=e_i(f_{i-1})$

Figure 8.6   Illustration of the Kaplan–Meier method showing on the left the study data and on the right the computed values; the lower part shows the method for calculating those values.

day 2 is also 100%. The cumulative probability of survival on day 2 is therefore 100% of the 100% who survived to see day 2. The third day is the same, and so on until day 30. On that day there was one death and therefore the probability of survival is not 100%. Only 20 of the 21 individuals at risk survived that day and thus the probability of survival on day 30 is estimated to be $20/21 = 95.2\%$. On day 30 we have that 100% of the subjects included in the study reached that day, but only 95.2% of these 100% survived to this day. Therefore, the probability of survival after 30 days is $95.2\% \times 100\% = 95.2\%$.

On day 31 no one died, so the survival probability on day 31 is 100% and, therefore, 95.2% of the initial subjects were still alive. On each of the days 31 to 49 the probability of survival was 100%, but on day 50 only 19 survived out of the 20 subjects still under study, that is, 95%. Thus, of the 95.2% subjects from the initial sample that survived to day 50, only 95% of them, that is, 90.5%, survived to day 51.

On day 51 there were only 18 subjects under study, because 1 was censored on day 50. One subject died on that day, so the probability of survival on day 51 was 17/18, or 94.4%. Therefore, 94.4% of the 90.5% initial subjects survived after 51 days.

The calculations are done in the same way for each day up to the day that corresponds to the maximum observation time. Because nothing changes on the days when there were no events, calculations need to be done only on those days when events occurred.

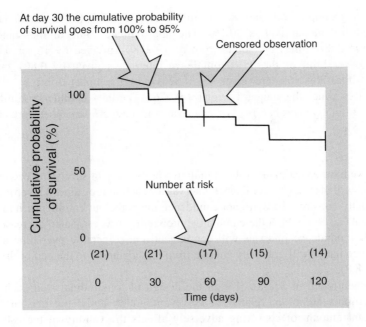

Figure 8.7   Kaplan–Meier curve.

The estimates of the cumulative probabilities at each time point are called the **product-limit estimates** (because they are the result of successive multiplications of survival probabilities as the time intervals are decreased toward zero). Together, these estimates are called the **survivor function**. They are usually presented in a graph called the **Kaplan–Meier curve**. Figure 8.7 shows the curve for the data in the example of Figure 8.6. In addition to presenting the cumulative probability at each time unit, these graphics should include an indication of the subjects at risk on various observation periods and the distribution of censored data over time. These are usually represented by vertical lines over the curve, as shown in Figure 8.7. In a well-managed study, censoring should have occurred randomly over time and in such cases the vertical lines will be distributed evenly along the entire curve.

The Kaplan–Meier method gives us, therefore, the cumulative probability of an event occurring, or of not occurring if one prefers, up to a given moment in time. Each of the computed probabilities, having been estimated from the data, is of course subject to sample variation. As is customary in statistics, we can obtain interval estimates of the true value of the cumulative probability for each unit of time.

We can get an approximate value of the standard error of the cumulative probability on a given day by multiplying the cumulative probability on that day by the square root of the quotient of one minus that probability and the number at risk on that day.

For example, on day 82 in the previous example, the cumulative probability of survival is 80.1%. The approximate value of the standard error is 0.801 times the square root of $1 - 0.801$ divided by 16, or 0.089. The lower limit of the 95% confidence interval is given by 0.801 minus 1.96 times the standard error ($= 0.626$) and the upper limit by 0.801 plus 1.96 times the standard error ($= 0.976$). The 95% confidence limits of the cumulative probability of survival on day 82 are therefore 62.6 to 97.6%.

As we have an estimate of the cumulative probability of survival in each unit of time, we will also have a confidence interval for each unit of time. Therefore, what we actually construct here are not confidence intervals, but **confidence bands**. As the sample size on which the cumulative probability was estimated decreases over time, the confidence intervals will consequently also increase over time and the confidence bands will gradually enlarge from the beginning to the end of the curve (Figure 8.8).

As mentioned in Section 4.9 on the design of longitudinal studies, because the drop-outs are often related to the event under study, the existence of a significant amount of censoring adversely affects the validity of the estimates. It is generally accepted that the drop-out rate should not exceed 15% of the initial sample size due to the risk of biasing the results. When presenting the results of a survival-time study, the median time of observation of individuals should also be given.

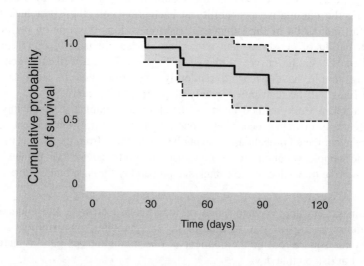

Figure 8.8   Kaplan–Meier curve with 95% confidence bands.

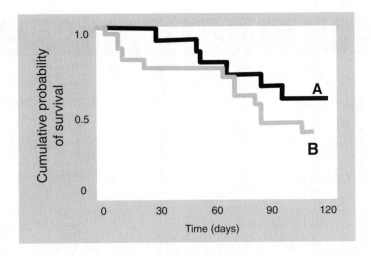

Figure 8.9   The logrank test is used to compare two or more Kaplan–Meier curves.

8.7 The logrank test

The logrank test allows us to test the null hypothesis that an event occurs with equal frequency in different populations. In other words, it tests the hypothesis of equality between several Kaplan–Meier curves (Figure 8.9). As we saw in the previous section, the Kaplan–Meier curve represents the survivor function, that is, the cumulative probabilities of survival at each unit of time. These probabilities are obtained from estimates of the rate of occurrence of the event at each unit of time. Taken together, these values are called the **hazard function**.

Thus, in the abstract, the logrank test compares the hazard function of an event between two or more groups. It is therefore a test of the association between an event and an attribute measured on a nominal scale defining different populations. It is possible to use the logrank test for attributes measured on ordinal or interval scales, but then it is necessary to group the values into classes, thereby converting these attributes into nominal attributes.

The logrank test has affinities with the chi-square test. It consists of comparing the number of events observed in each group to the number expected if the hazard function was the same for all. A measure of the departure of the observed from the expected numbers is used. Under the null hypothesis of equality of the hazard functions this measure follows approximately a chi-square distribution, so we can determine whether the observed departure is an infrequent occurrence under H_0.

Consider a study in which we intend to investigate the existence of an association between the occurrence of an event and the gender of the individuals. Suppose the time unit is months, the event is death, and gender is a binary attribute with values male and female.

Observation time	Males R	Males C	Males D	Females R	Females C	Females D	Total R	Total C	Total D	Expected M (TD/TR × MR)	Expected F (TD/TR× FR)
5	21			28		1	49		1	0.429	
6	21			27	1		48	1			
11	21			26		2	47		2	0.894	
13	21			24		1	45		1	0.467	
24	21			23		1	44		1	0.477	
30	21		1	22			43		1	0.488	
50	20		1	22			42		1	0.476	
50	19	1		22			41	1			
51	18		1	22			40		1	0.450	
63	17			22		1	39		1	0.436	
65	17		1	21			38		1	0.447	
66	16			21	1		37	1			
69	16			20		2	36		2	0.889	
79	16			18		1	34		1	0.471	
82	16		1	17		2	33		3	1.455	
92	15		1	15			30		1	0.500	
102	14			15		1	29		1	0.483	
115	14	14		14	13	1	28	27	1	0.500	
Total			6			13				8.861	10.139

M: male   F: female   R: at risk   C: censored   D: dead   TD: total dead   TR: total at risk
MR: male at risk   FR: female at risk

Figure 8.10   Illustration of the logrank test for the comparison of Kaplan–Meier curves: left, the study data; right, the computed values.

For the logrank test, we sort the observation times of all individuals in ascending order. Then, we write down the population at risk, the dead, and the censored in each group separately, as shown in Figure 8.10. If the null hypothesis is true and the hazard function is actually equal in both groups, the probability of an event occurring in a given month is the same in the two groups and can be estimated by dividing the total number of events in that month by the total number at risk.

For example, on month 5 there was 1 death and therefore the estimate of the probability of death on month 5 is 1/49 subjects at risk on that month, or 2.04%. As on month 5 there were 21 males and 28 females in the study, if the mortality rate is indeed equal in both genders, the expectation is that 2.04% of the 21 males have died, as well as 2.04% of the 28 females, or 0.429 males and 0.571 females. On month 11 there were 2 deaths, and therefore the estimate of the mortality rate in this month is 2/47 individuals at risk or 4.26%. The expected number of deaths on that month, if the mortality rate is identical in both groups, would thus be 4.26% of 21 males (= 0.894) and 4.26% of 26 females (= 1.106).

As shown in Figure 8.10, the total number of deaths observed among males was 6 and among females was 13. The total expected deaths is obtained by adding all the expected deaths on each month of observation. As the total number of observed deaths equals the sum of expected deaths in the two groups, we need only to calculate the deaths observed in one of the groups and obtain the value for the other group by subtracting the total number of expected deaths in one group from of all deaths observed in the two groups. Thus, under H_0 the number of expected deaths in males is 8.861 and in females is 10.139.

The test statistic, which quantifies the departure of observed to expected values under H_0, is given by the sum across all groups of the squared difference between observed and expected deaths divided by the number of expected deaths. In this example, its value is the square of 6 minus 8.861 divided by 8.861, plus the square of 13 minus 10.139 divided by 10.139. The result, 1.73, follows a chi-square distribution with degrees of freedom equal to the number of groups minus one. The 5% rejection limit of a chi-square distribution with 1 degree of freedom is 3.84, and thus the value we obtained is not large enough to enable us to reject the null hypothesis. If we read off the p-value in a statistical table of the chi-square distribution, we obtain the value 0.19.

We can obtain an estimate of the relative risk, in this context usually called the **hazard ratio**, by dividing the **relative death rates** (the ratio of observed to expected deaths) of the two groups. If the observed number of deaths in males and females were 6 and 13, and the expected number of deaths were 8.861 and 10.139, the hazard ratio of males to females would be 6/8.861 divided by 13/10.139, or 0.53. The risk of mortality among men is therefore about half of the risk among women. The logrank test tells us that the difference in mortality between genders is not statistically significant ($p = 0.19$). This means that if we constructed the 95% confidence interval for the hazard ratio, it would include the value 1 which, in turn, translates to equal mortality risk in both groups.

The determination of confidence limits for the hazard ratio is complicated. An alternative is to obtain them with Cox regression, a statistical method that can also replace the logrank test and which will be discussed later on in this book.

8.8 The adjusted logrank test

Consider now that we have explored the issue of an eventual association between gender and mortality a little further, and noticed that the two groups had a marked difference in the age distribution. Among the 21 males there were 11 (52%) over 50 years old, but among the 28 women only 9 (32%) were older than 50 years. As mortality is always associated with age, this observation leads us to speculate that, had the groups had the same age distribution, the excess mortality observed in women relative to the expected mortality could have been even greater, and perhaps we would have had evidence of a difference in mortality between genders. An attribute such as age, which is simultaneously related with the attribute that we are investigating and the event under study, is called a **confounding factor**.

The adjusted logrank test is intended to test the association between an event and a nominal variable, eliminating the confounding effect of one or more external variables. This test is an extension of the logrank test and, therefore, can be used only with nominal attributes or, as in this example, with interval attributes (or ordinal) previously converted into nominal attributes.

The adjusted test postulates that the hazard function is not necessarily the same in the groups defined by the values of the adjustment variable (confounding factor). We must therefore estimate a hazard function for the group of subjects under 50 years old and another for the group over 50 years old. Therefore we calculate the observed and the expected number of deaths in both groups, separately for each level of the confounding variable, to create a measure of the deviation of the observed to the expected number of deaths under H_0. Then we sum those measures to obtain a test statistic in the same way as we did for the logrank test. The test statistic follows the chi-square distribution with degrees of freedom equal to the number of groups minus one.

In our example, we begin by separating the two age groups and estimating the hazard function for each one of them. Under H_0, the function is the same for men and women within each age group, so we can calculate the expected number of deaths in men and women in each age group separately. The results are presented in Figure 8.11.

Thus, in the group under 50 years old, 1 subject died on month 5 among the 29 at risk on that month. The death rate in this group on month 5 is estimated as 1/29, or 3.45%. Under H_0 we expect the death of 3.45% of the 10 men, or 0.345, and 3.45% of 19 the women at risk on that month, or 0.655. In the group over 50 years old, the first death occurred on month 13, when there were 19 individuals at risk. The death rate on that month for that age group is thus estimated at 1/19, or 5.26%, and under H_0 the rate is equal in both genders. Consequently, the expected number of deaths is 0.579 in males and 0.421 in females.

The test statistic of the adjusted logrank test is constructed by adding the total expected and observed deaths for each gender. In Figure 8.11 we can see that, in men, there were 6 observed deaths and 8.663 expected deaths under H_0; in women, there were 13 observed deaths and 10.337 expected deaths under H_0. We obtain the test statistic in the usual way, by summing across groups the squared differences between the observed and expected deaths divided by the expected deaths. In this example it is the square of 6 minus 8.663 divided by 8.663, plus the square of 13 minus 10.337 divided by 10.337. The result (1.50) follows, as in the unadjusted test, a chi-square distribution with degrees of freedom equal to the number of groups minus one. The value obtained is below the 5% rejection limit of a chi-square distribution with 1 degree of freedom, so we cannot reject the null hypothesis. Finally, even after taking into account the age difference, we still cannot conclude on a difference between the survival curves for both genders.

Age <50 years old

Observation time	Males			Females			Total			Expected M (TD/TR × MR)	Expected F (TD/TR × FR)
	R	C	D	R	C	D	R	C	D		
5	10			19		1	29		1	0.345	
11	10			18		2	28		2	0.714	
24	10			16		1	26		1	0.385	
50	10		1	15			25		1	0.400	
51	9		1	15			24		1	0.375	
65	8		1	15			23		1	0.348	
69	7			15		2	22		2	0.636	
79	7			13		1	20		1	0.350	
92	7		1	12			19		1	0.368	
115	6	6		12	11	1	18	17	1	0.333	
Total			4			8				4.255	7.745

Age >50 years old

Observation time	Males			Females			Total			Expected M (TD/TR × MR)	Expected F (TD/TR × FR)
	R	C	D	R	C	D	R	C	D		
6	11			9	1		20	1			
13	11			8		1	19		1	0.579	
30	11		1	7			18		1	0.611	
50	10	1		7			17	1			
63	9			7		1	16		1	0.563	
66	9			6	1		15	1			
82	9		1	5		2	14		3	1.929	
102	8	8		3	2	1	11	10	1	0.727	
Total			2			5				4.408	2.592

R: at risk   C: censored   D: dead

Figure 8.11   Illustration of the adjusted logrank test.

The **adjusted hazard ratio** is obtained in the same manner as above, by dividing the relative death ratios of the two groups. In this example it would be 6/8.663 divided by 10/13.337, or 0.55.

8.9 The Poisson distribution

If the event under study can occur several times in the same individual during the period of observation, we cannot use the methods we have just discussed, unless we sacrifice information and use only the first occurrence of the event, for example. The methods that allow us to analyze events occurring repeatedly are collectively called **event-count analyses**.

In cohort studies designed to investigate this type of event we obtain for each individual the duration of the observation period and the number of events during this period. If the observations are independent and the occurrence of the event does not show a time trend, we can divide the total number of events observed on the entire cohort by the sum of observation times of all subjects in the cohort.

This gives us a measure of the frequency of the event called, as we saw earlier, **incidence density**, which is expressed as number of events per person-years.

Thus, in the case of events that can occur repeatedly in the same individual the quantity we want to estimate in the population is the incidence density. Naturally, the incidence density observed in a study is an observation from a random variable and is subjected to sampling variation. Therefore, we aim to find confidence limits for the population incidence density and in order to do that we need to know the distribution of sample incidence densities.

Suppose we monitored a cohort of 156 patients for the appearance of colon polyps with the aim of estimating the rate of occurrence of colorectal adenomas. Patients in this cohort were observed for variable lengths of time and, in all, these 156 patients were observed for a total of 15 295 months. During this time a total of 171 adenomatous polyps were identified and removed by colonoscopy.

If we look closely at the data we will find that during most months no polyps were identified, that on a few dozen months a single polyp was identified, and that, very occasionally, two polyps were identified in the same month. Additionally, the occurrence of a polyp in any given month is independent of whether or not polyps occurred in the previous or subsequent months. Therefore, whatever happens in each month is independent of what happens in any other months. Consequently, we may consider that the event is a binary variable with a defined probability and that the total number of observed polyps corresponds to the total number of hits in 15 295 independent trials. Thus, as we saw earlier, we expect the total number of hits (the polyps) observed on a cohort to have a binomial distribution, whose parameters π and n are the incidence rate of the event and the total observation time, respectively.

However, in situations such as those typically observed in incidence studies, where the probability of occurrence of an event per unit of time (the incidence rate) is very small and the number of trials (the total time of observation, or persons-time) is very large, the total number of events observed in a cohort is best described by the **Poisson distribution** than by the binomial distribution.

The Poisson distribution is therefore suitable for describing the probability distribution of rare events. This distribution counts, for a binary variable, the number of hits in a number of trials. This probability distribution has the peculiarity of possessing a single parameter, the mean, because its variance is equal to the mean.

For estimating the value of the standard error of the number of events, we use the estimate of the population variance obtained in our sample. In this example, the estimate of the variance of the number of events is equal to the total number of events observed, that is, 171. As the incidence density is a measure obtained by dividing the total number of events by persons-time, the estimate of the variance of the incidence density is equal to the number of events also divided by persons-time.

Therefore, we can estimate the standard error of the incidence density by the square root of the quotient of the sample variance of the incidence density and the number of trials (the persons-time). To find the confidence limits we refer to the Poisson distribution. However, there is an easier ways to obtain these limits.

We can take advantage of the convergence of distributions. We saw above that the Poisson distribution arises when, in a binomial distribution, the probability of a hit is very small. This means that when the probability of an event increases, and therefore also the number of observed events, the Poisson distribution converges to the binomial. In addition, we have also seen that the binomial, in turn, converges to the normal when the number of trials is large, as is typically the case in this type of problem. In conclusion, when the number of events is large (say, more than 10), we can use the normal approximation to find confidence limits using the square root of the number of events divided by the total observation time as an estimate of the standard error of the sample incidence density.

Here is how to construct 95% confidence limits for the incidence density (ID). The standard error of ID is the square root of the variance divided by the number of trials, in this case the persons-time (PT):

$$SE(ID) = \sqrt{\frac{var(ID)}{PT}}$$

On the other hand, the variance of ID is equal to the number of observed events (the incidence I of the event) divided by the persons-time

$$var(ID) = \frac{I}{PT}$$

Therefore, the first expression becomes

$$SE(ID) = \sqrt{\frac{I/PT}{PT}} = \sqrt{\frac{I}{PT^2}}$$

which can be written as

$$SE(ID) = \frac{\sqrt{I}}{PT}$$

The sample variance of the number of events is equal to the total number of observed events. In the example of the colon polyps, its value is 171. The sample variance of ID is equal to the sample ID and both are equal to 171 divided by PT, or $171/15\,295 = 0.0112$.

The standard error of ID is the square root of the total number of observed events divided by PT, or $0.000\,85$. Using the normal approximation, the 95% confidence limits are $0.0112 - 1.96 \times 0.000\,85 = 0.0095$ per person-month and $0.0112 + 1.96 \times 0.000\,85 = 0.0129$ per person-month.

Another way of constructing the confidence interval is by working on a linear scale using the natural logarithm of the incidence density, as we did before for the confidence limits of odds ratios. The standard error of the logarithm of the sample incidence density is equal to the square root of the reciprocal of the total number of events. Thus, the 95% confidence limits for the logarithm of the population incidence density are obtained by adding and subtracting 1.96 times the standard error from the logarithm of the incidence density. Then we return to the original scale by exponentiation of those limits.

In the example, the sample ID is 171/15 295, or 0.0112, and its logarithm is −4.494. The standard error of the logarithm of the incidence density is equal to the square root of 1/171, or 0.076 47. The 95% confidence limits of the logarithm of the population incidence density are −4.494 − 1.96 × 0.076 47 and −4.494 + 1.96 × 0.076 47, that is, −4.6439 to −4.3441. Exponentiation of these values gives 0.0096 and 0.0130, practically the same result we obtained above. Thus, the 95% confidence limits for the incidence density are 0.96 to 1.30 per 100 person-months. The interpretation of the incidence density is therefore that in a cohort of 100 people we would expect to observe about 1 to 1.3 polyps every month, or alternatively, on a person observed for 100 months, we would expect to observe about 1 to 1.3 polyps.

8.10 The incidence rate ratio

We saw earlier that differences between proportions are generally expressed as ratios. The same applies for rates, such as the incidence density, and the difference between two incidence rates can be expressed as the **incidence rate ratio**. Therefore, the incidence rate ratio (IRR) is a measure of association between a binary variable and an event. Of course it can also be used as a measure of association for attributes on any scale, as long as they are converted to a binary attribute. We can investigate the association of events with interval attributes without having to dichotomize them by a method called **Poisson regression**. In this section we will discuss the problem of comparing only two incidence densities.

We will expand the example of the previous section. Suppose we now want to investigate whether the incidence of colorectal polyps is associated to the gender of individuals. In other words, we want to find whether there is a difference between the incidence rates of colorectal polyps in the male and female populations.

In our sample, men were observed for a total of 11 165 months and during that time a total of 139 polyps were identified. Women were observed for a total of 4130 months and a total of 32 polyps were identified. The sample incidence density in men was 139/11 165 = 0.0125 and in women 32/4130 = 0.0077. The estimate of the IRR is, therefore, 0.0125/0.0077 = 1.6068. The 95% confidence limits for the population IRR are obtained, as before, using the logarithm of IRR, which we denote by ln(IRR).

In this example the logarithm of 1.6068, the incidence rate ratio, is 0.4742. The standard error of ln(IRR) is equal to the square root of the sum of the reciprocal of the events observed in each group. In this example, its value is equal to the square root of 1/139 + 1/32, or 0.1961.

Using the normal approximation we compute the 95% confidence limits of ln(IRR). The lower limit is $0.4742 - 1.96 \times 0.1961 = 0.0898$ and the upper limit is $0.4742 + 1.96 \times 0.1961 = 0.8586$. The 95% confidence limits in the original scale are obtained by the exponentiation of those values, giving 1.094 and 2.360.

The 95% confidence interval for the population IRR in our example is 1.09 to 2.36. This means that the incidence of colorectal polyps is, with 95% confidence, somewhere between 9 and 136% greater in men than in women. Since the value 1 is not within the 95% confidence interval, we can reject the null hypothesis of no difference in the population incidence densities between genders, at the 5% significance level.

Instead of calculating the confidence limits of ln(IRR) followed by exponentiation of the results, we can obtain the limits directly on the original scale by multiplying the observed IRR by the exponential of $-1.96 \times SE \ln(IRR)$ for the lower limit, and by the exponential of $+1.96 \times SE \ln(IRR)$ for the upper limit.

We can build a statistical test by dividing ln(IRR) by the standard error of ln(IRR) and reading in the table of the normal distribution the probability of obtaining such a high value under H_0. In this example, the z-value is 2.418, which corresponds to a p-value of 0.016. There is therefore evidence that male gender is associated with a higher incidence of colorectal polyps, which is estimated to be from 9 to 136% greater than the incidence in females.

Event-count analysis has a number of limitations due to the strong assumptions it makes. One assumption is that the variance of the number of events is equal to the mean, which may not be valid if some individuals have zero events and others have a large number of events. Another assumption is that the incidence rate is constant over time, which may not be the case in many clinical situations, such as the rate of deaths from cancer or the incidence of adverse events from medication. For this reason these methods should be used with caution.

9

Statistical modeling

9.1 Linear regression

The statistical method used for the investigation of associations between two interval variables is linear regression. This method is based on the same principles of the methods discussed so far, modeling the relationship between the variables and testing the plausibility of the model under the null hypothesis of independence of the variables.

Suppose we wanted to investigate the association between the body height of a person and the forced vital capacity (FVC), a measure of pulmonary function that measures the volume of air in liters expelled from the lungs during a forced exhalation. We would need a random sample of individuals and we would measure for each one of them the body height and the FVC. The study data would look like the data of Figure 9.1, which were collected on a sample of 20 healthy subjects. On the right, on the **scatterplot**, the most adequate form of visual presentation of an association between continuous variables, each point represents an individual whose FVC and height are read, respectively, on the vertical and horizontal axis.

When there is directionality in the relationship, that is, when the values of one attribute determine in some way the values of the other, the determinant attribute is placed on the horizontal axis and the determined attribute on the vertical axis. In this example, if the two attributes are related, then the height of a person may determine the FVC to some extent, but the FVC certainly does not determine the height. The putatively determined attribute is called the **explained or dependent variable**. The determinant attribute is called the **explanatory or independent variable**. Although it is generally recognized that the terms explained and explanatory are more adequate to designate the variables, in the literature there is a clear preference for the terms dependent and independent. We will just follow that trend and use the latter terms.

If now we want to represent the relationship between the two attributes, the simplest way is of course to draw a line running approximately through the center

---

Biostatistics Decoded, First Edition. A. Gouveia Oliveira.
© 2013 John Wiley & Sons, Ltd. Published 2013 by John Wiley & Sons, Ltd.

Height (cm)	FVC (liters)
171	3.44
167	3.52
170	3.63
159	2.19
165	3.52
180	4.23
170	4.14
166	3.87
170	4.84
162	2.78
160	2.86
163	4.03
160	2.05
164	3.02
168	3.64
171	4.25
177	4.66
167	3.93
162	3.79
167	4.14

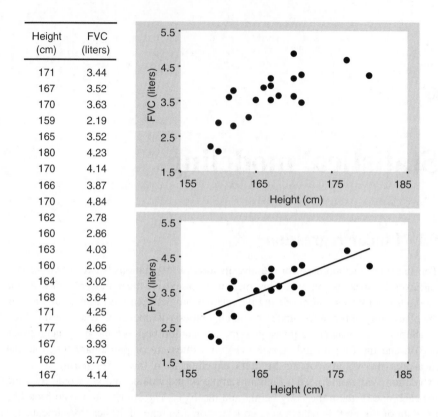

Figure 9.1   Representation of the relationship between two interval variables.

of the cloud of points. If we want to represent the relationship in the simplest way possible, without a doubt we will choose a straight line, the simplest of all lines. In the graph at the bottom of Figure 9.1, the line represents a model of the association between the two attributes that, although very simple, is an impressively powerful method for describing and analyzing relationships among attributes, as we will see in the following sections.

The name given to this line is the **regression line** of the dependent on the independent variable. Intuitively, we realize that a regression line will be horizontal when there is no relationship between the attributes, because it means that the average value of the dependent variable (FVC in this example) does not vary with the independent variable (height). In other words, a horizontal line means that the average value of FVC is the same whatever the height and, therefore, the two attributes are not related.

On the other hand, if the line is oblique, this means that FVC values tend to increase with height (as in Figure 9.1), or decrease if the line has a downward slope. That is, the values taken by the dependent variable are conditioned by the values of the independent variable and thus there is an association of the two variables.

Thus, a straight line is the simplest model of the relationship between two interval-scaled attributes, and its slope gives us an indication of the existence of an association between them. Therefore, an objective way to investigate an association between interval attributes will be to draw a straight line through the center of the cloud of points and measure its slope. If the slope is zero, the line is horizontal and we conclude that there is no association. If it is non-zero, then we can conclude on an association.

So we have two problems to solve: how to draw the straight line that best models the relationship between attributes and how to determine whether its slope is different from zero. We will begin by discussing the first problem.

9.2 The least squares method

Of all the straight lines we could draw through the cloud of points, the one that best represents the relationship between the two variables would logically be the line that runs exactly through the middle of all the points. But what does this mean, and how do we find that line?

If we think of a circumference, its center is defined by the point which is exactly at the same distance from all other points. In slightly more formal terms, we could say that the center is, of all the points we could choose inside a circumference, the one in which the variance of the distances to all the points on the circumference is the smallest of all (and equal to zero in this case). Generalizing this definition to any other shape, regular or irregular, we can define the center as the point in which the variance of the distances to all other points is the smallest of all. By the same token, we can define the line passing through the middle of all points as the one in which the variance of the distances of the points to the line is minimal.

As is well known, the position of a straight line in a two-way graph can be defined by its **slope** and its **intercept**. Thus, each value y^* of the variable Y, on the line shown in Figure 9.2, is defined by the slope, that is, a certain value b, which multiplies the value of X, to which is added the value a of Y corresponding to the value 0 of X. In other words, each value y^* is equal to $a + bx$. This is the **equation of the straight line**.

Thus, in a scatterplot in which we fitted a straight line to the cloud of points in order to model the relationship between the attributes, the location of each point on the graph can be expressed as a function of that line by $y = a + bx + e$ (see Figure 9.2), where e is the distance between the observed value y and the value y^* predicted by the regression line. Therefore, e is a measure of the distance of a point to the line. So now we can define the best straight line, the one that runs exactly through the middle of all the points, as the one line among all lines we can possible draw through a cloud of points for which the variance of e is the smallest.

We therefore need to find the line that minimizes the sum of squares of the differences of Y to the regression line, those differences being the quantities referred

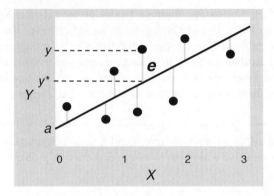

Figure 9.2   Rationale of the least squares method.

to above and in Figure 9.2 and as *e*. The method for determining this straight line is called the **least squares method**. Let us see what the foundation of this method is.

Figure 9.3 shows a scatterplot of three pairs of values (x, y) and we want to find the straight line that minimizes the sum of squared deviations of *y* to the line. Let us start by finding a point that, because of its position exactly at the center of the cloud of dots, certainly belongs to the line we want to find. That point is the centroid of the cloud of dots. From geometry we know it is defined by the means of X and Y, which we denote by $\bar{x}$ and $\bar{y}$. Now let us draw for every point (x, y) a straight line joining the point to $(\bar{x}, \bar{y})$.

We can now find the slope of each of these lines, using $(\bar{x}, \bar{y})$ as reference. By doing so we are, in essence, creating a new variable that condenses the information of each pair of observations (x, y) into a single value, the slope of a straight line passing through the point defined by the pair of values and by $(\bar{x}, \bar{y})$. If we now average the values of this new variable, what do we get?

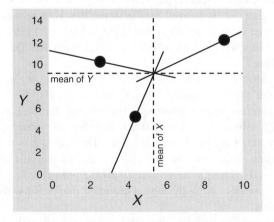

Figure 9.3   Rationale of the least squares method.

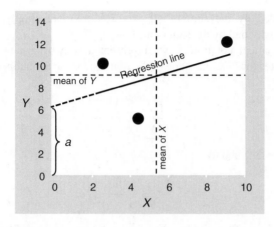

Figure 9.4   Rationale of the least squares method.

We get a quantity that is the average of the slopes of all the straight lines that we can draw connecting each point to $(\bar{x}, \bar{y})$. However, we need to consider that the points farthest from $(\bar{x}, \bar{y})$ are less frequent, and therefore we should include a weighting factor to compensate for the lower probability of observing the more extreme values in a sample. We can do this by assigning a weight to each value such that more extreme values have greater influence on average. Thus, we compute a weighted average of all the slopes to obtain the slope of the regression line we wanted to find.

What are the properties of this average? In any distribution, the quantity for which the sum of the squares of the differences to all other values in the distribution is the smallest of all is, precisely, the mean of the distribution. In other words, the mean is the value that minimizes the sum of squares of the deviations.

Accordingly, the weighted mean of the slopes of all straight lines that can be drawn in a sample of pairs of values (x, y) joining each pair with $(\bar{x}, \bar{y})$ is, by definition, the slope of the straight line that minimizes the sum of squares of the deviations, that is, of the least squares regression line.

Once we know the value of the slope b, obtaining the value of the intercept a in the equation of the straight line $Y = a + bX$ is straightforward: if $\bar{y} = a + b\bar{x}$, then $a = \bar{y} - b\bar{x}$ (Figure 9.4).

Let us now find what the value of the slope of the least squares line is. Using $(\bar{x}, \bar{y})$ as reference, the slope m of a straight line through point (x, y) is equal to $(y - \bar{y})/(x - \bar{x})$. Now, to compute the average of the slopes we will use as weights the distance of each value of X to the mean of X. As it is irrelevant for this purpose whether the difference is positive or negative, we will square the difference to remove its sign. The weights are, therefore, $(x - \bar{x})^2$.

A weighted average is obtained by the sum of the product of each value and its corresponding weight, divided by the sum of all the weights. Therefore, the weighted average of slopes is the sum of all products of $(y - \bar{y})/(x - \bar{x})$ by

$(x - \bar{x})^2$ divided by the sum of all $(x - \bar{x})^2$ or, more simply, the sum of $(y - \bar{y})(x - \bar{x})$ divided by the sum of all $(x - \bar{x})^2$.

In mathematical notation, if we represent a sum by a capital sigma $\sum$ (the Greek letter for 'S'), the slope of the regression line is obtained by

$$b = \frac{\sum \left[\frac{(y - \bar{y})}{(x - \bar{x})} (x - \bar{x})^2 \right]}{\sum (x - \bar{x})^2}$$

which can be simplified to

$$b = \frac{\sum (y - \bar{y})(x - \bar{x})}{\sum (x - \bar{x})^2}$$

In words, this expression says that the slope of the least squares line is equal to the sum of the products of the distances of the points to the means of the dependent and independent variables, divided by the sum of squares of the independent variable.

9.3   Linear regression estimates

In the example of the relationship between FVC and body height, the equation of the regression line obtained by the least squares method was FVC = $-12.92 + 0.099 \times$ height. The interpretation of this result is evident, taking into account the meaning of the coefficients a and b in the equation of the line. Thus, the value 0.099, which multiplies the independent variable, corresponds to the coefficient b, the slope of the line, here called the **regression coefficient**. Therefore, 0.099 liters is by how much FVC increases for each increment of 1 cm in height. If the relationship between FVC and height was the inverse, this coefficient would have a negative sign. It is also useful to think of the value of the regression coefficient as the difference of the mean values of the dependent variable between any two consecutive values of the independent variable.

The value -12.92, here called the **regression constant**, represents the inter-section of the line and the vertical axis, that is, the value of the dependent variable predicted by the regression when the independent variable takes the value 0. At first glance, this would mean that the predicted value of FVC is -12.92 liters for a height of 0 cm, obviously an impossible value for FVC. However, this interpretation is not legitimate because the regression line was determined only for height values between 159 and 180 cm. Outside this range, we cannot make predictions about the value of FVC because we do not know whether the slope of the regression line will be the same outside this range.

Let us find the predicted value of FVC for the minimum height in our sample, 159 cm. Substituting this value into the equation and solving $-12.92 + 0.099 \times 159$, we get 2.82 liters for FVC, not terribly different from the observed value of 2.19 liters.

This difference of $2.82 - 2.19 = 0.63$ liters, the error in the prediction of the value of FVC, corresponds to the value of e mentioned above and is called a **residual**. The residual encompasses the measurement error of height, the measurement error of FVC, and part of the variability of FVC that is not explained by height.

Clearly, individuals of the same height do not all have the same FVC. Therefore, if we repeated this study for a different sample of individuals with the same heights as those in the study, the values we would find for FVC would not be exactly the same. Consequently, the position of the regression line will vary slightly from sample to sample. More specifically, in each repetition of the study we would obtain slightly different values for the regression constant and coefficient.

The parameters of the regression line, the slope and intercept, are therefore subjected to sampling variation and the values of these parameters obtained from a sample are merely point estimates of the true values of the intercept and slope. This means that, if we know the sampling distribution of these parameters, we can construct confidence intervals for their true values.

As we saw earlier, the indication of the existence of an association between variables is given by a non-zero slope of the regression line. Thus, although very occasionally in clinical research it may be of interest to estimate the true value of the intercept, we are usually only interested in estimating the true value of the regression coefficient.

Let us consider first the probability distribution of the sample regression coefficients. We have seen that the coefficient b is the result of the sum of observations of random variables, the slopes of the straight lines connecting each point to the center of the cloud of points. Therefore, from the central limit theorem, if the number of observations is large the regression coefficient will have a normal distribution. In small samples its distribution will be normal only if the slopes of the straight lines connecting each point to the center of the cloud of points have a normal distribution. The numerator of the slopes is a random variable subtracted from a constant, and the denominator, if the independent variable is interval, is also a random variable subtracted from a constant. Therefore, in small samples we know that the regression coefficient has a normal distribution if the dependent variable has a normal distribution at each value of the independent variable, and if the independent variable (if it is interval) has also a normal distribution.

Now that we know that, under these assumptions, the sampling distribution of the regression coefficient is normal, in order to build confidence intervals we need only to estimate the **standard error of the regression coefficient**.

The same reasoning that we used earlier to find the standard error of sample means will allow us to find the standard error of the regression coefficient. We saw above that the regression coefficient b is the weighted average of the slopes m of all the straight lines that we can draw connecting each of the points defined by a pair of values (x, y) to $(\bar{x}, \bar{y})$, and that the weighting factor is $(x - \bar{x})^2$. Therefore, the squared standard error of b is the weighted variance of the slopes m divided by the sample size. This turns out to be equal to the variance of the residuals divided by the sum of squares of the independent variable.

We need to find what the variance of the slopes m is. A weighted variance is equal to the sum of the product of the squared differences to the weighted mean and the weighting factor, divided by the sum of all the weights. As shown in Figure 9.5, the difference of each slope m from the weighted mean of the slopes (the slope of regression line) is $(y - \bar{y})/(x - \bar{x}) - (y^* - \bar{y})/(x - \bar{x})$ or $(y - y^*)/(x - \bar{x})$.

Now, $(y - y^*)$ is the residual e, the deviation of the observed y to the value y^* predicted by the regression. So we can write the difference from the slope m to the weighted average of slopes as $e/(x - \bar{x})$. Thus, the squared deviations of each slope m from the weighted mean is $e^2/(x - \bar{x})^2$.

The weighted variance of the slopes m is the sum of $e^2/(x - \bar{x})^2$ multiplied by the weighting factor $(x - \bar{x})^2$, divided by the sum of the weights.

$$\text{var}(m) = \frac{\sum \dfrac{e^2}{(x - \bar{x})^2}(x - \bar{x})^2}{\sum (x - \bar{x})^2}$$

As the terms $(x - \bar{x})^2$ in the numerator cancel out, the weighted variance is equal to $\sum e^2 / \sum (x - \bar{x})^2$.

Therefore, the estimate of the standard error of b will be the weighted variance of the slopes m divided by the sample size. Actually, such an estimate will be biased, because the two parameters of the regression line a and b are not the true population values α and β, but were chosen to define a line through the exact center of the observed values. Therefore, we must divide the weighted variance of the m by $n - 2$ to account for the two parameters estimated from the sample and thus obtain an unbiased estimate of the standard error of b.

We can simplify this expression further. Let us write it first as

$$\text{SE}(b) = \sqrt{\frac{\sum e^2 / \sum (x - \bar{x})^2}{(n - 2)}}$$

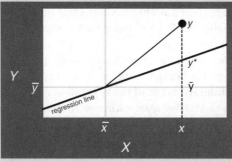

Figure 9.5   The slope m *of a line connecting* $(\bar{x}, \bar{y})$ *to an observation* (x, y) *is* $(y - \bar{y})/(x - \bar{x})$. *The weighted average of the slopes is* $(y^* - \bar{y})/(x - \bar{x})$. *Their difference is* $(y - y^*)/(x - \bar{x})$.

This is the same as

$$SE(b) = \sqrt{\frac{\sum e^2}{(n-2)\sum(x-\bar{x})^2}}$$

but in this expression $\sum e^2/(n-2)$, or $\sum(y^*-y)^2/(n-2)$ if you prefer, is the variance of the residuals since e has zero mean, which we call the **residual mean square**. Therefore, the standard error of b can be written as

$$SE(b) = \sqrt{\frac{\text{var}(e)}{\sum(x-\bar{x})^2}}$$

Figure 9.6 shows the procedure for obtaining the estimate of the standard error of the regression coefficient. In the example of the regression of FVC on height, the sum of squares of the residuals is 4.835. The estimate of the variance of the residuals is obtained by dividing 4.835 by $n-2=18$ degrees of freedom. The result is 0.269.

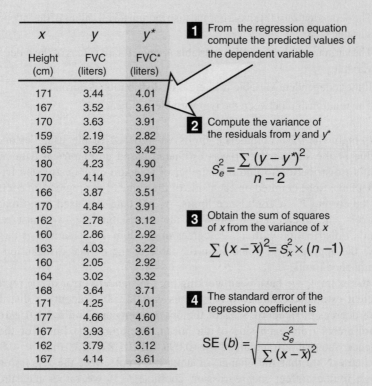

x	y	y^*
Height (cm)	FVC (liters)	FVC* (liters)
171	3.44	4.01
167	3.52	3.61
170	3.63	3.91
159	2.19	2.82
165	3.52	3.42
180	4.23	4.90
170	4.14	3.91
166	3.87	3.51
170	4.84	3.91
162	2.78	3.12
160	2.86	2.92
163	4.03	3.22
160	2.05	2.92
164	3.02	3.32
168	3.64	3.71
171	4.25	4.01
177	4.66	4.60
167	3.93	3.61
162	3.79	3.12
167	4.14	3.61

1 From the regression equation compute the predicted values of the dependent variable

2 Compute the variance of the residuals from y and y^*

$$s_e^2 = \frac{\sum(y-y^*)^2}{n-2}$$

3 Obtain the sum of squares of x from the variance of x

$$\sum(x-\bar{x})^2 = s_x^2 \times (n-1)$$

4 The standard error of the regression coefficient is

$$SE(b) = \sqrt{\frac{s_e^2}{\sum(x-\bar{x})^2}}$$

Figure 9.6   Steps in the calculation of the standard error of the regression coefficient.

The standard error of the regression coefficient is the square root of 0.269 divided by the sum of squares of height. An easy way to obtain the sum of squares is by multiplying the variance of height by $n - 1$. Thus, as the variance of height is 30.05, the sum of squares is given by $30.05 \times 19 = 570.95$.

The standard error of the regression coefficient is thus the square root of 0.269/570.95, or 0.022.

This estimate of the true standard error of b is unbiased on the condition that the dispersion of the points about the regression line is approximately the same along the length of the line. This will happen if the variance of Y is the same for every value of X, that is, if Y is homoscedastic. If this condition is not met, then the estimate of the standard error of b may be larger or smaller than the true standard error, and there is no way of telling which.

In summary, we can estimate the standard error of the regression coefficient from our sample and construct confidence intervals, under the following assumptions:

• The dependent variable has a normal distribution for all values of the independent variable.
• The variance of the dependent variable is equal for all values of the independent variable.
• If the independent variable is interval its distribution is normal.
• The relationship between the two variables is linear.

To obtain the 95% confidence limits we proceed in the usual manner. As the value of the standard error was estimated based on empirical data, we use Student's t distribution to find the number of standard errors estimated from the data that have to be counted on each side of the value of the regression coefficient to set the correct $1 - \alpha$ confidence limits. As our data was used to estimate two parameters, the intercept and slope of the regression line, we must refer to Student's t distribution with $n - 2$ degrees of freedom. If the sample is large, we can use the appropriate z-value from the table of the normal distribution to set the confidence limits.

In the example we have been working on, the standard error of the regression coefficient estimated from 20 observations is 0.022. In Student's t distribution with 18 degrees of freedom, 95% of the observations are within 2.101 estimated standard errors from each side of the mean. Therefore, the limits of the 95% confidence interval are obtained by $0.099 \pm 2.101 \times 0.022$, that is, 0.053 to 0.145 liters. Note that the change of units does not affect the regression line, although it does affect the regression coefficient. If we rather measured the height in meters instead of centimeters, the regression equation would be $FVC = -12.92 + 9.9 \times$ height (in meters) and the confidence interval would be 5.3 to 14.5 liters.

As usual, we can construct a test of the null hypothesis that the true (population) value of the regression coefficient is zero to test the association between the two variables. The null and alternate hypotheses of the test are H_0: $\beta = 0$ and H_A: $\beta \neq 0$.

For this purpose, we divide the sample regression coefficient by its standard error and we read in a table of Student's t distribution with $n - 2$ degrees of freedom the probability under H_0 of obtaining a value for b at least as large as the observed value. In this example, the value of the test statistic is $0.099/0.022 = 4.5$, far greater than 2.101, the 5% rejection limit based on a t distribution with 18 degrees of freedom. Consequently, we may conclude on an association between FVC and height.

9.4   Regression and correlation

We can transform a normal distribution into any other distribution with a different mean and variance just by multiplying its values by a constant and adding another constant.

If we look at the regression equation $y^* = a + bx$ we realize that when we predict the values y (denoted as y^*) from the values x we are applying a transformation of X on Y.

If the regression fits the data perfectly, the regression equation will transform X exactly into Y and the predicted values y^* will have exactly the same distribution as the values y of the dependent variable. On the other hand, if the regression is less than perfect, there will be a departure of the values y from those predicted by the regression equation and therefore the variance of y will be greater than the variance of y^*.

Therefore, the variance of y^* represents the variance of Y that is explained by the regression of Y on X. Thus, if we divide the variance of y^* by the variance of y, we will obtain the proportion of the variance of Y that is explained by the regression.

In the same example, the variance of the predicted values of FVC is 0.294 and the observed variance of FVC is 0.549. Dividing the former by the latter, we obtain 0.536. This result means that about 53.6% of the variance of FVC is explained by the body height. Therefore, in order to be able to explain (and predict) FVC completely, we need to find the variables that explain the remaining 46% of its variance.

This measure is called the **coefficient of determination** and it is an important measure of association between variables. It is usually represented as R^2 because its value is the square of another measure of association frequently used, called the **correlation coefficient**, which is represented by r. In this example, the correlation coefficient is therefore 0.73, the square root of 0.536.

Although we can obtain R^2 from r, the two measures are not completely equivalent. The coefficient of determination has values between 0 and 1, while the correlation coefficient ranges from -1 to $+1$. That is, the correlation coefficient, in addition to providing a measure of the strength of an association, also informs us of the type of association, a negative value meaning that the two attributes have

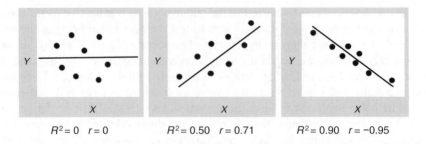

$$R^2 = 0 \quad r = 0 \qquad\qquad R^2 = 0.50 \quad r = 0.71 \qquad\qquad R^2 = 0.90 \quad r = -0.95$$

Figure 9.7   Scatterplots with regression lines illustrating different situations and the corresponding coefficients of determination and correlation.

an inverse relationship that is, one increases as the other decreases. In both cases, the greater the absolute value of the coefficient, the greater the strength of the association. However, unlike the coefficient of determination, the correlation coefficient is an abstract value that has no direct and precise interpretation, somewhat like a score. For this reason, it is being replaced by R^2.

These two measures are related to the degree of dispersion of the observations about the regression line. In a scatterplot, when the two variables are independent, the points will be distributed over the entire area of the plot. The regression line is horizontal and the coefficient of determination is zero (Figure 9.7). When an association exists, the regression line is oblique and the points are more or less spread along the line. The higher the strength of the association, the less the dispersion of the points around the line and the greater will be R^2 and the absolute value of r. If all the points are over the line, R^2 has value 1 and r value $+1$ or -1.

The importance of these measures of association comes from the fact that it is very common to find evidence of association between two variables, and it is the strength of the association that tells us whether it has some important meaning. In clinical research, associations explaining less than 50% of the variance of the dependent variable, that is, associations with R^2 less than 0.50 or, equivalently, with r between -0.70 and 0.70 are usually not regarded as important.

9.5   The *F*-test in linear regression

We can test the null hypothesis that $\beta = 0$ with a different test based on analysis of variance. Recall the reasoning we made when discussing anova for the problem of comparing several means, and let us make an analogy here. Figure 9.8 compares a situation where the null hypothesis is true, on the left, with a situation where the null hypothesis is false, on the right. When the two variables are independent, $\beta = 0$ and the slope of the sample regression line will be very nearly zero (not exactly zero because of sampling variation).

An estimate of the variance σ^2 of Y for fixed values of X can be obtained from the variance of the residuals, that is, the variance of the departure of each y from

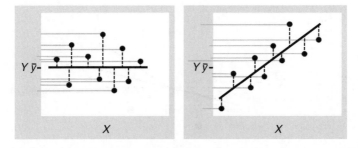

Figure 9.8   Illustration of the relationship between the total variance of Y and the residual variance in the case of true (left) and false (right) null hypotheses.

the value y^* predicted by the regression (dashed lines in Figure 9.8). We will denote this estimate by s_0^2 and call it the **residual mean square**.

Now, if H_0: $\beta = 0$ is true, s_0^2 will be very nearly equal to an estimate s_T^2 obtained from pooling together all the observed values y. However, if the null hypothesis is false (Figure 9.8, right), the regression line will be steep and the departures of the values y from the regression line will be less than the departures from $\bar{y}$. Therefore, the residual mean square will be smaller than the total variance of Y. We could compare the two estimates s_T^2 and s_0^2 by taking the ratio s_T^2/s_0^2. The resulting variance ratio would follow an F distribution if the two estimates of σ^2 were independent, and if the null hypothesis were false the variance ratio would have a value much larger than expected under H_0.

This is not the case, though. As both estimates were based on the same data, they are not independent. But the concept is appealing and so we need to look for a way of obtaining another estimate of σ^2.

Remember that in anova we obtained a third, independent estimate of σ^2 from the observed variance of sample means. Knowing that under H_0 the observed variance of sample means is equal to σ^2/n, we calculated the variance of sample means directly from the data as the sum of squared deviations of the sample means to the grand mean divided by the number of sample means. Then, by multiplying this estimate of the variance of sample means by n, we obtained an independent estimate of σ^2 under H_0.

We can do something equivalent in linear regression. Consider a set of observations on two uncorrelated variables X and Y. Figure 9.9 shows the regression line of Y on X fitted to four points (x, y). We have already seen that the sample regression line must pass through a point defined by the sample means of Y and X, that is, through $(\bar{x}, \bar{y})$. As H_0 is true, we know that the slope b of the regression line is due only to sampling variation and that the mean of the distribution of b is zero, the true slope of the regression line in the population.

As is shown in Figure 9.9, under H_0 the deviation of each slope m of the line connecting $(\bar{x}, \bar{y})$ to (x, y) from the true slope 0, that is, from $(y - \bar{y})/(x - \bar{x})$, is partly due to the variance of Y and partly due to the sampling variation of b. The variance of Y accounts for the difference between the slope m and the slope b of the

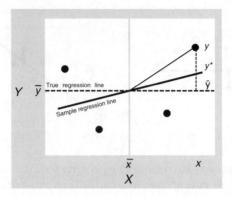

Figure 9.9   Under H_0 the variance of Y accounts for the difference $(y - y^*)$ and the sampling variance of the regression coefficient b accounts for the difference $(y^* - \bar{y})$.

regression line, and the sampling variance of b accounts for the difference between the slope b of the sample regression line and the slope 0 of the true regression line. The latter difference is the slope $(y^* - \bar{y})/(x - \bar{x})$ and consequently the variance of these slopes estimates the sampling variance of b.

We saw earlier in the discussion on the standard error of the regression coefficient that the sampling variance of b is equal to the variance of Y for fixed x divided by the sum of squares of x, that is, $\sigma^2 / \sum (x - \bar{x})^2$. So, if we calculate the sampling variance of b directly from our data and multiply it by $\sum (x - \bar{x})^2$ we will obtain an estimate of σ^2. This is exactly the same reasoning we used in anova. We call this estimate the **mean square due to regression** and denote it by s_1^2.

Remember that we are weighting the data, so the weighted variance is the product of $\sum (y^* - \bar{y})^2 / (x - \bar{x})^2$, the sum of squared differences to the mean, and the weight $(x - \bar{x})^2$, divided by the sum of all weights. As the $(x - \bar{x})^2$ cancel out, the expression becomes $\sum (y^* - \bar{y})^2 / \sum (x - \bar{x})^2$.

The product of $\sum (y^* - \bar{y})^2 / \sum (x - \bar{x})^2$ and $\sum (x - \bar{x})^2$ will give us the value of s_1^2, an estimate of σ^2 under H_0 that is independent of s_T^2 and s_0^2. As

$$\sum (x - \bar{x})^2 \times \sum (y^* - \bar{y})^2 / \sum (x - \bar{x})^2 = \sum (y^* - \bar{y})^2$$

we conclude that this new estimate of σ^2 is equal to the sum of squares of y^*.

Now, if the null hypothesis is true, s_1^2 and s_0^2 estimate the same quantity. But if the null hypothesis is false, the line will be steep and the points will lie near the line, so the departures of y^* from $\bar{y}$ will be large and the departures of y from y^* will be small. In that case s_1^2 will estimate a greater quantity than s_0^2.

So now we can compare the two estimates of σ^2 by taking the ratio s_1^2 / s_0^2 and looking on the F distribution to see whether the observed variance ratio is within the expected values when the null hypothesis is true. We have seen before that the residual mean square had $n - 2$ degrees of freedom, losing 1 degree of freedom to each parameter, a and b, of the regression line because in a dataset we may change all but the last two values in order to obtain a line with a given slope and intercept. So we will use $n - 2$ degrees of freedom for the denominator of the F distribution. For the numerator we will use 1 degree of freedom because, as the line must pass through $(\bar{x}, \bar{y})$, its position will be completely determined once we select one value for y^*. Therefore, we will refer to the F distribution with 1 and $n - 2$ degrees of freedom to test the null hypothesis.

We can obtain the sum of squares due to regression from the regression coefficient b. Actually, the product of b and the deviation of x about its mean, $b(x - \bar{x})$, will give us the difference of $\bar{y}$ to the regression line, $(y^* - \bar{y})$. Therefore, the sum of squares due to regression $\sum (y^* - \bar{y})^2$ is equal to $\sum [b(x - \bar{x})]^2$, which in turn is equal to $\sum b^2 (x - \bar{x})^2$. Since b is a constant, it comes out of the summation sign and the expression becomes $b^2 \sum (x - \bar{x})^2$. As $\sum (x - \bar{x})^2$ is the sum of squares of x, which can easily be obtained by the product of the variance of x and its degrees of freedom, the sum of squares due to regression is equal to $b^2 \times \mathrm{SSq}(x)$.

In the example of the regression of FVC on height, we saw that b was 0.099, the residual mean square with 18 degrees of freedom was 0.269, and $\mathrm{SSq}(x)$ was 570.95. The sum of squares due to regression is $0.099^2 \times 570.95 = 5.596$. The variance ratio is $5.596/0.269 = 20.80$. In the F distribution with 1 and 18 degrees of freedom, the 5% rejection boundary is 4.41 and, therefore, we may reject the null hypothesis. The exact p-value is 0.0002.

The F-test in linear regression is perfectly equivalent to the t-test of the regression coefficient. It tests the null hypothesis that the population regression coefficient of the independent variable is zero. We can also think of the F-test as a test of the null hypothesis that the coefficient of determination R^2 is zero.

9.6   Interpretation of regression analysis results

We are now able to read and interpret the results of linear regression analysis. We will use as illustration the output of a statistical package, in this case Stata. Figure 9.10 shows the results of the regression analysis of FVC on height.

On the upper left is displayed the analysis of variance table, from which the value of R^2 and the F-test are obtained. In the first column is the sum of squares (SS) due to regression (Model), the residual sum of squares (Residual), and the total sum of squares (Total). The next column shows the number of degrees of freedom (df) used to compute the mean squares (MS), that is, the estimates under H_0 of the variance of Y for each value of X.

```
  Source |       SS        df       MS                 Number of obs =       20
---------+----------------------------------            F( 1,     18) =    20.94
   Model |  5.61420097     1   5.61420097               Prob > F       =   0.0002
Residual |  4.82625418    18   .268125232              R-squared      =   0.5377
---------+----------------------------------            Adj R-squared =   0.5121
   Total |  10.4404552    19   .54949764               Root MSE       =  .51781


-----------------------------------------------------------------------------
     fvc |    Coef.    Std. Err.       t     P>|t|     [95% Conf. Interval]
---------+-------------------------------------------------------------------
  height |  .0991619   .0216705      4.58    0.000     .0536338       .14469
   _cons | -12.92858    3.61975     -3.57    0.002    -20.53339     -5.323771
-----------------------------------------------------------------------------
```

Figure 9.10   Regression analysis of FVC on height.

On the upper right is shown the number of observations (Number of obs), the value of the F statistic with the number of degrees of freedom (F(1, 18)), and the corresponding p-value (Prob > F).

Next is shown the value of R^2 (R-squared) and, below, the value of the **adjusted** R^2 (Adj R-squared). We will see later what the adjusted R^2 is.

In the last line of this group, Root MSE refers to the **root mean square error**. This quantity is the square root of the **mean squared error**. The mean squared error is the average of the squared differences between the observed and predicted values. Therefore, the MSE is similar to the residual mean square, the estimate of the variance of Y for fixed x, but with the n divisor. Thus, the root mean square error is the same to the MSE as the standard deviation is to the variance, and its units are the same as those of the dependent variable. A root mean square error value of 0 means that the estimate is perfect. The indicated value (0.517 81 liters) means that about two-thirds of the values of Y are less than 0.517 81 liters from the value y^* predicted by the regression. Besides being a measure of the goodness of fit of the model, along with R^2 and the F-test, the root mean square error may be useful for the comparison of two alternative models. The model with the lowest RMSE will have greater predictive ability.

The table at the bottom of the figure presents, for the independent variable (height) and the intercept (_cons), the value of the regression coefficients (Coef.), the standard error of the coefficients (Std. Err.), the statistics of the significance test of the coefficients (t), the corresponding p-values (P>|t|), and the 95% confidence limits of the regression coefficients ([95% Conf. interval]).

Therefore, the interpretation of the results of the linear regression of FVC on height is as follows. The model is based on 20 observations. The F-test is highly significant, meaning that height is associated with FVC, and we estimate that about 50% of the variability of FVC is explained by the body height. The model estimates that for each increase of 1 cm in height, the FVC increases from 54 to 145 mL. The evidence for this association is strong as indicated by a p-value of less than 0.001.

9.7   Multiple regression

The approach we have just discussed for the investigation of an association between two interval variables can be extended to problems where we want to study the relationship between one dependent variable Y and two independent variables X_1 and X_2. Try to visualize the variables arranged in a three-coordinate axis. Each observation would be represented by a point in a three-dimensional space at the intersection of the values of three variables. The association between the two explanatory variables and the dependent variable would also be modeled by a regression line passing through the middle of all the points, now in a three-dimensional space. The equation of the regression line would then be $y^* = a + b_1 x_1 + b_2 x_2$.

By the same token, we can extend this approach to more than two explanatory variables. There is nothing to prevent us from thinking of a space with as many dimensions as the number of explanatory variables (plus one for the dependent variable) and a cloud of points in this space being crossed by a line passing through the middle of all these points.

This is the concept of **multiple regression**, a technique called multivariate analysis because it allows the creation of statistical models with multiple variables simultaneously. This model has as its main objective the explanation of the variability of a dependent variable and the prediction of its values through a linear combination of independent variables $x_1, x_2, \ldots, x_n$ in the form of $y = a + b_1 x_1 + b_2 x_2 + \cdots + b_n x_n$.

As we will see, multiple regression has extensive application in clinical research. The principles of the method are exactly the same as were discussed for the case of simple regression, of which multiple regression is just a generalization. Thus, the regression coefficients, in this context called **partial regression coefficients**, are also the point estimates of the difference in mean values of the dependent variable between two consecutive values of the independent variable, but now at constant value of the other independent variables in the model. Standard errors and confidence intervals for each of the coefficients can be determined and each coefficient can be tested against the null hypothesis that its true value is zero. If the regression coefficient of an independent variable is not statistically different from zero, this means that the variable does not contribute to the explanation or prediction of the dependent variable and can be removed from the model without affecting the prediction.

The interpretation of the coefficient of determination R^2, here called **coefficient of multiple determination**, is also identical. However, it must be taken into account that the value of R^2 increases with the number of variables in the model. Therefore, if more variables are added to the model, even if they make no contribution whatsoever to the prediction of the dependent variable, the value of R^2 will always increase. The **adjusted R^2** is a correction of the coefficient of multiple determination, taking into account the number of independent variables on the model and, as such, should be the value of R^2 to be considered and reported.

A few remarks must be made about the F-test applied to multiple regression regarding the calculation of the degrees of freedom. For the residual mean square, the number of degrees of freedom is the number of observations minus 1 degree of freedom for each term in the regression equation, including the intercept. For the mean square due to regression, because we need one value y^* for each dimension to define the position of the regression line in the multi-dimensional space, the number of degrees of freedom is the number of terms in the equation excluding the intercept.

The F-test in multiple regression is still a test of the null hypothesis that $R^2 = 0$. Therefore, it is a test of the null hypothesis that all partial regression coefficients are zero in the population and, thus, it is basically a global significance test of the regression model.

One of the main uses of multiple regression is, of course, for finding the variables that explain the behavior of a particular attribute. For example, in the investigation of FVC discussed above we found an association between FVC and body height and estimated that FVC increases by about 100 mL per centimeter increase in height. Suppose we now want to investigate whether there is also an association between FVC and gender. So we run a regression of FVC on gender, coded 0 for females and 1 for males (yes, the independent variables in the regression can be binary, but not the dependent variable). The regression coefficient of gender, 0.865, is significantly different from zero ($p = 0.007$). The adjusted R^2 is 0.34. Therefore, we conclude that gender is associated to FVC, explaining about 34% of its variability, and we estimate that the mean FVC in men is 0.865 liters greater than the mean FVC in women.

However, we know that men are on average taller than women. Therefore, there is a possibility that the difference on average FVC between genders is due solely to the difference of heights and not actually related to the gender of individuals.

Multiple regression allows us to clarify the relationships among these variables. Figure 9.11 shows the results of a multiple regression analysis of FVC on both gender and height.

Source	SS	df	MS			Number of obs =	20
						F(2, 17) =	10.07
Model	5.66223037	2	2.83111519			Prob > F =	0.0013
Residual	4.77822478	17	.281072046			R-squared =	0.5423
						Adj R-squared =	0.4885
Total	10.4404552	19	.54949764			Root MSE =	.53016

fvc	Coef.	Std. Err.	t	P>|t|	[95% Conf. Interval]	
male gender	.148079	.358219	0.41	0.684	-.6076971	.9038551
height	.0891508	.0328451	2.71	0.015	.0198538	.1584479
_cons	-11.34607	5.328305	-2.13	0.048	-22.58782	-.1043324

Figure 9.11   Multiple regression analysis of FVC on gender adjusted for height to control for confounding.

From the results we conclude that FVC is associated with height ($p = 0.015$) but not with gender ($p = 0.684$) when height is taken into account. In other words, we have no evidence that mean FVC is different between men and women with the same height. We therefore conclude that the relationship between gender and FVC may be spurious and apparently due only to the association between gender and height. If we know the height of a person, knowing the gender does not allow us to make a better prediction of the value of FVC.

The effect of gender and height on FVC is therefore confounded. **Confounding** occurs whenever an external, often unrecognized and therefore unobserved, variable is associated with both the dependent and independent variables. In this example, height was the confounder and, had we not entered height into the model, we would have erroneously concluded that men had on average greater lung volumes, just because they were men.

Confounding is a terrible nuisance in clinical research because every time an association is discovered between two variables there is always the possibility that the association is spurious and due to a hidden confounder. This is the reason why it is so hard to establish causality in observational research. The corollary is that, in virtually any investigation of an association between variables in an analytical study, one needs to consider all the potential confounders and should include them in a multiple regression analysis.

The ever-present possibility of confounding and the need to control for confounders is what makes multivariate methods so important in biostatistics. It is also the reason why the simple statistical tests for comparison of means and proportions described in the previous chapters have little application in analytical research and are usually reserved for experimental studies. This applies to unadjusted confidence intervals as well. However, as we will see, because experimental research is also not immune to confounding, multivariate methods have great importance in those studies.

Besides controlling for confounders, the other major use of multiple regression is in the investigation of variables that are associated and explain the dependent variable in order to develop explanatory or predictive models of a variable of interest.

If we look at the example in Figure 9.12 we will see that a new variable, age, has been added to the model. The F-test is highly significant ($p < 0.0001$) meaning that at least one independent variable is associated with FVC. The tests of the partial regression coefficients of height and age are both significant, so we conclude that both variables are associated with FVC. According to the model, mean FVC decreases about 38 mL (18 to 58 mL) for each year of life, at constant value of height. In turn, among people of the same age, mean FVC increases by 100 mL (67 to 134 mL) for each centimeter increase in body height. The value of the adjusted R^2 increased to 74% and the root mean square error decreased from 0.53 to 0.38, as expected, since we have improved the predictive ability of the model. Therefore, we have identified a second variable that, together with height, explains almost 75% of FVC.

Naturally, these results based on a sample of only 20 subjects may not be very robust. In addition, we have not verified whether the assumptions of multiple

```
   Source |      SS          df      MS                Number of obs =       20
----------+----------------------------------         F(  2,    17) =    27.56
    Model | 7.97936972        2  3.98968486            Prob > F      =   0.0000
 Residual | 2.46108544       17  .144769732            R-squared     =   0.7643
----------+----------------------------------         Adj R-squared =   0.7365
    Total | 10.4404552       19  .54949764             Root MSE      =   .38049


---------------------------------------------------------------------------------
     fvc  |    Coef.    Std. Err.       t     P>|t|      [95% Conf. Interval]
----------+----------------------------------------------------------------------
   height |  .1003373   .0159262      6.30   0.000       .066736     .1339387
      age | -.0380223   .0094069     -4.04   0.001      -.0578692    -.0181755
    _cons | -11.0469    2.700231     -4.09   0.001     -16.74388    -5.349905
---------------------------------------------------------------------------------
```

Figure 9.12   Multiple regression analysis of FVC on two predictors.

regression were true. Multiple regression is, without doubt, a very powerful technique of data analysis but, on the other hand, relies on a significant number of assumptions. We will now review those assumptions and discuss how we can check that at least the most important ones were not violated.

9.8   Regression diagnostics

We saw earlier that the determination of the standard error of the regression coefficient was based on the assumptions of a normal distribution and of homoscedasticity of the dependent variable in all combinations of values of the independent variables.

In addition, for the values of partial regression coefficients to make sense, that is, they actually estimate the change in the value of the dependent variable for a unit change in the value of each independent variable, it is necessary for the relationship between the dependent and each independent variable to be linear because only then will the change in the dependent variable be constant across all values of the independent variable.

For estimating the standard errors of the partial regression coefficients and for several statistical tests of multiple regression, the assumption of normality of the independent variables has also been made. The independent variables may be random as well as controlled; that is, the investigator may elect to collect data from fixed values of an independent variable, but their distributions must be normal.

It is not possible to verify the validity of several of these assumptions since, in practice, for each combination of values of the independent variables we usually have only a single observation. This makes it impossible to confirm the normal distribution and the equal variance of the dependent variable for each combination of the values of the independent variables.

However, if the above assumptions are true, the residuals should have a normal distribution, zero mean, and should not be correlated with the values of the independent variables. Verification of the regression assumptions can thus be made indirectly by the **analysis of residuals**.

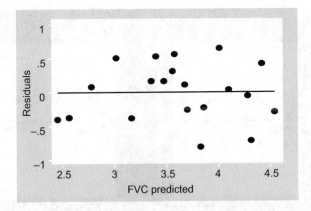

Figure 9.13   *Residual plot of the regression of FVC on height and age.*

An essential tool for the analysis of residuals is the **residual plot**. This is a scatterplot of the residuals on the vertical axis with the predicted value of the dependent variable on the horizontal axis. Figure 9.13 shows the residual plot of the regression of FVC on height and age of the previous section.

Visual inspection of the graph reveals several indications about the possibility of violation of the assumptions of regression and may help in the identification of specific problems with the data, as is illustrated in Figure 9.14.

In a residual plot no discernible pattern should be observed (Figure 9.14a). If the points are more to one side of the horizontal line that corresponds to the value 0 of the residuals, this means that the distribution of residuals is not normal (Figure 9.14b). If the spread of the residuals is not constant over the predicted values and increases for smaller or larger values of the prediction, or both, this is an indication of heteroscedasticity (Figure 9.14c). A transformation of the dependent variable (e.g., a logarithmic transformation) may correct these two problems. If the residuals are larger in the region near the center of the chart and decrease to each side (Figure 9.14d), this means that the relationship between the dependent variable and one or more independent variables is not adequately represented by a straight line. The residual plot of Figure 9.13 does not present any of these problems.

For the verification of the assumption of linearity one can inspect a scatterplot of the dependent variable with each of the independent variables.

Besides residual plots, there are several statistical tests available that help us identify violations of the assumptions. For example, to test the hypothesis of zero mean of the residuals we can use Student's t-test to compare the mean of the distribution of residuals to the theoretical value 0. Examples of other tests available for identifying violations of the assumptions of multiple regression are the Shapiro–Wilk test of the normality of the residuals, the Cook–Weisberg test of heteroscedasticity, and the Ramsey RESET test of linearity

Another useful analysis is **DfBeta**, which helps in the identification of outliers that exert a great influence, called **leverage**, on the estimates of the regression

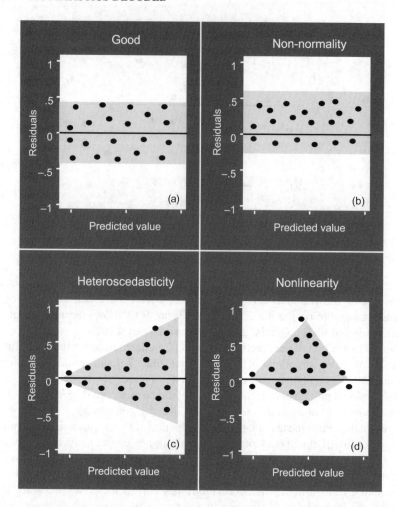

Figure 9.14   Patterns in residual plots suggestive of non-normality of the residuals (b), heteroscedasticity (c), and nonlinearity (d).

coefficients. Figure 9.15 illustrates the effect of an outlier on the estimation of the regression coefficient. In this example, there is obviously no correlation between the two variables, but because of an extreme value the least squares regression line has a significant slope. DfBeta is a measure of the change in regression coefficients when each observation is removed in turn from the data. If there are outliers the regression coefficients will change markedly.

The least squares method is fairly tolerant to some deviations from the assumptions, such as non-normality of the variables in the model (especially in large sample sizes) and heteroscedasticity of the dependent variable, but is markedly sensitive to extreme values. However, this does not mean that any suspect outlier should be excluded from the analysis, because it could lead to the truncation

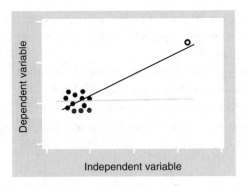

Figure 9.15   Influence of an outlier (open circle) in the regression estimates. Without the outlier the slope of the regression line would be nearly zero (gray line).

of the dependent variable. Instead, outliers should be reviewed and excluded only if there is evidence that they result from an error in the data that cannot be corrected.

Although in regression analysis the variables in the model are usually interpreted as independent predictors of the explained variable, the method does not select predictors based on their mutual independence. Rather, the independence of the predictors is an assumption of the model and, as such, it is the responsibility of the investigator to select variables for the model that are mutually independent, as well as to verify this assumption. Correlated predicted variables have an impact on the estimation of the regression coefficients and their standard errors, which will decrease the predictive ability of the model to a degree that cannot be assessed.

In practice, though, some degree of correlation between the variables in the model virtually always exists and that can be tolerated. However, the model does not tolerate perfect or important **multicollinearity**.

Perfect multicollinearity exists when the value of one independent variable is fully determined by the value of another independent variable or by a set of independent variables. When perfect or important collinearity is present, the slope of the regression line will vary erratically from one sample to another and therefore the standard errors of the coefficients will be so wide that showing an association will be unlikely.

Finally, it is important to mention other assumptions of multiple regression which are related to the study design. First, the data has to come from a random sample and the observations must be independent. Second, the dependent variable cannot be limited (allowed to vary only between certain values), censored (values below a certain lower limit, or above a certain upper limit, are given the same value), or truncated (values below or above a certain limit, or both, are excluded). We will discuss additional restrictions on multiple regression later in this book, but for now let us review the assumptions of multiple regression:

- The dependent variable has a normal distribution.
- The variance of the dependent variable is the same for all combinations of values of the independent variables (homoscedasticity).

- The dependent variable cannot be limited, censored, or truncated.
- The independent variables are either binary attributes or interval attributes with a normal distribution.
- The independent variables may be random or controlled.
- The independent variables must not be correlated.
- There can be no important or perfect multicollinearity.
- The relationship between the dependent variable and each independent variable must be linear.

9.9   Selection of predictor variables

Two other assumptions of multiple regression that have not yet been mentioned are, first, that the model should not contain unnecessary variables and, second, that all important variables are in the model. These assumptions mean that the selection of variables to be included in the model requires common sense and domain knowledge.

If we include in the model several variables that are not associated with the dependent variable, the model will try to fit to all independent variables, leading to a decrease in the influence of variables truly associated with the dependent variable and to the distribution of their influence by the redundant variables. In addition to obtaining inaccurate estimates of regression coefficients and increasing the type II error, this process will result in an excessive adjustment to the sample data, a problem called **overfitting**. The consequence of overfitting is a loss of reliability of the model, which results in a considerable increase in the error of the predicted value of the dependent variable (and consequent decrease of R^2), when the same model is applied to a different sample of the same population.

On the other hand, if important independent variables are omitted, their effect on the dependent variable will be distributed by other variables in the model, biasing their coefficients and possibly assigning significance to variables that are associated with the omitted variables but not with the dependent variable, thus increasing the type I error. As seen in the example on FVC, the omission of height in a regression model that included gender led to the erroneous conclusion that there was an association between gender and FVC.

As a consequence of the assumptions referred to above, during the specification of a multiple regression model designed to explain the behavior of a particular attribute, all decisions regarding the inclusion or exclusion of variables should be duly weighed and justifiable. The model must include all variables that are known to be associated with the dependent variable, even if their regression coefficients are not, in the sample, significantly different from zero. On the other hand, the model should include only those variables for which there is a theoretical explanation for an eventual association with the dependent variable.

After the set of candidate predictors of the dependent variable are selected and we run a computer program for multiple regression, we will typically find that the coefficients of some of the variables included in the model are not statistically different from zero. Of these, the variables that we know to be associated with the dependent variable should nevertheless stay in the model. For the remainder, as their coefficients are not significant, we should be able to delete them from the model without affecting the quality of the adjustment.

This is not always the case, though. In many models there are associations between independent variables, that is, there is multicollinearity among a group of predictors. As a result, the elimination of one variable can affect the coefficients of others, to the extent that a non-significant variable may become significant upon elimination of a highly correlated variable. Or, the other way around, the elimination of one variable may change one or more coefficients from significant to non-significant.

For this reason, the elimination of variables with non-significant coefficients after a regression must be done one variable at a time. There are two main approaches to this procedure. One is to run the regression with all variables, then remove the variable whose coefficient has the highest p-value, and run the regression again with the remaining variables. The procedure is repeated until all the regression coefficients are significant. This procedure can be performed manually or automatically by some statistical analysis programs.

In the other approach, called **stepwise regression**, the selection is automatic and is based on an algorithm. Different programs may use different algorithms, so results may vary, but the general principle is the same. In the **backward elimination** method the fit of the current model is compared to the fit of the various models obtained by removing one variable at a time, thereby identifying the variables that can be removed without significantly influencing the fit of the model. In the **forward elimination** method, the algorithm starts with the null model, that is, the model that contains only the constant, and in each step includes the variable that contributes the most for the fit of the model. With both methods, the elimination or inclusion of variables in the model stops when a certain criterion is reached. Some algorithms combine the two methods, evaluating at each step which variables should be removed and which should be included.

Some stepwise regression programs allow the specification of variables that have to be in the final model and cannot be removed, while others treat all variables equally. Because of the lack of control of the investigator on the process of selection of variables, stepwise regression is mainly used in problems where there are a large number of independent variables and no underlying theory guiding the selection of variables. Examples of such problems are the development of purely predictive models in which the goal of accurately predicting the dependent variable largely outweighs the explanation of the direction and magnitude of the associations. Therefore, most often the selection of variables after a regression is done by manual selection with the method described above.

9.10   Regression, *t*-test, and anova

We saw earlier that it is allowed to have binary independent variables in regression. As any coefficient represents the difference between the average values of the dependent variable between two consecutive values of the independent variable, if a binary attribute is encoded by two consecutive numbers (typically 0 and 1), its coefficient can be interpreted as the point estimate of the difference in the population means of the dependent variable between the groups defined by values of the attribute. The confidence interval of the coefficient therefore represents the 95% confidence interval of the difference between the population mean values of the dependent variable. And the test of significance of the coefficient is equivalent to a test of the null hypothesis of equality of means. Everything is completely analogous to what we discussed earlier on the estimation of differences between population means and Student's *t*-test. Even the assumptions of normality and homoscedasticity are identical.

Actually, Student's *t*-test for two independent samples can be replaced by linear regression, with fully coincident results. Regression, however, has the additional advantage of allowing us to adjust the analysis for external variables and thereby correct statistically any differences between the groups in the distribution of those external variables.

Multiple regression can also compare the means of more than two groups and thus replace anova. This procedure is a little more complicated and requires some explanation.

If we want to compare the means of the dependent variable between, say, four groups, we cannot simply create a categorical variable that encodes the groups with the values 1, 2, 3, and 4, and run the regression of the dependent variable on this variable. The regression does not differentiate between nominal variables with three or more categories and interval variables, and therefore would assume erroneously that the classes of the categorical variable represent increasing and equally spaced values.

Therefore, the only possibility is to convert the categorical variable into several binary variables. For example, suppose we wanted to investigate an association between FVC and education and we considered education to be a categorical attribute with values elementary, secondary, middle, and higher corresponding to the highest education level attained by an individual. We could replace this categorical attribute by four binary variables called Elementary, Secondary, Middle, and Higher, each variable taking the value 1 if an individual belongs to that group and 0 otherwise. These variables that exist only to allow an analysis are called **dummy variables**.

We cannot, however, have a regression model of FVC on those four variables because we would violate one of the conditions of application of multiple regression, namely, the absence of important or perfect multicollinearity. In multiple regression it is not allowed for an independent variable to be completely determined by another or a group of other independent variables. However, this would happen if we were to include the four variables in the model. For example, if we know that the variables Elementary, Secondary, and Middle are all 0 in a given individual, then we immediately know that the value of the variable Higher must be 1. The value of each dummy variable is therefore fully determined by the values of the other three, in clear violation of the assumption of no perfect multicollinearity.

The solution is simply to omit one of the dummy variables. Then, knowledge of the value of three of them does not necessarily tell us the value of the fourth, yet the group to which each observation belongs will remain perfectly encoded.

Let us first check whether three dummies are sufficient to encode the four categories. For example, say we omit the variable Elementary. Then, an individual belongs to the group where the value of the corresponding dummy variable is 1 and to the Elementary group if the values of all three dummy variables are 0. Therefore, a nominal variable with n categories can indeed be encoded by $n - 1$ dummies.

Let us now check whether we remove collinearity, again omitting one of the dummies, say Elementary. For example, if Secondary and Middle have the value 0 the value of Higher may be 1, if the group to which the individual belongs is Higher, or 0 if the individual belongs to the Elementary group. Therefore, the value of a dummy is no longer fully determined by the values of the other three.

This encoding method of nominal variables is called **binary encoding** or **encoding by reference**. The omitted category becomes the reference category, because the partial regression coefficients of the remaining dummy variables will estimate the difference of the average value of the dependent variable between each of the dummy variables and the reference category. Figure 9.16 shows the result of regression analysis of FVC on the set of dummies encoding the attribute Education.

The regression coefficient of the constant (intercept) is the mean FVC in the group Elementary. The coefficients of each dummy variable are the point estimates of the difference in mean FVC between the group defined by a dummy and the reference category. The significance tests of the regression coefficients of the dummy variables also test only the difference in the mean of the dependent variable between each dummy and the reference category. In other words, the analysis

Source	SS	df	MS			
Model	4.92595667	3	1.64198556			
Residual	4099.02104	5222	.784952325			
Total	4103.947	5225	.785444402			

	Number of obs =	5226
	F(3, 5222) =	2.09
	Prob > F =	0.0991
	R-squared =	0.0012
	Adj R-squared =	0.0006
	Root MSE =	.88598

fvc	Coef.	Std. Err.	t	P>|t|	[95% Conf. Interval]	
secondary	.0214739	.0266201	0.81	0.420	-.0307125	.0736603
middle	.1242445	.0574457	2.16	0.031	.0116268	.2368621
higher	.069435	.0442156	1.57	0.116	-.0172461	.1561162
_cons	3.860886	.0195441	197.55	0.000	3.822571	3.8992

Figure 9.16   Results of the regression of FVC on three dummy variables encoding a nominal variable with four categories.

results do not give us an unequivocal answer to the question of whether FVC is associated with education.

In order to get the answer, it is necessary to test globally the set of dummy variables that encode a categorical variable. This amounts to saying that we want to test the null hypothesis that the coefficients of the dummies encoding a categorical attribute are all equal to 0. The F-test of multiple regression tests precisely this hypothesis, and this is the test that must be considered if the dummies are the only explanatory variables in the model. In the example, the p-value of the F-test is 0.099 and, therefore, we do not conclude on an association between FVC and education. If a model includes additional independent variables the test is more complicated, but generally available in statistical analysis programs.

Thus, we have seen how to include nominal variables in multiple regression models and that anova can be replaced by multiple regression. The results of the two methods are absolutely equivalent.

9.11 Interaction

Imagine now that we continue our study of respiratory function parameters and intend to investigate the relationship between the peak expiratory flow rate (PEF) and gender. In a quick analysis, we find that the average value of PEF is significantly higher in men (580 liters/minute with standard deviation 55 liters/minute) than in women (486 liters/minute with standard deviation 72 liters/minute). However, we must take into account the average difference in height between genders, as men are about 12 cm taller than women. If PEF is related to height, then the difference in the mean PEF between genders is perhaps attributable solely to the difference in heights.

To clarify the issue, we do a multiple regression analysis of PEF on gender and height, the results of which are shown in Figure 9.17.

The results are clear: the model is statistically significant by the F-test ($p < 0.0001$) and the two attributes explain 53% of the variance of PEF. According

```
    Source |       SS        df        MS              Number of obs =      116
-----------+--------------------------------          F(  2,    113) =    64.68
     Model | 390373.892       2    195186.946          Prob > F      =   0.0000
  Residual | 340985.936     113    3017.57465          R-squared     =   0.5338
-----------+--------------------------------          Adj R-squared =   0.5255
     Total | 731359.828     115    6359.65067          Root MSE      =   54.932


--------------------------------------------------------------------------------
       PEF |     Coef.    Std. Err.         t     P>|t|       [95% Conf. Interval]
-----------+--------------------------------------------------------------------
male gender |   43.85963   12.62565      3.474     0.001      18.84594    68.87331
     height |   4.117939    .6115718      6.733     0.000      2.906306    5.329573
      _cons |  -203.0461   102.5968     -1.979     0.050     -406.3089    .2167124
--------------------------------------------------------------------------------
```

Figure 9.17   Results of the regression of PEF on gender and height.

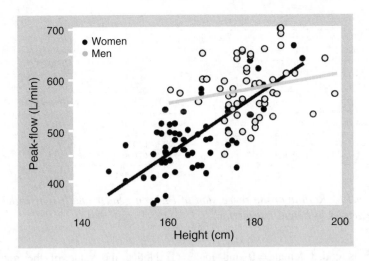

Figure 9.18 Scatterplot of the data of the regression of PEF on gender and height with regression lines fitted separately for women (black) and for men (gray).

to the partial regression coefficients, keeping the height constant the PEF is on average 44 liters/minute higher in men than in women ($p = 0.001$) and, in the same gender, PEF increases by 4 liters/minute for every centimeter increase in height ($p < 0.001$).

However, in a scatterplot of the values of PEF differentiating the two genders we obtained the result shown in Figure 9.18. In the graph we have also drawn separate regression lines for men and women.

The data is not in agreement with the results of the analysis. We can see that PEF increases more in women than in men for every centimeter of height, so that in women more than 185 cm tall the PEF is on average greater than in men. Thus, the data contradicts the model, which predicted that PEF was always higher in men regardless of height.

The model is wrong because the relationship between PEF and height is not the same in both genders and, conversely, the relationship between PEF and gender is not the same for all heights. This situation is called **interaction** between two variables: that is, the effect of an explanatory variable on the dependent variable is not the same for all values of another explanatory variable. When there is interaction, the model is not simply additive – for assessing the effect of a variable on the dependent variable it is necessary to take into account the value of the other variable in the interaction.

To properly analyze this data, we need to represent the interaction between gender and height in the model. For this purpose, a term for the interaction is included in the model, whose value is the product of the two variables. The results of the multiple regression model with interaction are shown in Figure 9.19.

The term for the interaction is statistically significant ($p = 0.001$). The regression equation is PEF $= -447 + 5.6 \times$ height $+ 802 \times$ gender $- 4.3 \times$ height $\times$ gender.

```
    Source |      SS         df        MS                Number of obs =     116
-------------+------------------------------             F(  3,   112) =   51.53
      Model |   424110.143     3   141370.048            Prob > F      =  0.0000
   Residual |   307249.685   112   2743.30076            R-squared     =  0.5799
-------------+------------------------------             Adj R-squared =  0.5686
      Total |   731359.828   115   6359.65067            Root MSE      =  52.377


-----------------------------------------------------------------------------
        PEF |    Coef.    Std. Err.      t     P>|t|     [95% Conf. Interval]
-------------+---------------------------------------------------------------
male gender |   802.1632   216.5725    3.704   0.000     373.0525    1231.274
     height |   5.578493   .7165818    7.785   0.000     4.158678    6.998308
interaction |  -4.323519   1.232894   -3.507   0.001    -6.766341   -1.880698
       _cons |  -447.5004   120.1193   -3.725   0.000    -685.5014   -209.4994
-----------------------------------------------------------------------------
```

Figure 9.19   Results of the regression of PEF on gender and height with a term for the gender by height interaction.

Gender is coded female $= 0$ and male $= 1$. Notice the value of the regression coefficient of gender, which is now 802 liters/minute. Clearly, this coefficient no longer represents the difference in mean PEF between the two genres – now it is necessary to take the height into account.

The interpretation of models with interaction is always more complicated, especially if there are two or more pairs of interacting variables or if the interaction involves three variables (called **second-order interactions**) or more. If one of the variables in the interaction is binary, one way of interpreting the model is to run regressions of the dependent on the interval variable separately for each value of the binary variable, as illustrated in Figure 9.18.

Another possibility consists of solving the regression equation for each value of the binary variable in the interaction. The regression equation with a first-order interaction and the main effects is $y = a + b_1x_1 + b_2x_2 + b_3x_1x_2$. Therefore, if x_2 is the binary variable coded $(0, 1)$, when its value is 0 the equation becomes $y = a + b_1x_1$ and y increases by b_1 for each unit increase in x_1. When its value is 1, the equation becomes $y = a + (b_1 + b_3)x_1 + b_2$ and y increases by $b_1 + b_3$ for each unit increase in x_1.

Accordingly, in the example the increase in PEF for each centimeter increase in height is 5.6 L/min in women and 1.3 L/min in men.

When the two variables in the interaction are continuous, it is often too complicated to understand the nature of the relationship between them and the dependent variable. In such cases it may be preferable to dichotomize one of the interacting variables and do separate regressions.

Although a model with interaction is more difficult to interpret, interactions often provide interesting information about the dynamics of the relationships between variables. Figure 9.20 shows how the relationship between PEF, height, and gender would look without interaction (top) and with interaction (bottom) in

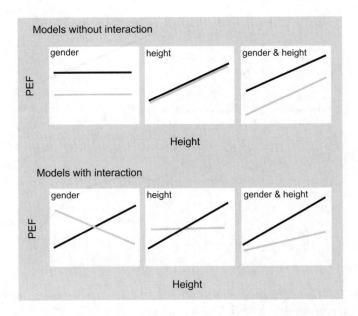

Figure 9.20   Exemplification of an interaction effect on the relationship of two variables with one dependent variable. Each graph indicates which of the explanatory variables is significantly associated with the dependent variable.

different settings where an association existed between the dependent variable and one interacting variable only, or with the two interacting variables.

As illustrated in this figure, the possibility of including terms to model interactions allows the evaluation of a much greater variety of relationships between the dependent variable and independent variables.

In addition, the existence of an interaction is sometimes the analysis of primary interest, such as when one wants to compare differences in means between two populations. For example, suppose we wanted to investigate whether the difference in mean PEF between genders is the same in smokers and non-smokers. The correct analysis would be to test the coefficient of the interaction of gender with smoker.

The analysis of interactions is also the method by which one may test the equality of the slopes of two regression lines, a problem encountered with some frequency in basic research, but only very occasionally in clinical research. As illustrated in Figure 9.20, the two lines are parallel only when the interaction is non-significant.

9.12   Nonlinear regression

Sometimes the change in the dependent variable is not constant over the range of values of the independent variable, that is, the relationship between the two is nonlinear. For example, it may happen that the dependent variable increases in the

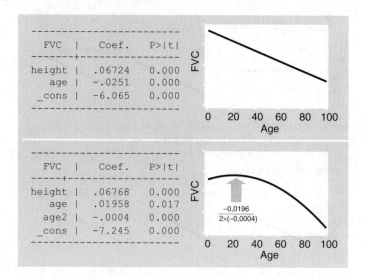

Figure 9.21   Comparison of models obtained by linear regression (top) and by polynomial regression (bottom).

same direction as the independent variable up to a certain value, showing thereafter a tendency to decrease rather than increase. An example is the daily production of growth hormone, which increases with age until puberty and gradually decreases from then on. The relationship between growth hormone and age is not linear; rather, it has the shape of a curve.

This relationship can be modeled by including a new term in the regression, the square of the independent variable. Such a term is called a **quadratic term**. The regression equation will have the form $y = a + b_1 x + b_2 x^2$, hence its name of **polynomial regression** or **curvilinear regression**. The regression equation is no longer the equation of a straight line, but the equation of a parabola. Figure 9.21 illustrates the models obtained with the regression of FVC on age, adjusting for height (top), and introducing the quadratic term age-squared (bottom).

To interpret the relationship between the two variables, it is useful to estimate the value of the independent variable at the inflection of the curve. This value corresponds to the vertex of the parabola $y = a + b_1 x + b_2 x^2$ and is given by $-b_1/2b_2$. In Figure 9.21, the inflection point is thus $-0.019\,58/(-0.0004 \times 2)$, or 24.5. According to the model, FVC increases with age until about the age of 24 years and then decreases, evermore rapidly as age increases.

The usual technique for evaluating a curvilinear regression is to enter a quadratic term into the model and test its coefficient for significance. If the term is non-significant then there is no evidence of a curvilinear relationship and the quadratic term is dropped from the model. Otherwise, the term is kept in the model and the independent attribute will be represented by the two variables.

9.13   Logistic regression

We saw above that multiple regression allowed binary attributes as independent variables, but that the dependent variable had to be interval. The explanation for this is illustrated in Figure 9.22, which shows a scatterplot of the relationship between a binary variable (gender, coded female = 0, male = 1) and an interval variable (body height). The points on the graph represent the proportion of males in each value of height from a random sample of about 1900 individuals, as well as the regression line estimated by the least squares method.

The result is clearly unsatisfactory. The straight line does not fit the points well, which are distributed in the form of an S-shaped curve. The residuals are almost all negative on the left half of the regression line and nearly all positive on the right half and, therefore, their distribution is certainly not normal. Moreover, the distribution of the proportion p of men is binomial, not normal as assumed by the least squares method, and being binomial its variance $p(1 - p)$ depends on the value of proportion and cannot therefore be equal for all values of height. In addition, the dependent variable can take values only between 0 and 1, violating the assumption that the dependent variable is not limited.

In summary, the application of linear regression to the case of a dependent binary variable leads to the violation of many assumptions of the least squares method and thus the estimates of the regression coefficients are not valid. Another

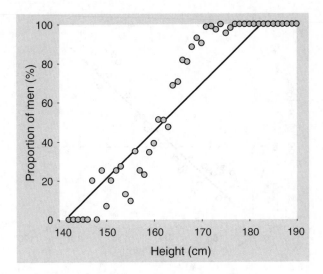

Figure 9.22   Illustration of the problems of fitting a least squares line when the dependent variable is binary.

important problem is that, according to the equation of the linear model in Figure 9.22, for heights exceeding 182 cm the predicted proportion of men exceeds 100%, which is an impossibility.

To solve these problems we need to find a transformation of the dependent variable such that the dependent variable can have a linear relationship with the independent variables and will no longer be limited to vary between 0 and 1.

We saw in Section 5.9 that the logit transformation satisfies these requirements. Recall that if p is a probability, then $p/(1 - p)$ are the odds and $\ln(p/1 - p)$ is the log-odds or logit, that is, the natural logarithm of the odds. When p has the value 0, the odds are 0 and the logit is $-\infty$. When p has the value 1, the odds are $+\infty$ and the logit is $+\infty$.

Consider now Figure 9.23, which shows the regression line of gender on height after the logit transformation of the dependent variable.

The result is clearly superior to the previous one, as the fit of the regression line to the data is much better. The equation of that regression line is, therefore,

$$\ln\left(\frac{p}{1 - p}\right) = a + bx$$

Naturally, we can extend this principle to multiple dimensions defined by so many independent variables, resulting in a model with the general form

$$\ln\left(\frac{p}{1 - p}\right) = a + b_1 x_1 + b_2 x_2 + \cdots + b_n x_n$$

In words, this expression says that the probability of a binary attribute, expressed as logit, is modeled by the linear combination of a set of independent variables.

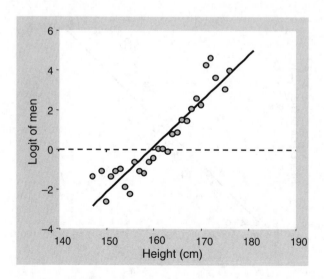

Figure 9.23   The least squares line fitted to the data of Figure 9.22 but with the logit transformation of the dependent variable.

As we usually prefer to think in terms of proportions or probabilities rather then logits, we can transform the values predicted by the equation into proportions. As $\ln(p/1-p) = \text{logit}(p)$, after exponentiation and solving for p the regression equation becomes

$$p = \frac{e^{a+bx}}{1 + e^{a+bx}}$$

In this expression, e is the base of natural logarithms (approximately 2.72).

There is no complex mathematics involved in obtaining the above equation. The equation of the regression line is

$$\ln\left(\frac{p}{1-p}\right) = a + bx$$

which by exponentiation becomes

$$p/(1-p) = \exp(a+bx)$$

Then we remove the fraction

$$p = (1-p)\exp(a+bx)$$

and collect together terms in p

$$p = \exp(a+bx) - p \times \exp(a+bx)$$
$$p + p \times \exp(a+bx) = \exp(a+bx)$$
$$p[1 + \exp(a+bx)] = \exp(a+bx)$$

to get the regression equation

$$p = \frac{\exp(a+bx)}{1 + \exp(a+bx)}$$

The result of this transformation is shown in Figure 9.24, with an obvious improvement in the interpretation of the model. We might, for example, estimate that among individuals with a height of 170 cm, about 90% are male or, equivalently, that the probability of an individual of 170 cm being a male is about 90%.

This is the **logistic regression** model, one of the most important analytical methods in clinical research. The affinities of this method with linear regression are manifold, as we have just seen, but some problems still persist. That is, for some values of height, there was only one individual in the sample and in these cases the proportion of men could only be 0 or 1, with corresponding logit values $-\infty$ or $+\infty$. Of course, no least squares line could possibly predict such values. This effect is clearly noted at the most extreme values of the independent variable, typically those where there are fewer observations. Consequently, these observations have vanished from the graphs of Figures 9.23 and 9.24.

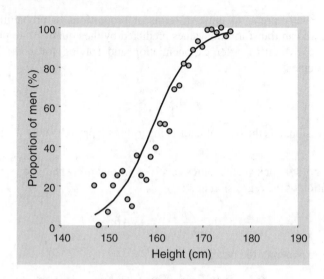

Figure 9.24   Transformation of the dependent variable back into proportions after fitting a least squares line to the data of Figure 9.23.

Therefore, the equation of the regression line cannot be obtained with the least squares method. Fortunately, there is an alternative approach for obtaining this equation, the method of maximum likelihood.

9.14   The method of maximum likelihood

We will begin the discussion of the method of maximum likelihood with a simple problem of estimation. Suppose we wanted to estimate, in a given population, the proportion of individuals with a certain binary attribute. To this end, we obtained a random sample of 50 individuals of this population, in which we observed the presence of this attribute in 30 of them.

As the attribute is binary, we know that the sample proportion p is from a binomial distribution and we want to estimate the mean of this distribution, which is equal to the proportion π with the attribute in the population. Now, the likelihood of the result we obtained in our sample depends on the proportion π with the attribute in the population. For example, if π were 50%, the probability P of observing 30 subjects with the attribute in a sample of 50 is 4.19%, and if π were 75% the probability P is 0.77%. We obtained the probabilities P from the formula of the binomial distribution.

Thus, we can compute the probability P of observing 30 individuals with the attribute in a sample of 50 for different assumptions about the population proportion π. We can calculate P in a systematic manner, starting at a value for π of 0% and going up to 100% in intervals of, say 1 percentage point. If we display on a graph the probability P of the result we obtained by sampling for each hypothetical value

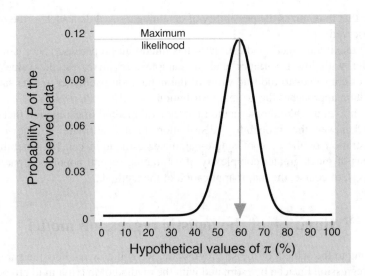

Figure 9.25   Likelihood function. The plotted values P *are binomial probabilities of observing 30 hits in 50 trials when the probability* π *of a hit in one trial ranges from 0 to 100%. The maximum likelihood of the observed result is 60%, corresponding to the maximum value of the likelihood function.*

of the population proportion π, we will get the result shown in Figure 9.25. In the graph we can see that the probability P of observing 30 individuals with the attribute in a sample of 50 is maximum (about 11.5%) if π, the proportion of the population with the attribute, is 60%.

The probability P of the data observed in the sample is a probability conditional on a given value for the probability π of the attribute. To distinguish the former from the latter probability, the former is given the name of **likelihood** and the curve relating the likelihood with the unknown parameter π is called the **likelihood function**. The value of 11.5% that in this example corresponds to the maximum value of the likelihood function is therefore the **maximum likelihood**.

In other words, 60% is the population proportion π that best explains the observed results and is thus the best estimate of the true value of the proportion of the attribute in the population. From our calculations we can also see that if the population proportion is less than 45%, the likelihood of the observed result is 2.5%. Therefore, 45% is the lower bound of the 95% confidence interval. Similarly, if the population proportion exceeds 74%, the likelihood of our data is 2.5%. The maximum likelihood estimate of the population proportion of the attribute is therefore 60% and the (exact) 95% confidence interval is 45 to 74%.

This method is therefore based on intuitive principles that we discussed earlier regarding the interpretation of data obtained by sampling and, in essence, just says that whatever is observed in the sample must be consistent with what exists in the population. Thus, the most likely value of the parameter we want to estimate is the one that maximizes the possibility of obtaining a set of data such as the one that was observed.

The maximum likelihood estimator has important properties. There is no other estimator with lower variance and, as sample size increases, the value of the estimator converges to the true value of the population parameter and its sampling distribution approaches the normal distribution.

The maximum likelihood method is a general method of estimation that can be used whenever the probability distribution of an attribute is known. This demonstration of the case of single-parameter estimation can be generalized to problems of much greater complexity of estimating several population parameters and this is, of course, the major application of the method.

9.15   Estimation of the logistic regression model

Returning to the subject of logistic regression, we will now see how the parameters of the regression line can be estimated with the method of maximum likelihood.

Recall that the logistic model is

$$\text{logit}(p) = a + b_1x_1 + b_2x_2 + \cdots + b_nx_n$$

where p is the probability of an individual having the dependent binary attribute, x_1 to x_n are the independent variables, b_1 to b_n the respective regression coefficients, and a the constant of the regression line. The model can also be written as

$$p = \frac{e^U}{1 + e^U}$$

if we prefer the prediction to be expressed as the probability of an individual having the attribute instead of the logit of having the attribute. In this equation, $U = a + b_1x_1 + b_2x_2 + \cdots + b_nx_n$ and e is the base of natural logarithms.

Maximum likelihood estimation begins by creating a tentative regression model by assigning a random value to each of the coefficients and to the intercept. Using this equation of the regression line, we calculate for each individual the predicted probability of having the dependent binary attribute A by multiplying the values of the independent variables by their (tentative) partial regression coefficients, summing these products overall, and adding the regression constant also multiplied by its (tentative) coefficient. The result is the predicted logit of A, which we transform into the predicted probability of A using the formula above.

Now let us pretend that this tentative model is the right one. If it actually is, then the observed values of A should be consistent with the predictions of the model.

This is how we evaluate the plausibility of the regression coefficients. We arrange the observations in groups having the same set of values of the independent variables, that is, observations that have the same predicted probability of A. We will denote the predicted probability of A in group i by p_i.

We know that the predicted value is a binomial probability. Therefore, we use the formula of the binomial distribution to compute, for each group i of subjects with the same characteristics, the likelihood of observing the number of subjects with A that was actually observed, given the number of subjects in that group and assuming that the true probability of the attribute in that group is p_i. For example, suppose that in a group of five similar individuals the predicted probability p_i of A was 75% and that two of them had the attribute A. The binomial probability of two hits in five trials when the probability of a hit is 75% is about 9%.

After we have done these calculations for all groups, we compute the joint probability (likelihood) of our dataset, under the assumption that the tentative model is correct, by multiplication of the probabilities calculated for all groups.

In practice, it is preferable to work with logarithms, because then we can add all the computed probabilities instead of multiplying them, making the calculations easier. The value obtained for the joint probability of the data is thus the logarithm of the likelihood, or the **log likelihood**. As any of these names is tricky to pronounce, hereafter we will refer to the log likelihood as LL.

After this initial evaluation, an algorithm computes the changes that must be made in the value and sign of the regression coefficients to increase the likelihood. The process is repeated for the new tentative values of the coefficients and the likelihood of the dataset is computed. The process continues over and over again in multiple iterations. At each step the algorithm makes corrections to the values of the coefficients until no further increase in LL can be achieved, at least to an appreciable extent, and the process stops. It is said that **convergence** has been reached. The maximum likelihood estimates of the regression coefficients are the latest values tested, those that maximized the likelihood.

The result of the method of maximum likelihood applied to the data in our example is shown in Figure 9.26. The model seems to fit the data well, but before we can accept the model we must verify that the assumptions of logistic regression were not violated.

Like any other statistical method, logistic regression is based on a number of assumptions. Compared to multiple linear regression, though, the assumptions of logistic regression are extremely reasonable. For example, as the dependent variable is the proportion of individuals who present a binary attribute, and that proportion has a binomial distribution, the assumptions that existed in multiple regression of a normal distribution of the dependent variable, homoscedasticity for all combinations of the values of the independent variables and the normal distribution of the residuals are consequently removed. The independent variables need not have a normal distribution and may be binary, ordinal, and interval. A linear relationship between the dependent variable and each explanatory variable is also not assumed. However, logistic regression assumes a linear relationship between the logit of the dependent variable and all independent variables.

As in multiple linear regression, the observations must be independent, the model should not omit important variables (underfitting), and it should not contain irrelevant variables (overfitting). There can be no perfect or important multi-collinearity between independent variables and all outliers responsible for the leverage of estimates should be investigated and explained. Basically, these are all problems that can be easily prevented with good study design.

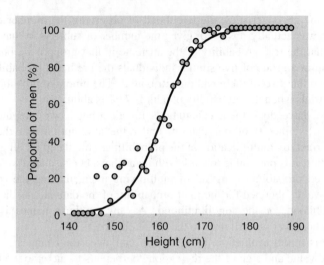

Figure 9.26   Logistic regression line obtained with the method of maximum likelihood fitted to the data of Figure 9.22.

In addition to verifying the assumptions of the method, the adequacy of the model should be assessed with formal methods. We saw in the discussion on multiple linear regression that the decision on the adequacy of the model was based on a test of the null hypothesis of all population regression coefficients being equal to 0, and that this test was based on the analysis of variance. In logistic regression the same principle applies, but the test is based on the **analysis of deviance**.

9.16   The likelihood ratio test

Suppose we could obtain several samples of the same population with the subjects matched for the same independent variables. The coefficients of the regression equation estimated with the maximum likelihood method will not be exactly the same. Because of sampling variation, the observed proportions p_i of the dependent binary attribute in identical groups i from the several samples would be different. Consequently, the maximum likelihood of the observed data would vary from sample to sample.

If the null hypothesis that all regression coefficients are zero is true, the model reduces to $\text{logit}(p) = a$. This model, consisting only of the constant, is called the **null model**. Thus, under H_0 the maximum likelihood of our model should not be much different from the maximum likelihood of the null model.

Therefore, we can construct a statistical test of the null hypothesis that all regression coefficients are zero by comparing the maximum likelihood of the null model to the maximum likelihood of our model. We do this by computing the maximum likelihood of a null model and taking the ratio of the maximum likelihoods of the null model to the current model. If our model explains the data better than the null model, the denominator will be greater than the numerator and the likelihood ratio will be less than 1. The smaller the ratio, the stronger the evidence against the null hypothesis.

All that remains to do now is to decide which value of the likelihood ratio statistic is large enough to allow us to reject the null hypothesis with known probability. The problem is that the distribution of the likelihood ratio statistic is very difficult to determine in most situations. Fortunately, there is a way around this.

We define a measure called the **deviance**, which is twice the difference between the LL of a perfectly fitted model and the LL of the current model. The likelihood of a perfectly fitted model is, of course, one. That is, the observed data is completely explained by the model. Therefore, the LL of a perfectly fitted model is the logarithm of one, that is, zero. The deviance of the current model is thus $2 \times (0 - \text{LL of the current model})$ or, more simply, -2LL of the current model.

The quantity -2LL is called the **log-likelihood ratio**, because the difference between the LLs is equal to the logarithm of the ratio of the two likelihoods. Naturally, the better the current model fits the data, the smaller the deviance and the log-likelihood ratio.

This quantity has the interesting property of following a chi-square distribution with degrees of freedom equal to the difference between the number of parameters in the two models. We can thus use the distribution of the log-likelihood ratio to reject with known probability the null hypothesis that the likelihood of our model is no different from the likelihood of the null model.

All we need is to find the difference between -2LL of the null model and -2LL of the current model. The result, called the **log-likelihood ratio statistic**, follows, as we have seen, a chi-square distribution with degrees of freedom equal to the difference between the number of parameters of the current model (the constant plus the number of independent variables) and the number of parameters of the null model (the constant). In summary, the number of degrees of freedom is equal to the number of independent variables in the model.

This test is called, appropriately enough, the **likelihood ratio test**. In the same way as the F-test in multiple regression, it tests the hypothesis that all regression coefficients (except the constant) are zero. A low p-value (i.e., less than 0.05) rejects the null hypothesis and is evidence that the model has some predictive power. Later on we will discuss methods for assessing the predictive ability of the model and to compare models according to their predictive capabilities.

9.17   Interpreting the results of logistic regression

We are now able to interpret the output of a statistical analysis program. Figure 9.27 shows the results of the logistic regression example we have been using.

In the top left corner, we see the results of the various steps of the convergence algorithm. We can see that the LL will be increasing (becoming less negative) until the difference from the previous iterations is minimal. The end of the iterations is determined by a convergence criterion included in the algorithm.

```
Iteration 0:    log likelihood = -1048.2766
Iteration 1:    log likelihood = -692.93092
Iteration 2:    log likelihood = -638.18855
Iteration 3:    log likelihood =  -631.5008
Iteration 4:    log likelihood = -631.33458
Iteration 5:    log likelihood = -631.33444

Logit estimates                          Number of obs    =        1853
                                         LR chi2(1)       =      833.88
                                         Prob > chi2      =      0.0000
Log likelihood = -631.33444              Pseudo R2        =      0.3977

------------------------------------------------------------------------
male gender |    Coef.    Std. Err.      z     P>|z|     [95% Conf. Interval]
------------+-----------------------------------------------------------
     height |  .2674934   .0133891    19.98    0.000    .2412513    .2937355
      _cons |   -43.07    2.190711   -19.66    0.000   -47.36371   -38.77628
------------------------------------------------------------------------
```

Figure 9.27   Results of logistic regression analysis of gender on height.

Displayed on the right is the number of observations (Number of obs), the log-likelihood ratio statistic of the likelihood ratio test with the number of degrees of freedom (LR chi2(1)) and the *p*-value (Prob> chi2) corresponding to the probability of the log-likelihood ratio statistic exceeding the value 833.88 in a chi-square distribution with 1 degree of freedom (less than 0.0001). The null hypothesis of all coefficients being equal to 0 can be excluded with a risk of less than 1:10 000.

In the following line there is another statistic, called **Pseudo R^2** (Pseudo R2). This statistic is analogous to the coefficient of determination R^2 of linear regression and has a similar interpretation, although it is not strictly, as it is in linear regression, a measure of the proportion of the variance of the dependent variable explained by the model.

Finally, the logistic regression table displays, for each independent variable, the partial regression coefficient (Coef.) and its standard error (Std. Err.), the result of the significance test of the regression coefficient (z) and its *p*-value (p>|z|), and the limits of the 95% confidence interval of the regression coefficients ([95% Conf. Interval]).

We can now use this model to estimate the probability that an individual has the attribute simply by inputting a value for height. For example, if we want to know how likely it is that an individual 170 cm tall is a male, we calculate

$$P = \frac{2.72^{-43+0.27\times170}}{1 + 2.72^{-43+0.27\times170}} = 0.948$$

Unless one is accustomed to think of probabilities in terms of logits, the coefficients themselves have no direct interpretation, which is frustrating at first glance when we think about how important the information contained in the coefficients was in linear regression. That is just at first glance, though, because actually we can obtain a measure of great clinical interest from the coefficients with a simple operation. We will see how in the next section.

9.18   Regression coefficients and odds ratios

Just as in linear regression, each logistic regression coefficient represents the difference in the average values of the dependent variable between two consecutive values of the independent variable. Thus, in the example we have been working on, the value of the coefficient of height means that the logit of being a male increases by 0.267 for each centimeter increase in height. In other words, 0.267 is the value of the difference between the logit of being a male for a given height and the logit of being a male for that height minus 1 cm.

Recall that the logit is the natural logarithm of the odds and that the difference between the logarithms of two quantities is equal to the logarithm of the ratio between these quantities, that is, $\ln(a) - \ln(b) = \ln(a/b)$. Hence, if 0.267 is the difference between the logarithms of the odds of the male gender at two consecutive values of height, it is also the logarithm of the ratio of the odds of male gender at two consecutive values of height. In other words, the logistic regression coefficient is the natural logarithm of the odds ratio. Therefore, exponentiation of its value will give us the odds ratio.

Thus, the exponential of the logistic regression coefficients is an estimate of the odds ratio of the dependent variable for each unit increase of the independent variable. In our example, the probability (or, more precisely, the odds) of an individual being male increases by $e^{0.267} = 1.30$ times per centimeter in height. Alternatively, we could also say that the probability increases 130%, or that the probability increases by 30%, per centimeter of height. The 95% confidence limits of the odds ratio are obtained in the same way. In this example they are 1.27 to 1.34.

Suppose that, for some reason, we wanted to express the association between gender and height as the odds ratio for each 5 cm increase in height. We could multiply the regression coefficient by 5 followed by exponentiation of the result. We can also work directly with the odds ratios, but as we are no longer in the logarithmic scale the odds ratios must be multiplied, not added. Therefore, if the odds ratio is 1.3 per centimeter in height, then per 5 centimeters in height the odds ratio will be 1.3^5, or 3.7. Similarly, when we want to express the combined effect of two independent variables in a logistic regression, their odds ratios should be multiplied.

In the case of multiple logistic regression, the odds ratio of an independent variable measures the degree of association between that variable and the dependent variable, controlling for the effects of all other independent variables in the model. For this reason the odds ratio is often called the **adjusted odds ratio**.

9.19   Applications of logistic regression

The applications of logistic regression in clinical research are manifold. Logistic regression is often used for modeling purposes. The dependent variable can encode the occurrence of a particular clinical outcome (e.g., death, complications, hospital discharge), and the model applied to identify patient factors associated with the outcome and their relative importance. This is the typical **study of prognostic factors** often found in the scientific literature.

These studies are based on case–control or one-sample cohort designs. Data on all attributes suspected or known to be associated with the outcome is collected from a random sample of subjects, and controls when applicable. In order to avoid including unnecessary variables in the model, a preliminary selection of the attributes is made by testing the association of each attribute with the outcome using the two-sample tests described in previous sections. Attributes that show some evidence of an association with the outcome, plus the attributes that are already known to be associated with the outcome, are included in the logistic model. Then, the model is perfected with a stepwise procedure and finally the assumptions of logistic regression are checked for possible violations.

We can also apply logistic regression to problems of prediction, where we want to be able to estimate the probability of some specific outcome in a particular individual and thereby develop predictive systems of individual risk, or **risk stratification systems**.

The dependent variable may identify two groups, as in case–control studies, and the logistic regression model used for the identification of the important attributes for the **discrimination between groups**, enabling the development of clinical classification instruments that may be used, for example, for the rapid screening of a particular disease or even for diagnostic purposes.

Logistic regression can also be used to test differences in proportions between two groups, being equivalent to the chi-square test for the comparison of proportions, or to test these differences, controlling the effects of nominal covariates being, in this case, equivalent to a test often used in epidemiology, the **Mantel–Haenszel test**.

Like multiple regression, logistic regression allows the use of dummy variables and, consequently, the analysis of independent categorical variables. Likewise, logistic regression allows analyses adjusted for the values of covariates, the inclusion of dummies for modeling nonlinear relationships (e.g., quadratic terms), and testing of interactions between independent variables.

The selection of variables is based on the same principles that were discussed in regard to multiple linear regression. For the determination of the variables that must be dropped from the model in **stepwise logistic regression**, the likelihood ratio of the current model is compared to the likelihood ratio of that model with one of the independent variables removed. If the likelihood ratio test is not significant, then the reduced model has a similar performance and therefore that variable can be removed.

Although the analysis of studies of prognostic factors is basically the same as in studies of predictive factors, there are some minor differences that should be noted. Studies of prognostic factors have above all an explanatory purpose, while predictive studies have the very pragmatic aim of producing excellent predictions. Thus, the former have the requirement of interpretability of the models from the biological and clinical standpoints, which limits the parameterization of the model to simple variables and relationships. As we saw previously, the presence of an interaction or a quadratic term complicates the interpretation, so these models are typically developed without those terms.

Rather, models that are intended primarily for prediction need explain neither why certain variables are in the model, nor how their effects are exerted. These models can therefore be rich in interactions and quadratic terms as long as they contribute to improving the adjustment. Interactions should be the product of standardized variables (divided by the standard deviation) to reduce the variance. However, these models are much more susceptible to overfitting than the explanatory models. The validation of a predictive model on a separate sample of individuals is therefore of utmost importance. We will see later on how to validate a predictive model.

For some time, logistic regression was the method of choice for the study of prognostic factors. However, due to the characteristics of these studies, many of them requiring the observation of individuals over sometimes prolonged periods of time, with the consequent difficulty in obtaining complete data on the outcome, logistic regression has been replaced in these problems by another regression method that can deal with censored data. We will cover this subject right after we discuss the method for model validation.

9.20   The ROC curve

The demonstration by a significant likelihood ratio test that a given logistic model fits the data well and has good explanatory properties by a high value of the pseudo R^2 does not necessarily mean that its performance in the classification of individuals or in the prediction of individual outcomes is clinically interesting. Therefore, in addition to those test statistics, in studies of prognostic factors for risk stratification, and in the construction of clinical instruments for patient classification, it is of great importance to actually measure the performance of the model in the setting where it was meant to operate.

The evaluation of the performance of the model begins with the definition of a classification rule based on the probabilities predicted by the model. A natural choice for this rule is to adopt a threshold of 50%, that is, if the predicted probability of an individual having the attribute is greater than 50% then the individual is classified as having the attribute.

After we have set this rule, we may arrange the observations in a **classification table**. Figure 9.28 displays a classification table with data drawn from the problem of predicting the gender of an individual from body height. The table shows how individuals with and without the attribute were classified by this rule.

From this table we can obtain important measures of model performance. In the group of 1377 individuals who had the attribute (man), 1283 (93.2%) were correctly classified. In the group of 466 who did not have the attribute (woman), 274 (58.8%) were correctly classified. These two measures are called, respectively, **sensitivity** (percentage of correctly classified among those with the attribute) and **specificity** (percentage of correctly classified among those without the attribute). The sensitivity is thus the **true positive rate** and specificity is the **true negative rate**. The

Observed			
Classified	Man	Woman	Total
Man	1283	192	1475
Woman	94	274	368
Total	1377	466	1843

Figure 9.28   Classification table of observed to predicted.

false positive rate is thus 192 classified as man among 466 women (41.2%) and the **false negative rate** is 94 classified as woman among 1377 men (6.8%).

In this example, such measures mean that the adopted classification rule will identify, from body height, 93% of the men and 59% of the women. The total number of individuals correctly classified by this rule was 1557 from the total of 1843 in the sample (84.5%).

The percentage of correctly classified must be interpreted with caution, because it is not always generalizable. It depends on the design of the study from which the data was obtained. Whereas in a one-sample study of the target population the percentage of correctly classified can be generalized to the population from which the sample came (being also advisable to present the confidence interval), in a case–control study there is no population for which that result can be generalized, as these studies are by definition based on samples from two distinct populations. Therefore, in case–control studies, the proportion of correctly classified depends on the relative size of the two samples. In this example, if we reduce the size of the sample of men, the percentage of correctly classified will decrease and, conversely, it will increase if we increase the number of men.

In one-sample studies, the proportion of individuals with the attribute is a point estimate of the probability of the attribute in the population. Thus, in such studies we can divide individuals into two groups according to their classification by the decision rule, and estimate in each of the groups the proportion of individuals with the attribute. If the data from the example we have been discussing was obtained in a one-sample study (as it actually was), we can estimate that the proportion of men in the group classified as man is 1283/1475 (87%), and that the proportion of women in the group classified as woman is 274/368 (74.5%). But if the study had been a case–control study, in which we established at the outset the number of men and women who would be included in the study, these estimates would be distorted by the relative number of subject in the two groups. In the example, if we increase the size of the sample of men the first proportion would increase, and if we increased the size of the sample of women then that proportion would decrease.

These two measures of model performance are called, respectively, the **positive predictive value** and the **negative predictive value**. The first is the probability of an individual actually having the attribute when the model predicted him or her to have the attribute, and the second is the probability of an individual not having the attribute when the model predicted him or her not to have the attribute.

Therefore, while predictive values give us a measure of performance and an individual estimate of the probability of the attribute, the sensitivity and specificity give us only a measure of performance. However, since the latter statistics can be determined both from one-sample and from case–control studies, the evaluation of model performance is based mainly on sensitivity and specificity.

Sensitivity and specificity are related and vary inversely. If the decision rule were to classify all individuals as men, the sensitivity would be 100% and the specificity 0%. Conversely, if the rule were to classify all individuals as women, the sensitivity would be 0% and the specificity 100%. Therefore, the sensitivity and specificity depend on the cut-off point defined by the classification rule and neither measure when isolated will tell us anything about the performance of the model.

Therefore, a better way of evaluating the performance of classification and predictive models would be by assessing how those two statistics are related through a graph of the sensitivity and specificity of each cut-off point of the classification rule. Such a graph is called a **ROC curve** and an example is presented in Figure 9.29.

The name of this curve stands for Receiver Operating Characteristics, an analytical tool used in telecommunications that relates the signal to noise for different settings of a receiver. By analogy with the problem of the accuracy of classification systems, the signal reflects the true positive rate, and the noise the false

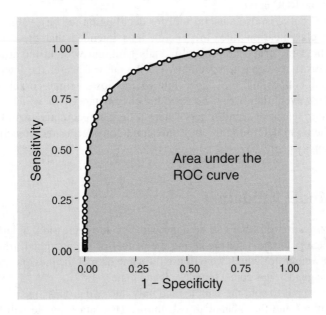

Figure 9.29   The ROC curve.

positive rate. The former is the sensitivity of a classification rule and the latter is one (or 100%) minus the specificity of the rule. In a ROC curve the sensitivity corresponds to the vertical axis and 1 − specificity to the horizontal axis. Each point in the graph corresponds to a possible cut-off value, that is, to the value of the logit or, willingly, to the probability of the attribute predicted by the model.

In a perfect system that correctly classifies all individuals the curve would pass through the upper left corner of the graph (100% sensitivity and specificity). The more the curve departs from that corner, the worse the performance of the classification system. The **area under the ROC curve** reflects thus the performance of the classification system, being 100% of the area of the chart in a perfect system, 50% if the system does not perform better than classifying individuals strictly by chance, and below 50% if the system performs worse than classifying by chance.

We can think of the area under the ROC curve as reflecting the proportion of individuals correctly classified by the system, regardless of the cut-off chosen for the decision rule (although an exact interpretation of the meaning of the area under the ROC curve is another). In general, a system for risk stratification or an instrument for the classification of patients will have the potential for clinical application only if the area under the ROC curve exceeds 80%.

The area under the ROC curve thus represents an additional measure of the performance of a classification system. It has an advantage over the other measures discussed so far because it integrates information about sensitivity and specificity into a single measure that is independent of the chosen cut-off. This means that we can compare the performance of two different classification systems through their areas under the ROC curve.

The ROC curve is also a useful tool for deciding on the best value of the cut-off for the classification rule. In general, it is intended that the decision rule identifies the same proportion of individuals with and without the attribute, that is, it has sensitivity equal to specificity, so the cut-off nearer the intersection of the diagonal connecting the upper left to the lower right corner should be selected. However, there may be specific circumstances in which one might prefer greater sensitivity, and others where a higher specificity is preferred, and the curve helps to determine the best threshold considering the objectives of the classification system.

9.21   Model validation

As a demonstration of what can be done with logistic regression, assume that we want to develop a system for predicting the gender of an individual, from the age, height, and weight, for which we will use the same data of the previous example. In the analysis, gender was coded female = 0 and male = 1.

The first step is to split the sample randomly into two groups, one with a third of the observations and the other with two-thirds. This latter sample will be used to develop the logistic model and is called the **training set**. The other, smaller sample

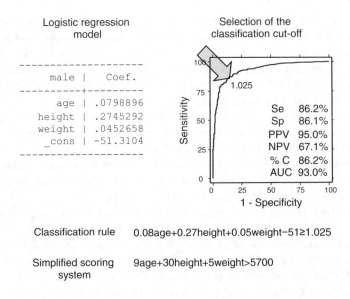

Figure 9.30 *Development of a risk stratification system.*

will be used to verify whether the performance of the predictive system we are about to develop is maintained when it is applied to a different set of data belonging to the same sample. This is called the **testing set**. We need the testing set to make sure that our model is generic enough and is not overfitted to the dataset from which it was developed.

Then, we develop a logistic regression model of gender on age, height, and weight, using the training set. The result is presented in the table of Figure 9.30. All partial regression coefficients were significant.

Next, we compute for each individual in the training set the predicted logit of being a male from the regression equation $0.08 \times$ age(years) $+ 0.27 \times$ height (cm) $+ 0.05 \times$ weight(kg) $- 51$. If our goal is to evaluate the overall performance of the model, the area under the ROC curve is probably the best single measure. If our goal is to develop a risk stratification system, then with the assistance of the ROC curve we select the cut-off value of the classification rule as the value of the predicted logit that defines a rule with equal sensitivity and specificity, as is shown in Figure 9.30. In the example, the cut-off point corresponds to a value of 1.025 of the predicted logit and we adopt the following rule: an individual whose predicted logit is equal to or greater than 1.025 will be classified as male, otherwise as female.

At this point we have defined the classification rule of our system as $0.08 \times$ age (years) $+ 0.27 \times$ height(cm) $+ 0.05 \times$ weight(kg) $- 51 \geq 1.025$ predicts   a   male, otherwise a female. The performance of a classification system with this cut-off seems adequate. The sensitivity and the specificity are 86% and the area under the ROC curve 93%. As this is a one-sample study, we may compute the positive predictive value (95%), the negative predictive value (67%), and the proportion correctly classified (86%).

If we plan to use this system in a clinical setting, perhaps it will be better accepted if the coefficients are rounded numbers easier to memorize, so we divide all the coefficients by the one with the lowest value and multiply them by an integer (5 in this example) to obtain rounded scores. The simplified scoring system has the form $9 \times$ age(years) $+ 30 \times$ height(cm) $+ 5 \times$ weight(kg) and an individual whose score exceeds 5700 is predicted to be male.

We then evaluate the performance of the system on the testing sample. If our aim is to verify the overall performance of the model, then the areas under the ROC curve should be compared. If our aim is to develop a risk stratification system, then we need to compute the score of each subject and calculate from a classification table the sensitivity and specificity of the adopted classification rule. In this example, these values were, respectively, 89% and 85%.

The testing sample is not meant to show that our system is generalizable. Rather, its purpose is mainly to show that our system is not generalizable if its performance for a different set of data is below acceptable standards. In practice, when the sample test results are unsatisfactory, we should try to understand the reasons for the degradation in performance. Possible reasons include violation of model assumptions, missed important variables in the model, multi-collinearity, overfitting, and poor model parameterization with failure to consider interactions and nonlinear relationships. If these problems can be corrected, a new model is estimated and reevaluated until eventually a reasonably robust model is obtained.

If the system passes this test, as was the case in our example, it is necessary to show that its performance is reproduced for a completely different sample of individuals, called the **validation set**. Some degradation of performance of the classification system almost always happens in validation sets, as is shown in Figure 9.31 when the model was applied to a sample of 663 subjects from a study conducted on the same population but in a different setting.

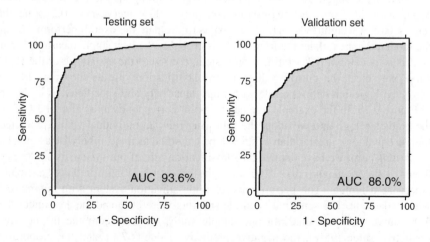

Figure 9.31   ROC curves on the testing set and validation set.

The important point is that the performance measures estimated in the validation set remain at a level that makes the system relevant for the purpose it was meant to have.

9.22   The Cox proportional hazards model

We will now see how we can apply the same regression principles to longitudinal studies where we are interested in the time to the occurrence of a specific event or outcome (e.g., death) and in identifying a set of attributes that will allow us to predict the occurrence of that outcome.

The appropriate study design for this type of research is the cohort study, both the one-sample and the two-sample designs. For each individual we need to collect data on the values of the attributes that we are interested in assessing as predictors and on the time elapsed from entry into the study to the occurrence of the event of interest. As is usual in this type of study, at its conclusion we will have complete information about the time to the event in the subset of subjects in whom the event did occur, and censored data for those in whom the event did not occur during the observation period.

What would we do to analyze these data with multiple linear regression? We could select as the dependent variable the time to the event and as independent variables the attributes of the subjects. The coefficients would then estimate the difference of mean time to the event between two consecutive values of each respective independent variable.

Unfortunately, this solution would not work because we would not have complete data for many observations. Censored data on time to the event would be a violation of the assumptions of multiple regression.

What if we selected as dependent variable the occurrence of the event, ignoring the time to the event, and fitted a logistic model? That would not work either. Because of censoring and varying observation times, the outcome of a subject censored after a very short follow-up would be considered as having the same outcome as a subject censored after a very long follow-up. This, of course, would be inappropriate.

Remember that we had a similar problem when discussing the comparison of survival rates when the time to an event was censored. The solution was based on estimating the risk of the event in each day (or whatever unit of time was used) among those surviving up to that day. If censoring was non-informative, that is, not related to the outcome, an unbiased estimate of the risk of the event in each day could be based on the number at risk up to that day.

Likewise, when we want to investigate the relationship between time to an event and a set of patient attributes we can adopt a similar approach. We cannot estimate the time to an event because the data is censored, but we can estimate from our observations the rate of occurrence of the event As the time to an event is inversely related to the rate, that is, the higher the rate, the shorter the time to the event, instead of trying to predict the mean time to an event from the explanatory

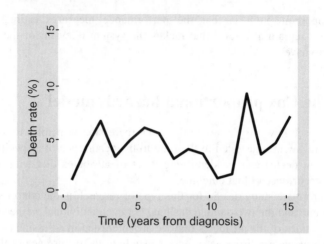

Figure 9.32   Hazard rate in women diagnosed with breast cancer.

variables, we will predict the rate of occurrence of an event on time. We will now see how we can develop such a model.

Let us first define the **hazard rate**. Consider, for example, that we have followed up a cohort of women with breast cancer for a long time and that we are searching for patients variables related to an event that occurs only once in each subject, say, death.

If we look at Kaplan–Meier estimates we will realize that women in our cohort are exposed each day to a certain risk of death which is not the same every day. For example, Figure 9.32 shows the yearly death rate after diagnosis of breast cancer in a cohort of 683 women. We can see that the death rate is high in the years following diagnosis because of disease progression, then it de-creases progressively to a minimum at about 10 years, increasing thereafter mostly because of death from old age.

We call the description of the event rate over time the **hazard rate** or **hazard function**. A suggestive name also used sometimes is the **force of mortality**. The hazard rate at a given moment t, denoted by $h(t)$, is thus the event rate at moment t among the subjects in whom it did not occur up to that moment.

Now consider that we wish to investigate whether a certain patient attribute is associated with decreased survival, for example, the finding of lymph nodes positive for the tumor at the time of diagnosis. If the attribute is actually related to survival time, then the hazard rate will be higher among those with positive nodes than in those with negative nodes. We call the hazard rate of those without the attribute, the **baseline hazard rate**, and denote it by $h_0(t)$.

Accordingly, if we estimate that the hazard rates are different in the two populations, then we can conclude on an association between positive lymph nodes and survival after diagnosis in breast cancer patients.

Clearly, it makes no sense to estimate the difference between the hazard rates at every time t. However, if the difference in the hazard rates between the two populations is assumed constant over time, then we can obtain a single estimate of that difference.

If two conditions are met, the difference between the hazard rates is indeed a constant. One condition is that the baseline hazard rate is the same for all women without the attribute. The other condition is that the difference between the hazard rate $h(t)$ of those with the attribute and the hazard rate $h_0(t)$ of those without the attribute does not change over time. Later we will look deeper into the implications of these conditions, but for now let us write down our model.

Recall from the discussion on relative risks that it was said that the best way to express differences between two rates or proportions is by their ratio. It was also said that it is easier to work with natural logarithms than with ratios, because then we can work with subtractions instead of divisions. Therefore, our problem comes down to estimating the model

$$\ln h(t) - \ln h_0(t) = bx$$

which is assumed to be the same for all t.

If we want to estimate a model adding another attribute, say, positive hormonal receptors, and if the two attributes are independent, then each attribute will change the difference $\ln h(t) - \ln h_0(t)$ by a certain amount b. If we call the two attributes x_1 and x_2, the expression above becomes

$$\ln h(t) - \ln h_0(t) = b_1 x_1 + b_2 x_2$$

We can generalize this equation for any number of explanatory attributes and write a general expression for this model as

$$\ln h(t) - \ln h_0(t) = b_1 x_1 + b_2 x_2 + \cdots + b_n x_n$$

Then we can write that equation in regression fashion as

$$\ln h(t) = \ln h_0(t) + b_1 x_1 + b_2 x_2 + \cdots + b_n x_n$$

Its interpretation is that the logarithm of the hazard rate can be expressed as a function of the baseline hazard rate, analogous to the intercept in multiple regression, and of a linear combination of explanatory variables.

As a final touch, we take the antilogarithm to express the hazard rates as proportions and the result is

$$h(t) = h_0(t)\exp(b_1 x_1 + b_1 x_2 + \cdots + b_n x_n)$$

In words, this model postulates that the hazard rate at a given moment in time can be split into a baseline hazard rate $h_0(t)$ and a hazard rate that is determined by the values of a set of independent variables. The independent variables need not be binary, but can also be interval if a linear relationship between the logarithm of the hazard rate and each variable can be assumed. In other words, this model requires the hazard rate to increase or decrease by a fixed value across the values of each independent variable on a logarithmic scale and, consequently, the effect of each variable on the hazard rate is exponential on a linear scale.

This model is called the **Cox model**. It belongs to a group of survival models called **proportional hazard models** and has some interesting and useful properties that make it preferable to other models in many clinical research problems.

There are several proportional hazard models, and they are distinguished from each other by the assumptions they make about the distribution of the baseline hazard (e.g., exponential, Weibull). The innovation of the Cox model is that it makes no assumptions at all about the distribution of the baseline hazard. Actually, this model does not even estimate the value of $h_0(t)$, it only estimates the values of the partial regression coefficients.

Now we can understand the reason for the wide application of this model in clinical research. The fact that the distribution of baseline risk is irrelevant increases the robustness of the method and allows its application to the study of any type of event. Of course, this advantage comes at a price.

Elimination of the estimation of $h_0(t)$ in Cox regression is obtained at the cost of doing without the possibility of estimating the value of the dependent variable $h(t)$ from the values of the independent variables, unlike the usual case in other regression models. What the Cox model does estimate is the **hazard ratio**. The hazard ratio on a particular day is thus defined for the set of individuals in whom the event has not yet occurred (called the **risk set**) as the ratio of the hazard rate of the event among those with the attribute to the hazard rate of the event among those without the attribute.

Take the case of a binary independent variable x. If we represent this variable by x_0 when it takes the value 0 and by x_1 when it takes the value 1, the hazard ratio (HR) between the two values of this variable is

$$HR = \frac{h_0(t)\exp(bx_1)}{h_0(t)\exp(bx_0)}$$

If we assume that the baseline hazard rate $h_0(t)$ is the same for all subjects we can drop that term from the expression so it becomes

$$HR = \frac{\exp(bx_1)}{\exp(bx_0)}$$

It follows that, under the assumption that the baseline hazard function is the same for all subjects, its knowledge is not necessary for estimating the hazard ratio.

For the estimation of partial regression coefficients, the method of maximum likelihood we used in logistic regression is not feasible because the population at risk is not the same every day – it decreases because of censoring and because patients have reached the outcome under study. To take account of the changing sample size, the method for estimating the regression coefficients used in the Cox model is the **method of maximum partial likelihood**.

Analogous to the method of maximum likelihood, this method is also based on the determination of the probability of a particular dataset for various values assigned to the regression coefficients. The difference to the method of maximum likelihood is that here a partial likelihood is computed for each occurrence of the event, therefore taking into account the decreasing sample size.

This method assumes that time is continuous and, therefore, there cannot be two or more individuals with exactly the same time to the event. As in practice time is measured on a discrete scale, that is, in number of days, months, or years, this may result in several ties in elapsed time to the event. In this case there are a number of methods that deal with ties, perhaps the most popular being the **Breslow method**. This is the simplest method but also one of the least accurate, whereas exact methods exist but have the drawback of being computationally very expensive.

9.23   Assumptions of the Cox model

We have already mentioned several assumptions of the Cox model, but further discussion is useful in order to better understand their full meaning and the consequences when they are not met.

In that model, each regression coefficient represents by how much each variable increases or decreases the hazard rate above or below the baseline hazard. This is illustrated in Figure 9.33, representing on the left the baseline hazard function, which is unknown but assumed equal for all subjects. An attribute with a negative regression coefficient decreases the hazard rate below the baseline hazard and is therefore a factor of good prognosis. This is represented in the middle graph of Figure 9.33. Conversely, a positive coefficient increases the hazard rate above the baseline hazard and is therefore a factor of poor prognosis (Figure 9.33, right graph). Needless to say, if the coefficient is zero, that variable does not change the baseline hazard and therefore has no predictive interest.

The assumption of equality of the baseline hazard function for all subjects implies that the entry point in the study must be a clearly defined moment in the life of a subject, so that all follow-ups are synchronized from that moment on. Sometimes this is not perfectly achieved, such as when study entry is defined as the time of diagnosis because a varying amount of time may have elapsed between the onset of a disease and diagnosis. Good study design is therefore essential for this assumption to hold.

As the hazard ratio is assumed to be constant over time, this means that in this model it is assumed that the effect of an attribute on hazard rate is constant over time, as illustrated in Figure 9.33. Thus, for example, if positive nodes at diagnosis double the death rate at 1 year, then the rate at 5 and 10 years after diagnosis, and for the entire lifetime of the subject, is also doubled. This is a key assumption of the Cox model and is called the **proportionality assumption**. This assumption must always be checked for each independent variable in the model.

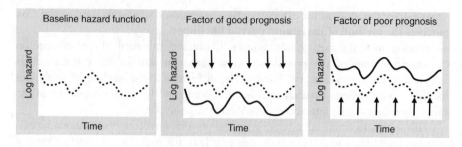

Figure 9.33   Illustration of the proportionality assumption.

There are several methods for testing the proportionality assumption, some of them graphical and some based on statistical tests. One of the most popular, although subjective, is the **log–log plot**. This is a graph of the logarithm of the cumulative hazard for each value of an independent variable against the logarithm of time. While the hazard rate is the instantaneous probability of the event, the **cumulative hazard** at time *t* is the probability of the event from entry into the study up to time *t*. The cumulative hazard is the logarithm of the reciprocal of the survivor function and can be obtained from Kaplan–Meier estimates. If the proportionality assumption is valid, the log–log plot will contain several lines, one for each value of the independent variable, separated by constant vertical distances (Figure 9.34, left graph). If the lines converge (Figure 9.34, right graph), this is an indication that the proportionality assumption does not hold.

Violation of the proportionality assumption can be readily suspected by inspecting Kaplan–Meier curves. If the curves cross twice, then the assumption does not hold.

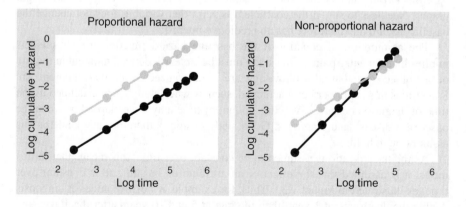

Figure 9.34   Visual assessment of the proportionality assumption with log–log plots: left, the expected plot when the proportionality assumption holds; right, the converging lines indicate violation of the proportionality assumption.

In addition to the assumptions of proportionality and of equality of the baseline hazard function for all subjects, the Cox model also assumes that censoring is not informative, that the hazard rate increases or decreases proportionally across the values of the independent variables, and that the observations are independent, as well as the other conditions common to other regression models with respect to multicollinearity, overfitting, and underfitting.

9.24   Interpretation of Cox regression

In the Cox model the partial regression coefficients b_i estimate the difference in the logarithm of the hazard rate between two consecutive values of an independent variable. Therefore, exponentiation of the coefficients will give the hazard ratio between two consecutive values of the independent variable because $\ln(a) - \ln(b) = \ln(a/b)$ and $\exp[\ln(a/b)] = a/b$.

The hazard ratio has an interpretation identical to the relative risk. For explanatory binary attributes the hazard ratio reflects the proportional increase in the hazard rate in subjects with the attribute to subjects without the attribute. In the case of explanatory interval attributes it reflects the proportional increase in the hazard rate for each unit increase in the value of the attribute. Thus, for example, if the regression coefficient of the independent variable X has the value 0.5, the hazard ratio is $\exp(0.5) = 1.65$. This means that for each increase of one unit in X, the rate of the event increases 1.65 times. Likewise, if the coefficient is -0.17, then $\exp(-0.17) = 0.84$, meaning that for each increase of one unit of X, the risk of the event decreases by 16%.

Now we should be able to interpret the results of an analysis using the Cox proportional hazards model. Figure 9.35 shows the results of a Cox regression of overall survival after the diagnosis of breast cancer on three patient attributes, namely age in years, T (tumor size greater than 5 mm), and N (tumor spread to four or more lymph nodes). Below the results of the iterations of the maximum partial likelihood method is the information that the Breslow method for ties of survival times was used. Next is information on the composition of the sample, where we are told that a total of 158 deaths were observed in 683 women during an observation period of 4105 person-years.

On the right, the significant ($p < 0.0001$) result of the likelihood ratio test means that the overall model is significant, that is, at least one independent variable is associated with survival. In the table below are the estimates of the partial regression coefficients, their standard errors, the statistics of the significance tests of the regression coefficients and the associated p-values, and the 95% confidence limits. All coefficients are significant and therefore we conclude that all three attributes are associated with overall survival.

Exponentiation of the coefficients and of the 95% confidence limits will allow easier interpretation of the regression coefficients. The coefficient of age is 0.015 971 4 and $\exp(0.015\,971\,4) = 1.0161$. Therefore, the death rate increases by 1.61% for each year increase in age, at a constant value of the other explanatory

```
Iteration 0:     log likelihood = -934.65558
Iteration 1:     log likelihood = -919.71675
Iteration 2:     log likelihood =  -919.701
Iteration 3:     log likelihood =  -919.701
Refining estimates:
Iteration 0:     log likelihood =  -919.701

Cox regression -- Breslow method for ties

No. of subjects =          683              Number of obs   =        683
No. of failures =          158
Time at risk    =         4105
                                            LR chi2(3)      =      29.91
Log likelihood  =     -919.701              Prob > chi2     =     0.0000

-----------------------------------------------------------------------------
      _t |     Coef.    Std. Err.      z     P>|z|     [95% Conf. Interval]
---------+-------------------------------------------------------------------
     age |   .0159714    .0070531    2.26    0.024     .0021476     .0297952
       T |   .4580184    .1775713    2.58    0.010     .1099851     .8060516
       N |    .610401     .162813    3.75    0.000     .2912935     .9295086
-----------------------------------------------------------------------------
```

Figure 9.35   Example of the result of a Cox regression.

variables. In other words, among women with the same T and N, each year of age increases the death rate by 1.61%.

The hazard ratios for the other attributes are obtained the same way. Thus, for T the hazard ratio is $\exp(0.458\,018\,4) = 1.580\,938$ and for N is $\exp(0.610\,401) = 1.841\,17$. Accordingly, among women of the same age and with the same N, a tumor size greater than 5 mm at diagnosis increases the death rate by 58%. Among women of the same age and tumor size, the presence of four or more positive nodes increases the death rate by 84%.

The Cox proportional hazards model has been used extensively in medical research for modeling, prediction, and classification. In many respects, Cox regression is similar to logistic regression in the modeling techniques and in its applications. Thus, we may use dummy variables to model and test interactions, and to represent categorical variables. We can also model curvilinear relationships. We can use Cox regression for the same problems where the logrank test is used. However, in situations where the estimation of hazard ratios is not of primary interest, the logrank test may be preferable because it is free of assumptions.

To conclude, a brief reference to some extensions of the Cox proportional hazards model is in order. One of the most important is **Cox regression with time-varying covariates**. The proportional hazards model assumes that the value of the independent variables is the same throughout time. For example, if we enter the baseline hemoglobin as an independent variable into a model, the method will estimate the hazard ratio as if in each individual the hemoglobin level was constant over time until the occurrence of the event. To obviate this condition, a model with time-varying covariates allows the inclusion of multiple values of one or more independent variables recorded at different points in time. The analysis is based on splitting each subject into several observations, one for each value of the covariate.

This model assumes that the time-varying variables are exogenous, that is, their value does not depend on elapsed time. One example of an endogenous variable, and therefore inadequate for this model, is patient age, which, obviously, increases with elapsed time.

Other extensions to the model that are occasionally used are the **Cox model with multiple failure data**, which allows the analysis of events that may occur several times in the same individual, and the **Cox model with competing risks**, used in situations where several distinct outcomes are studied simultaneously, which may occur in the same individual, but the occurrence of an event prevents the occurrence of others. A simple example might be to assess the response to therapy in cancer patients, in whom death due to drug toxicity prevents evaluation of the time to progression.

10

Measurement

10.1 Construction of clinical questionnaires

In the opening section of the book it was mentioned that biostatistics was also involved in measurement. It was also said that biostatistics provides methods for measuring things that are known to exist but cannot be measured by conventional instruments, like pain or anxiety. This is surely intriguing for many people, and some may even think this is suspicious, because common sense tells them that we can only measure what is observable. Of course, that is not true. Everybody is familiar with measurements of things that we cannot observe, for example, time and the force of gravity.

Much more intriguing, though, is the possibility of measuring things that are abstract concepts that exist only in our mind, like quality of life, satisfaction with the received care or, for that matter, anything we might think of. Again, this is not impossible. For example, we have seen in linear regression that we can find the coordinates of the center of a cloud of points in a scatterplot, which itself is an imaginary point, through the means of the two variables in the graph. The center of the cloud of points is what is called a **construct**, that is, an unobservable entity that cannot be measured directly but can be derived from the measurement of other directly observable variables.

Actually, there is nothing strange about this – we have seen before that we are able to measure things even though we do not observe them. In regression analysis we were able to predict the value of an attribute of a subject through the values of a set of variables that were related to that attribute. For example, we could derive the respiratory peak flow of any subject from the values of height and gender. Therefore, we do not actually need to observe something in order to measure it, as long as we can measure a set of attributes that are related to it.

So we may begin by considering that the construct we want to measure is a continuous variable like any other continuous variable, but with the particular

Biostatistics Decoded, First Edition. A. Gouveia Oliveira.
© 2013 John Wiley & Sons, Ltd. Published 2013 by John Wiley & Sons, Ltd.

characteristic that it cannot be observed. The difficulty, then, lies in the creation of a model that will allow us to predict the value of the unobservable variable from a set of observed attributes. In regression, the models are developed with knowledge of the actual value of the dependent variable, so this method applies only to observable attributes. The method of **factor analysis** allows us to define a set of independent predictors of an unobservable attribute.

Actually, factor analysis goes beyond that. The constructs of interest are often complex entities which involve several **dimensions**, and factor analysis helps us discover the hidden dimensions underlying a dataset, thereby helping us understand the relevant factors that determine the expression of a set of variables that can be measured.

However, factor analysis would not tell us explicitly what the dimensions explaining the observed data stand for. Factor analysis will only tell us which observed variables are associated with each dimension, and it will be up to the investigator to figure out what each dimension is measuring, guided by the pattern of relationships between the observed variables and each dimension. Thus, identification of the dimensions is a subjective exercise and consequently may not always be reproducible. This is the reason why factor analysis is not generally regarded as an adequate method for scientific work.

Nevertheless, this approach can be taken to develop clinical instruments for measuring virtually anything. Suppose we want to be able to measure some construct, say, happiness. We could start by collecting observations on measurable variables that in some way may be related to the construct, representing for example opinions, beliefs, attitudes, behaviors, values, feelings, sensations, and so on. These variables need to be quantified, at least in an ordinal scale, and Likert scales are often used for this purpose. Factor analysis would extract from the data the main dimensions, called **factors** in this context, that explain the variability in the data. The investigator would then name the factors guided by the meaning of the variables most related to each factor. For example, the investigator could identify factors such as wealth, professional success, family support, social relationships, health, and so forth. Therefore, happiness would be a state with several dimensions and each dimension can be assigned a score. If the factors are assumed to be independent, a total score for happiness can be compounded by adding up the partial scores of each dimension.

There are many other ways of developing clinical questionnaires but factor analysis, being a formal statistical method, deserves an explanation in this book and we will discuss briefly the underpinnings of the method in the next section. However, once a questionnaire is developed it needs to be validated, and the process of validation is as important as the development. Therefore, we will also cover methods of validation.

10.2 Factor analysis

Assume that we want to develop a scale to measure, say, happiness and that we asked 15 random subjects to respond to five questions presented as 10-point Likert scales. Usually at least 100 subjects will be necessary for the initial development of

Subj.	x1	x2	x3	x4	x5
1	5	6	2	8	0
2	9	9	2	4	5
3	9	8	7	3	9
4	1	3	8	0	6
5	0	3	3	5	2
6	3	4	7	2	7
7	5	6	0	6	0
8	1	1	7	2	8
9	6	6	3	6	5
10	7	6	9	3	9
11	7	3	5	4	8
12	5	5	7	0	7
13	0	0	0	6	3
14	6	6	8	0	7
15	5	4	3	9	3

$x1$-I have a great deal of energy
$x2$-I feel in good health
$x3$-I have a positive influence on other people
$x4$-people do not find me attractive
$x5$-I am optimistic about the future

Correlation matrix

	x1	x2	x3	x4	x5
x1	1.00				
x2	0.85	1.00			
x3	0.11	0.05	1.00		
x4	0.04	0.00	-0.79	1.00	
x5	0.30	0.05	0.84	-0.70	1.00

Figure 10.1   Dataset and correlation matrix of the factor analysis example.

a questionnaire, but for illustrative purposes we will consider only 15. We will name those questions $x1$ to $x5$. Suppose the questions were formulated as follows: $x1$, I have a good deal of energy; $x2$, I feel in good health; $x3$, I have a positive influence on other people; $x4$, people do not find me attractive; and $x5$, I am optimistic about the future. Subjects were requested to state their degree of agreement with each sentence on a 0 (totally disagree) to 9 (totally agree) Likert scale.

Figure 10.1 shows our artificial dataset along with the five questions and a correlation matrix displaying the correlation coefficients between the five variables. On a first inspection of the correlation matrix we find high correlation coefficients between $x1$ and $x2$, and between $x3$ and $x4$, $x3$ and $x5$, and $x4$ and $x5$. Therefore, there seems to be two patterns of relationships in the data, one involving variables $x1$ and $x2$, the other variables $x3$, $x4$, and $x5$. Of course, these two patterns were easily identified with only a correlation matrix because there are just five variables and their values were selected for this illustration to give clear relationships. On a real dataset with several dozen variables it will be much harder to discern patterns of relationships between variables through a correlation matrix. This is where factor analysis comes into play.

Factor analysis usually begins by standardizing the variables, that is, by transforming each one into a variable with mean 0 and variance 1. This is accomplished by taking the difference of each value to its mean and dividing by the standard deviation.

In order to understand factor analysis we will need to stretch our imagination to great lengths. In our example, we will need to imagine a space in 15 dimensions with 15 coordinate axes at right angles. Each axis represents one subject in the sample and each variable is defined in this space as a point located according to the values observed in each subject. Of course, this is impossible to illustrate in two dimensions, so we will imagine those points projected onto the first two observations. The variables would be located as shown in Figure 10.2.

Each axis corresponds to a subject and the points to each variable of the questionnaire.

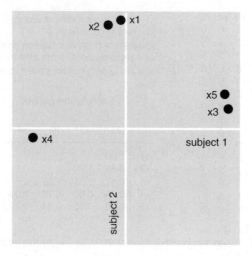

Figure 10.2   Data points representing the value of the questionnaire items defined by the observed values in each subject.

The first factor will be a straight line running through the center of the cloud of points corresponding to the x variables, in the same way that a least squares line runs through the center of a cloud of points. As all variables were previously centered, they all have zero mean, and thus the first factor must pass through the origin. Thus, the first factor would be positioned as is shown in Figure 10.3.

The projections of the five variables on the factor, represented by the dashed lines in Figure 10.3, define on that factor quantities called the **factor loadings**. The factor loadings are the correlation coefficients between the variables and the factor. Therefore, the square of a factor loading represent the proportion of the variance of a variable that is explained by the factor. We can see in Figure 10.3 that variables $x3$, $x4$, and $x5$ load high, and variables $x1$ and $x2$ load low, on the first factor.

Factor analysis proceeds by placing the second factor through the origin and at a right angle from the first factor. By placing the second factor at a right angle from the first, the two factors are made uncorrelated. This is shown in Figure 10.4. Again, try to imagine that Figure 10.4 represents the projection in a two-dimensional space of the 15 coordinate axes. Now variables $x1$ and $x2$ load high on factor 2, while variables $x3$, $x4$, and $x5$ load low on that factor.

The third factor is then positioned through the origin and at right angles from the first two factors, and loadings on this factor are determined. The process goes on until as many factors as variables are positioned.

The next step in the analysis is to establish how many factors should be retained. Intuition tells us that the important factors are those that explain most of the variance in the questionnaire items. As the squared factor loading represents the proportion of the variance of a questionnaire item that is explained by a factor, and as, due to the standardization of all the questionnaire items, all items have variance 1, if we sum all the squared factor loadings on a factor we will obtain a measure of

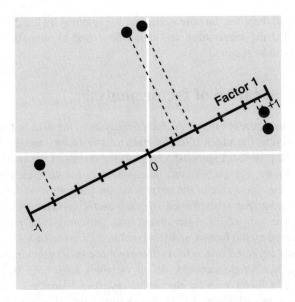

Figure 10.3   Factor analysis. The first factor is positioned through the center of the points representing the questionnaire items The projection of each variable on the factor defines the factor loading of a variable on the factor and is a measure of the correlation between each item and the factor.

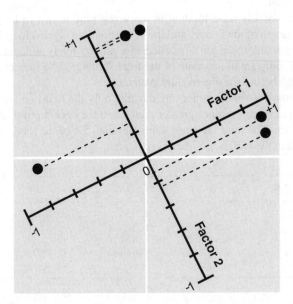

Figure 10.4   Factor analysis. The second factor is positioned at a right angle from the first factor, and is therefore uncorrelated. The projections of the questionnaire items on the second factor are represented by the dashed lines.

the variance in all the questionnaire items that is explained by the given factor. This quantity is called the **eigenvalue** and it is often used to suggest the number of factors that should be retained.

10.3 Interpretation of factor analysis

We will now look at the results of the factor analysis of the data in Figure 10.1. We begin by inspecting the eigenvalues of each of the factors to decide how many factors should be retained. Figure 10.5 presents the results of the analysis.

The table shows, for each factor, its eigenvalue, the difference between each eigenvalue and the eigenvalue of the next factor, the proportion of the variance in all questionnaire items that is explained by each factor (obtained by dividing each eigenvalue by the sum of all eigenvalues), and the cumulative proportion of the variance explained by the factors up to and including a given factor.

A commonly accepted rule is to retain only those factors that explain at least as much variance as a single variable. As all variables have been standardized and have variance 1, a factor with an eigenvalue less than 1 does not explain the data better than a single variable and is therefore irrelevant. This is called the **Kaiser criterion**. By the same reasoning, an alternative rule is to retain those factors explaining the same proportion of the total variance as an average single item. As there are five items in the example, each one accounts on average for 20% of the total variance, and thus those factors explaining more than 20% of the variance should be retained.

Another commonly used method is the **scree plot** shown in Figure 10.6. The scree plot is a graph of the factors and the corresponding eigenvalues. When a factor analysis is successful, the eigenvalues decrease progressively and then plateau when factors explain only small amounts of the total variance. The factors to be retained are the ones just before the eigenvalues plateau.

In this example it seems rather obvious that only the first two factors should be retained. Both have eigenvalues greater than 1 and they account for about 98.5% of the total variance. The third factor explains less than 5% of the total variance, which is much less than the variance of the average item.

```
---------------------------------------------------------------------
    Factor  |  Eigenvalue   Difference   Proportion    Cumulative
------------+--------------------------------------------------------
   Factor1  |    2.43041     0.73335       0.5800        0.5800
   Factor2  |    1.69706     1.50254       0.4050        0.9850
   Factor3  |    0.19452     0.23756       0.0464        1.0314
   Factor4  |   -0.04304     0.04553      -0.0103        1.0211
   Factor5  |   -0.08856        .         -0.0211        1.0000
---------------------------------------------------------------------
```

Figure 10.5   Factor analysis. Eigenvalues for the five factors. The eigenvalue represents the variance in all the questionnaire items that is explained by each factor.

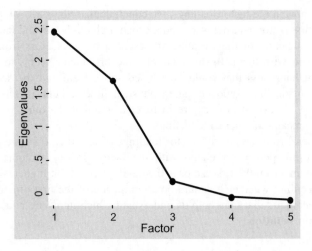

Figure 10.6   Scree plot.

Once we have decided on the number of factors to be retained we then look at the factor loadings of the questionnaire items on those factors. Figure 10.7 shows the factor loadings of the items $x1$ to $x5$ on the first and second factors, as well as a statistic called uniqueness. The **uniqueness** of a variable is the proportion of the variance of the variable that is not explained by the factors. A complementary measure is the **communality**, that is, the proportion of the variance of a variable that is explained by the factors. The latter quantity is obtained for each variable by summing the squared factor loadings for all factors. This quantity is analogous to R^2 in multiple regression and is equal to one minus the uniqueness. In the table of Figure 10.7 we can see that variable $x4$ is not well explained by the two factors, which account only for about 70% of its variance, while over 85% of the variance of any other items is explained by the factors. This information may help us refine our model by rephrasing the item and applying the questionnaire again, or sometimes by just dropping that item.

```
-------------------------------------------------
Variable |  Factor1    Factor2  |  Uniqueness
---------+--------------------+---------------
     x1  |   0.3702     0.8736  |    0.0998
     x2  |   0.2557     0.8799  |    0.1604
     x3  |   0.8847    -0.2274  |    0.1656
     x4  |  -0.7808     0.3068  |    0.2962
     x5  |   0.9141    -0.1179  |    0.1505
-------------------------------------------------
```

Figure 10.7   Factor loadings of the item questionnaires on the two retained factors.

Looking now at the factor loadings, we conclude that items $x3$, $x4$, and $x5$ load high on the first factor, and items $x1$ and $x2$ load high on the second factor. We can interpret the meaning of the variables represented by the factors from the items most correlated with them. In this example, the first factor seems to represent a dimension of happiness that could be labeled self-esteem and the second factor another dimension that could be called physical fitness. Item $x4$ has a negative factor loading because its values are in reverse order of the other items, that is, higher scores mean more negative feelings.

As a general rule, items with factor loadings of less than 0.30 are disregarded because they are poorly correlated with the factor. In our example, the factor loading for item $x1$ on the first factor is 0.37 and, were it a bit higher, we would be in trouble deciding whether it should be a composite of the first dimension or the second. A technique that may help us understand which items are related to which factor is **factor rotation**.

10.4 Factor rotation

Factor rotation is a technique for improving the interpretation of factors. As shown in Figure 10.8, the factors can be rotated around the origin so that those items that loaded high on a factor will have higher loadings and those that loaded low will have loadings near zero. This may offer a clearer insight into which items correlate with each factor and, therefore, may improve interpretability of the meaning of the factors.

The rotation illustrated in Figure 10.8 is an **orthogonal rotation** and the method used is known as **varimax rotation**, one of several methods of orthogonal rotation. Varimax rotation maximizes the sum of the variances of the squared loadings,

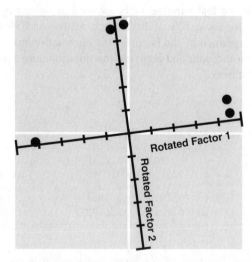

Figure 10.8   Orthogonal rotation of the factors shown in Figure 10.4.

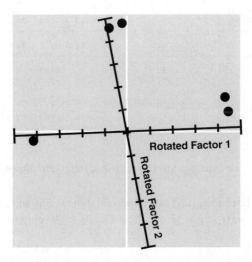

Figure 10.9   Oblique rotation of the factors shown in Figure 10.4.

resulting in high loadings for a few variables, and the remainder will load near zero. It is named orthogonal because the factors are kept at right angles to each other, meaning that the rotated factors are also uncorrelated. Depending of the rotation, the first factor may no longer explain the largest amount of variation, as is the case in unrotated factor analysis. The factor loadings, eigenvalues, and proportion of variance explained will also change, but not the uniqueness.

If after an orthogonal rotation the pattern of correlations is still not meaningful, an **oblique rotation** may help. In an oblique rotation the factors are rotated individually in order to obtain an optimal solution and they may no longer be orthogonal. Figure 10.9 shows an oblique rotation with the method known as **promax rotation**.

The main distinction between the two methods is that in oblique rotation the factors may no longer be uncorrelated. This may be seen as a limitation or an improvement, depending on the objectives of the analysis. If the purpose of the research is to develop an instrument for measuring a given construct which produces a single score, then the orthogonal rotation may be preferable because a composite score obtained by summing the individual dimensions will be easier to interpret. If the primary purpose is to understand which dimensions exist and how they interrelate, then an oblique rotation will fit the data better and will provide additional information on the relationships among factors. Oblique rotations are also adequate when only the scores of the subscales measuring distinct dimensions are relevant and no single score for the construct is sought.

Figure 10.10 shows the factor loadings after orthogonal and oblique rotation of the two retained factors in our example. In this case there are no dramatic changes in the factor loadings and, as seen in Figure 10.9, after an oblique rotation the factors are almost at right angles. This means that the two dimensions are indeed uncorrelated.

```
Orthogonal rotation                      Oblique rotation
--------------------------------         --------------------------------
 Variable |  Factor1    Factor2           Variable |  Factor1    Factor2
----------+---------------------         ----------+---------------------
      x1 |   0.1013     0.9433                 x1 |   0.0638     0.9412
      x2 |  -0.0101     0.9163                 x2 |  -0.0469     0.9192
      x3 |   0.9126     0.0386                 x3 |   0.9138    -0.0036
      x4 |  -0.8362     0.0675                 x4 |  -0.8413     0.1064
      x5 |   0.9091     0.1520                 x5 |   0.9056     0.1102
--------------------------------         --------------------------------
```

Figure 10.10   Factor loadings after orthogonal (left) and oblique rotation (right).

Up to now we have identified two important dimensions and the items related to each of them. The next step is to use these results to develop a scoring system for happiness.

10.5 Factor scores

We can think of a factor as a new variable, one that cannot be observed but whose values are obtained from a linear combination of the variables associated with it. We call this a **latent variable**. Therefore, in this example we could compute for each subject how much the happiness score was, as well as how much the subject scored on the self-esteem and the physical fitness dimensions.

A simple method is to sum the values of the questionnaire items of a given subject weighted by the loading of each variable on a given factor, thereby giving more weight to the variables most correlated with the factor. As the variables usually have different distributions, it is convenient to standardize them to mean 0 and variance 1. This is called the **factor score** for a given factor. For example, the standardized values of items $x1$ to $x5$ in subject 1 in the dataset of Figure 10.1 are, respectively, 0.1335, 0.5465, −0.9007, 1.4631, and −1.7355. Therefore, the factor score of this subject for factor 1 (self-esteem) is the sum of (0.1335×0.3702), (0.5465×0.2557), (-0.9007×0.8847), $(1.4631 \times (-0.7808))$, and (-1.7355×0.9141), which is equal to −3.3365. Likewise, the factor score for factor 2 (physical fitness) is the sum of (0.1335×0.8736), (0.5465×0.8799), $(-0.9007 \times (-0.2274))$, (1.4631×0.3068), and $(-1.7355 \times (-0.1179))$, which is equal to 1.4558. The factor scores should also be standardized so that their means and variances are the same. Then, the value of the **standardized factor score** of each subject represents the **Z-score**, that is, the departure of the factor score from the sample mean expressed as standard deviations. For the first subject in the dataset, the standardized factor score for self-esteem is −1.3249 and for physical fitness 0.8275. This subject has below-average self-esteem and above-average physical fitness. A score for happiness can be obtained by summing, or averaging, the standardized factor scores of the retained factors.

There are alternate methods for obtaining factor scores which are model based, providing more refined population estimates, like the least squares regression method and the Bartlett method. Whatever the method used, standardized factor scores can be used as a variable in hypothesis testing and in statistical modeling. Particularly in multiple regression, factor analysis may be useful by reducing a large number of explanatory variables to a few factors, thereby decreasing the sample size requirements and, possibly, simplifying the interpretation of the results.

However, factor analysis is not commonly used in medical research for several reasons. It has already been mentioned that the number of retained factors and their interpretation is highly subjective, but there are still other problems. There are several methods for extracting the factors, there are several methods for rotation of the factors, and there are several methods for obtaining factor scores, each method giving different results. These features of factor analysis open the door to the possibility of experimenting with several methods until the results are in agreement with the prior beliefs and expectations of the researcher. Furthermore, factor analysis assumes that the variables are in an interval scale (although it is often used with ordinal variables), that their distribution is normal, and that they are linearly related to each other.

Nevertheless, factor analysis may be a useful technique for understanding the structure of the data and the interrelationships between a large set of variables, and this may be hypotheses generating. The method described in these sections is a type of factor analysis called **exploratory factor analysis** and the factoring approach is called **common factor analysis** or **principal factor analysis**. The other type is called **confirmatory factor analysis** and the difference is that in this case the method is used to verify whether the number of factors and the loadings of the variables on each factor correspond to what was expected based on an underlying theory.

10.6 Reliability

Many instruments used in clinical research are based on items scored on Likert scales and developed using factor analysis. As stated above, factor analysis is used as a data reduction technique and, for that purpose, commonly used criteria include retaining only factors with eigenvalues greater than 1 and dropping items that load less than 0.30 on all factors. Items displaying high uniqueness, which therefore are poorly related to the factor solution, may also be deleted. Factor analysis may also uncover hidden dimensions in the items, allowing the creation of multi-dimensional scales in the case where no underlying theory has already proposed those dimensions. If a theory exists, or if previous research has already suggested some dimensions, then factor analysis will help identify the items related to each dimension specified by theory or in the literature.

In clinical questionnaires based on Likert scales, the total score is usually obtained by the simple sum of the values of all items. If the scale is multi-dimensional and dimensions are uncorrelated, partial scores for each dimension, or subscale, are obtained in the same way and a composite score is obtained from the sum of the partial scores. Thus, scores are usually obtained giving the same weight to all items, regardless of their loadings on the factors. This approach is based on **classical test theory** and is supported by the assumption in Likert scales that all items are considered to be replicates of each other. In contrast, **modern test theory** (also called **item response theory**) takes into account the importance of each individual item when forming the scores. However, the latter approach is rather complex for multi-dimensional scales.

Once an instrument is developed, it must be evaluated and possibly further improved by analyzing its reliability. **Reliability** is the ability of a measurement to give consistent results when measuring a construct in different persons, in different situations, and at different times. Basically, a reliable instrument is one that is free from error.

Reliability can be assessed with several statistics, the most popular being **Cronbach's alpha**. This statistic is widely considered to be a measure of the **internal consistency** of an instrument, that is, to what degree the different items on an instrument designed to measure a certain construct produce similar scores. As we have seen, the underlying assumption in instruments measuring a single dimension and based on Likert scales is that all items are measuring the same construct on the same scale, and this is what justifies creating a score based on the average, or the sum, of all the items. If this assumption holds, then all items should be strongly correlated, and the greater the correlation between the items, the greater the internal consistency of the instrument.

Cronbach's alpha is a measure of the degree of correlation of the items in a questionnaire. As all items are on the same scale, when they are all measuring exactly the same construct, then the items are identical variables and Cronbach's alpha will have the value 1. Conversely, when each item is measuring a different construct, then the items have zero correlation between them and Cronbach's alpha will have the value 0. Therefore, Cronbach's alpha is a scale between 0 and 1 that evaluates the degree of correlation between the items in a questionnaire; the closer it is to 1, the greater the confidence that all items are measuring the same quantity, that is, that all the items produce the same scores. Cronbach's alpha is thus a measure of **internal consistency reliability** of a questionnaire.

Cronbach's alpha is a statistic is based on the properties of variances. We have already seen that, when we sum k independent random variables, the variance of the resulting variable is equal to the sum of the variances of the k variables. For example, for three variables this can be expressed as

$$\text{var}(y) = \text{var}(x_1) + \text{var}(x_2) + \text{var}(x_3)$$

If the variables are not independent, the variance of the resulting variable is not equal to the simple sum of the variances. It will be equal to

$$\text{var}(y) = \text{var}(x_1) + \text{var}(x_2) + \text{var}(x_3) + \text{cov}(x_1, x_2) + \text{cov}(x_2, x_1) + \text{cov}(x_1, x_3)$$
$$+ \text{cov}(x_3, x_1) + \text{cov}(x_2, x_3) + \text{cov}(x_3, x_2)$$

As $\text{cov}(x_i, x_j) = \text{cov}(x_j, x_i)$, the expression can be simplified to

$$\text{var}(y) = \text{var}(x_1) + \text{var}(x_2) + \text{var}(x_3) + 2\text{cov}(x_1, x_2) + 2\text{cov}(x_1, x_3) + 2\text{cov}(x_2, x_3)$$

We can write this expression more generally as

$$\text{var}(y) = \sum \text{var}(x_i) + 2 \sum \text{cov}(x_i, x_j)$$

with i different from j.

The terms $\text{cov}(x_i, x_j)$ are the **covariance** of two variables, a measure of how two variables change together. If two variables are independent, their covariance is zero, and if they are identical their covariance is equal to their common variance.

Therefore, if each item in a questionnaire is measuring a different construct, then they are all independent and the covariance of all pairs of items will be zero. Conversely, if all items are identical, then the covariance of all pairs of items are the same, and equal to the common variance because the covariance of two identical variables is equal to their common variance, that is, $\text{cov}(x, x) = \text{var}(x)$.

We can easily find the value of $2\sum\text{cov}(x_i, x_j)$ in the above expression. If we subtract the sum of the variances of the items in the questionnaire from the variance of the total score of the questionnaire, the result is twice the sum of the covariances between all the items in the questionnaire.

For example, consider our happiness questionnaire again. Item 4 of the questionnaire was reversed, so we need to reverse its values so that all items give greater scores to positive attitudes. If we compute the total score of the questionnaire in the 15 subjects, the variance of the total score is 103.11. The sum of the variances of the individual items $x1$ to $x5$ in those 15 subjects is $8.97 + 5.95 + 9.21 + 7.98 + 9.21 = 41.32$. The difference $103.11 - 41.32 = 61.79$ is thus the added variance of the total score that results from the intercorrelation of the items. This quantity corresponds to the term $2\sum\text{cov}(x_i, x_j)$ in the expression above.

We would get the same result by obtaining the covariance matrix and summing all the off-diagonal elements of the matrix. The **covariance matrix** displays the covariance of all combinations in a set of variables. Figure 10.11 shows the covariance matrix of the five items in our questionnaire.

Our questionnaire items are correlated, otherwise their covariances would be zero and we would get a value of 0 for $2\sum\text{cov}(x_i, x_j)$. The value 41.32 is a measure of the intercorrelation of the questionnaire items, but just by looking at that value we have no clue whether it means a large or a small intercorrelation.

So the question now is how to interpret that value. We may approach this issue by constructing an index that ranges from 0 (no correlation between the questionnaire items) to 1 (identity of all questionnaire items). For this, we will need to know what the variance of the total score would be if all items were perfectly correlated.

	x1	x2	x3	x4	x5
x1	8.97	6.21	1.03	-0.30	2.76
x2	6.21	5.95	0.33	-0.02	0.38
x3	1.03	0.33	9.21	6.75	7.72
x4	-0.30	-0.02	6.75	7.98	6.03
x5	2.76	0.38	7.72	6.03	9.21

Figure 10.11   Covariance matrix. The diagonal elements are shaded.

If all items were perfectly intercorrelated but with different distributions, the variance of the total score obtained with the sum of the k questionnaire items would be equal to the sum of the variances of the k items (which are the same as the values in the diagonal of the covariance matrix) because the covariance of two identical variables is equal to their common variance, plus the sum of all the off-diagonal covariances.

In any matrix the number of off-diagonal elements is $k^2 - k$ and the total number of elements in the matrix is k^2. Thus, the proportion of off-diagonal elements in a matrix is $(k^2 - k)/k^2$, which can be written as $k(k-1)/k^2$ and simplified to $(k-1)/k$.

Thus, if all the items are perfectly correlated, the sum of the off-diagonal covariances, $2\sum \text{cov}(x_i, x_j)$, must account for $(k-1)/k$ of the variance of the total score, S_T^2. So, when the items are perfectly correlated, if we divide $2\sum \text{cov}$ (x_i, x_j) by $(k-1)/k$ we should get a quantity equal to S_T^2. Therefore, the ratio of the two quantities

$$\frac{2\sum \text{cov}(x_i, x_j)}{(k-1)/k} \quad \text{and} \quad S_T^2$$

for i different from j, will be equal to 1.

When the correlations between the items are less than perfect, the sum of the off-diagonal covariances will decrease and consequently the ratio of the two quantities will be less than 1. When all the items are uncorrelated, their covariances, and their sum, are zero and the ratio of the two quantities will be zero. This index will thus inform us about the degree of inter-item correlation: the greater their correlation, the closer to 1 the index will be.

Accordingly, the procedure for computing this index is as follows. First, we obtain the value of $2\sum \text{cov}(x_i, x_j)$, for i different from j, that is, the sum of all off-diagonal covariances, by subtracting the sum of the k item variances, S_i^2, from the variance of the total score. In other words, we obtain the term $2\sum \text{cov}$

(x_i, x_j) from the difference $S_T^2 - \sum S_i^2$. If all items are perfectly correlated this quantity must correspond to $(k - 1)/k$ of the variance of the total score. So we obtain the estimate of the variance of the total score under the assumption that all items are perfectly intercorrelated by

$$\frac{S_T^2 - \sum S_i^2}{(k - 1)/k}$$

which can be written as

$$\frac{k}{k - 1} \times \left(S_T^2 - \sum S_i^2 \right)$$

To obtain the index we divide the result by the variance of the total score:

$$\alpha = \frac{k}{k - 1} \times \frac{S_T^2 - \sum S_i^2}{S_T^2}$$

This simplifies to

$$\alpha = \frac{k}{k - 1} \times \left(1 - \frac{\sum S_i^2}{S_T^2} \right)$$

which is the usual formula of Cronbach's alpha.

In our example of the happiness questionnaire $\sum S_i^2 = 41.32$ and $S_T^2 = 103.11$, so Cronbach's alpha for that questionnaire is

$$\alpha = \frac{5}{4} \times \left(1 - \frac{41.32}{103.11} \right) = 0.75$$

It is a matter of considerable debate what the minimum value of Cronbach's alpha should be and that indicates an adequate internal consistency reliability. It is widely assumed that a questionnaire intended for use in research should have a value of Cronbach's alpha greater than 0.80, while some advocate that it should be as high as 0.90.

However, it must be noted that, as Cronbach's alpha measures the degree of inter-item correlation, its value will be inflated if a questionnaire contains several items that are essentially identical, such as rephrased questions. It will also increase with the number of items in the questionnaire. Figure 10.12 simulates question-naires composed of independent items and shows how Cronbach's alpha increases with the number of items in the questionnaire and with the proportion of identical items in the questionnaire, leaving the remaining items uncorrelated.

Therefore, before claims are made about the internal consistency reliability of a questionnaire, it is important to assert that the high alpha coefficient has not been inflated by eventually redundant items. A set of statistics may be helpful in this regard by giving some insight on whether the questionnaire has the properties expected of a measuring instrument. In one-dimensional instruments, as well as for

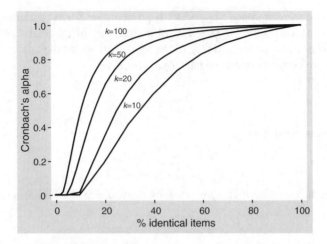

Figure 10.12   Relationship of Cronbach's alpha with the number of k *uncorrelated items in a questionnaire and the proportion of identical items.*

each subscale of a multi-dimensional instrument, inter-item correlations should not be too high, say, greater than 0.70, otherwise the items could be redundant. On the other hand, inter-item correlations should not be too low, say, less than 0.30, as this may indicate that they are measuring something different. These situations can be readily identified by inspection of the correlation matrix of all items.

Another property expected of an adequate measuring instrument is for each item to have a high correlation with the total score in one-dimensional instruments, and with its own subscale score in multi-dimensional instruments. Therefore, the correlation between each item and the total score in one-dimensional instruments, called the **item-test correlation**, or in multi-dimensional instruments the correlation of each item with its own subscale score, called the **item-to-own-dimension correlation**, should be high. On the other hand, as each item should be measuring a different facet of a construct, its correlation with a total score formed with the remaining items of the questionnaire should be moderately low. This is called the **item-rest correlation**. Furthermore, in multi-dimensional questionnaires, as each dimension is usually assumed to be independent, the correlation between subscale scores should be moderately low. This can be verified by inspecting the **interdimension correlation matrix**. Figure 10.13 shows some of these statistics for our example of the happiness scale. Additionally, the last column shows the alpha coefficient of a scale formed by excluding the referenced item. Thus, if item $x2$ was dropped, the alpha coefficient would increase from 0.75 to 0.77, suggesting that item $x2$ is not adequately measuring the construct, which is further suggested by its low (0.55) item-test correlation.

Accordingly, when reporting a new clinical questionnaire, at least the following statistics should be presented: in one-dimensional scales, Cronbach's alpha of the full questionnaire, median item-test correlation, median item-rest correlation,

Item	Obs	Sign	Item-test correlation	Item-rest correlation	Average Inter-item correlation	Alpha
x1	15	+	0.64	0.41	0.40	0.73
x2	15	+	0.55	0.31	0.45	0.77
x3	15	+	0.79	0.64	0.31	0.64
x4	15	-	0.70	0.50	0.37	0.70
x5	15	+	0.83	0.69	0.29	0.62
Test scale					0.34	0.75

Figure 10.13   Item analysis.

median inter-item correlation, and median Cronbach's alpha excluding one item in turn; and in multi-dimensional scales, Cronbach's alpha of each dimension, median item-to-own-dimension correlation, median item-rest correlation within each subscale, median inter-item correlation within each dimension, and the interdimension correlation matrix.

It should be noted that we used the data in our example only for illustrative purposes. In a real setting, internal consistency reliability should be assessed in studies on a different sample with a much larger size, usually greater than 400 subjects.

10.7 Concordance

We have discussed a widely used method for the evaluation of internal consistency reliability, but other types of reliability also need to be evaluated. A good instrument is supposed to produce consistent scores between two applications in the same subject. This is called **test–retest reliability**. It is evaluated by comparing the scores obtained in the same subjects when the questionnaire is administered at two different moments in time.

Test–retest reliability has some issues, though. First, the second application of an instrument must occur only after an elapsed period of time, allowing for the subjects to forget their answers on the first administration of the questionnaire. Second, the construct that the instrument is measuring should be relatively stable over the span of time between the two administrations.

At first glance, one might think that a measure of test–retest reliability could be obtained by computing the correlation coefficient between the two measurements. Thus, Pearson's correlation coefficient r or, perhaps more appropriately, the nonparametric equivalent **Spearman's correlation coefficient** ρ (rho) given that the scores are in an ordinal scale, would provide an easily interpretable quantity for describing reliability.

This would not be appropriate, though, as Figure 10.14 shows. The figure displays three situations where the correlation coefficient is 1, but only in the left graph are the two measurements in the same subjects equal. In the middle graph the

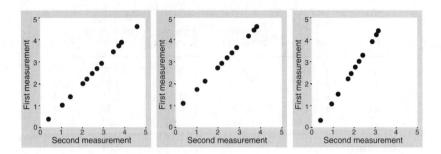

Figure 10.14   Three situations with equal correlation coefficients but different reliability.

first score is systematically greater than the second by a fixed amount, and in the right graph the first score is systematically higher than the second by a fixed proportion. In these latter two cases the instrument is unreliable, although the correlation coefficient between the two measurements is 1.

There is another reason why the correlation coefficient is not appropriate for the analysis of test–retest data. The correlation coefficient assumes an order in the measurements. In that case, one of the measurements is labeled x and the other y. When there are replicates of the same measurement, as is the case in test–retest studies, there is no way of telling which is x and which is y.

Therefore, the correlation coefficients mentioned above are not adequate for measuring test–retest reliability. The **intraclass correlation coefficient** method conveniently addresses those two issues and is the most commonly used method of analysis of test–retest data. The intraclass correlation coefficient ranges from 0 to 1, higher values indicating greater concordance between replicates. A value of at least 0.7 for the intraclass correlation coefficient is usually considered as indicative of adequate test–retest reliability. As it is expected that a questionnaire has high test–retest reliability, studies aimed at its demonstration require sample sizes of only 15–20 subjects.

The intraclass correlation coefficient estimates how much of the variance in the score of a questionnaire administered to the same subjects on two occasions is due to the variability of the scores between the two administrations. If the scores completely agree, the variance of the score between the two administrations will be zero and the coefficient will be 1. As the concordance between administrations decreases, that variance will increase and the coefficient will decrease.

The total variance of the score is due, on one hand, to the variance due to differences in scores between subjects, σ_B^2, and, on the other hand, to the variance due to differences in scores within the same subjects, σ_W^2. Therefore, if we want to evaluate the magnitude of σ_W^2 we can divide σ_B^2 by the sum $\sigma_B^2 + \sigma_W^2$. If the scores given by the subjects are totally concordant, σ_W^2 will be zero and the ratio will be one. The lesser the concordance between administrations, the lesser the ratio will be.

The quantities σ_B^2 and σ_W^2 can be estimated from an anova table, which will give us the between-subjects and the within-subjects mean squares. The within-subjects mean square represents the variation from one administration to the other, that is, σ_W^2. The between-subjects mean square is the sum of the between-subjects variance, σ_B^2, times the two administrations, plus the variance of the two administrations in the same subject, σ_W^2. Thus, the between-subjects mean square estimates $2\sigma_B^2 + \sigma_W^2$.

Therefore, the difference of the between-subjects mean square and from the within-subjects mean square is $(2\sigma_B^2 + \sigma_W^2) - \sigma_W^2 = 2\sigma_B^2$. The sum of the between-subjects mean square and the within-subjects mean square is $(2\sigma_B^2 + \sigma_W^2) + \sigma_W^2 = 2\sigma_B^2 + 2\sigma_W^2$.

If we now divide the two quantities we will get the intraclass correlation coefficient (ICC). Representing symbolically what was just said, the population intraclass correlation coefficient is

$$ICC = \frac{\sigma_W^2}{\sigma_W^2 + \sigma_B^2}$$

which can be estimated by using the components of variance in the anova table by

$$ICC^* = \frac{MS_B - MS_W}{MS_B + MS_W}$$

Returning to our example of the happiness questionnaire, suppose that the questionnaire was retested in the first four subjects and that the results were as shown in Figure 10.15.

The estimate of the intraclass correlation coefficient is computed from the components of variance obtained with anova as $(109.46 - 21.38)/(109.46 + 21.38) = 0.67$.

Subject	Scores	
	1st	2nd
1	14	18
2	30	25
3	39	28
4	27	30

ANOVA table

Source	SS	df	MS
Between subjects	328.38	3	109.46
Within subjects	85.5	4	21.38
Total	413.88	7	59.13

Figure 10.15   Illustration of the results of a test–retest study with anova table.

Sometimes, clinical questionnaires are not based on Likert scales and are not developed with factor analysis. Such instruments are often used for the diagnosis, grading, or prediction of a condition or outcome. For example, we saw previously that risk stratification instruments could be developed with logistic regression. Some clinical questionnaires used for classification or prediction are even developed without resorting to any statistical methods and consist of a combination of items with high sensitivity and items with high specificity, with small weights given to the former and large weights to the latter. Typically, those instruments are not meant for self-administration; it is the investigator who scores the questionnaire. As different researchers may have different criteria for scoring each item, the scores are subject to inter-observer variability. Therefore, instruments administered by a third-person should be evaluated for **inter-rater reliability**.

Consider a dichotomous item, say the presence of lung metastasis. Two raters evaluating the chest CT scan of a patient may not always agree as to the presence or absence of metastases. We would be led to think that an adequate measure of inter-rater reliability would be the percentage agreement between the raters.

This could be misleading, though. The reason is that, if the raters were guessing at random whether a patient has lung metastases, on a number of occasions, which could be significant, they might agree by chance alone. Therefore, chance agreement inflates the true degree of concordance in the ratings.

Cohen's kappa is a measure of the agreement between two raters that accounts for agreement occurring by chance. This statistic is used only for items in a nominal scale. A test of the null hypothesis that the true kappa is different from 0 is also available, but it only tells us that some degree of concordance exists, not whether the inter-rater agreement is within a value suitable for research. Indicative cut-offs for kappa values commonly accepted are <0.2 poor, 0.2–0.4 fair, 0.4–0.6 moderate, 0.6–0.8 good, and >0.8 excellent agreement.

Cohen's kappa calculates the difference between the observed agreement and the agreement expected purely by chance, and creates an index with a maximum of 1. If an item has k classes, a contingency table of the ratings of two observers will have $k \times k$ cells and the observed agreement, O, is the total number of observations where the two raters agree, that is, the sum of the diagonal cells of the table running from top left to lower right. The agreement expected by chance alone, E, can be estimated from the row and column marginal totals as follows. For each cell in the diagonal, the expected number of observations is the product of the row total and the column total, divided by the total sample size, n. Then, the agreement expected by chance alone is discounted from the observed agreement by subtracting one from the other, that is, $O - E$. To create an index with a maximum of 1, we divide that result by the maximum possible agreement, discounting chance agreement. The maximum possible agreement is verified when all n observations lie on the diagonal and so the maximum possible agreement discounting chance agreement is $n - E$.

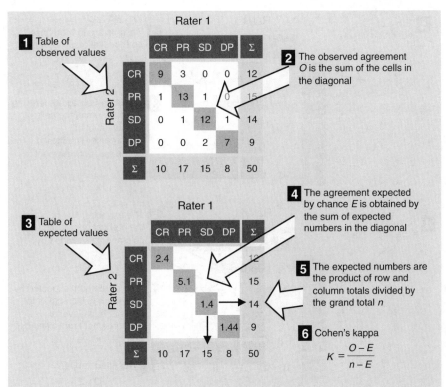

Figure 10.16   Cohen's kappa.

For example, imagine an inter-rater reliability study of two raters classifying cancer patients into four grades according to the patients' response to chemotherapy: complete response (CR), partial response (PR), stable disease (SD), and disease progression (DP). The results are presented in Figure 10.16.

The observed agreement between raters, O, is $9 + 13 + 12 + 7 = 41$. The agreement expected by chance alone, E, is the sum of $(10 \times 12/50)$, $(17 \times 15/50)$, $(15 \times 14/50)$, and $(8 \times 9/50)$, which is equal to 13.14. Thus, there were $41 - 13.14 = 27.86$ more concordant observations than expected by chance alone. The maximum number of concordant observations is 50, the total sample size n, and discounting those due to chance we get $50 - 13.14 = 36.86$. Therefore, the observed agreement is $27.86/36.86 = 0.76$ of the maximum possible agreement. This is the kappa statistic.

When the item classes are ordered, the disagreement is less serious if the raters differ by only one class than if they differ by two or more classes. Therefore, the degree of disagreement should be included in the calculations by assigning weights to that degree. This is known as the **weighted kappa** and its interpretation is analogous to the intraclass correlation coefficient.

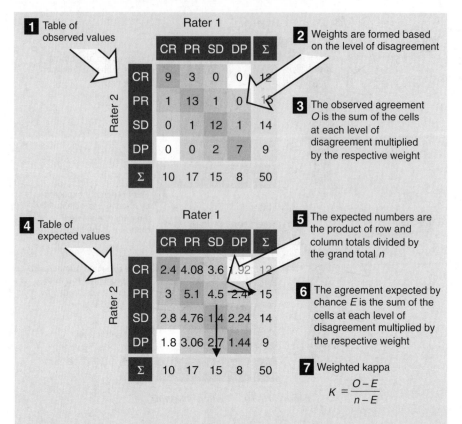

Figure 10.17   Weighted kappa.

We could consider the response to chemotherapy to be ordered from complete response to disease progression, representing decreasing success of the therapy. Thus, there would be lesser disagreement if a patient classified as complete responder by one rater were classified by the other as partial response than as disease progression, and we want to incorporate this judgment into the calculations. Therefore, the weighted kappa will be used in the analysis (Figure 10.17).

Weights are usually made equally spaced from 0 (maximum disagreement) to 1 (maximum agreement). As there are four classes, the weights will be multiples of $1/3 = 0.33$. These weights will be multiplied by the observed and expected number of observations at each level of disagreement. The calculation of the weighted kappa statistic is then the same as for Cohen's kappa.

Using the data from the table in Figure 10.16, for zero disagreements there were 41 observations and the expected number by chance alone is 13.14. The weight for this level is 1, so the weighted observed and weighted expected numbers are also 41 and 13.14. For disagreements in one class, there were

$1+3+1+1+2+1=9$ observations. The number expected by chance is obtained from the respective row and column totals: $(10 \times 15 + 17 \times 12 + 17 \times 14 + 15 \times 15 + 15 \times 9 + 8 \times 14)/50 = 21.28$. The weight for this level of disagreement is 0.66, so the weighted observed and weighted expected numbers are $9 \times 0.66 = 5.94$ and $21.28 \times 0.66 = 14.04$. For disagreements separated by two classes there were zero observations. The number expected by chance is $10 \times 14 + 17 \times 9 + 15 \times 12 = 9.46$ and as the weight for a departure of two classes is 0.33, the weighted expected number is $9.46 \times 0.33 = 3.12$. For a discrepancy in rating of three classes there were zero observations and as the weight for this level of disagreement is 0, so are the weighted observed and weighted expected numbers.

Summing all the weighted observed numbers we get $41 + 5.94 = 46.94$, and summing all the weighted expected numbers we get $13.14 + 14.04 + 3.12 = 30.3$. The difference between the weighted observed and weighted expected numbers is thus $46.94 - 30.3 = 16.64$. The maximum number of concordant observations discounting the number expected by chance is $50 - 30.3 = 19.7$. The weighted kappa is thus $16.64/19.7 = 0.84$.

Reliability can also be measured when there are more than two raters by using a different method called **Fleiss's kappa**. This statistic is used only with nominal items.

10.8 Validity

Reliability is not synonymous with validity. **Validity** is the property of an instrument of actually measuring what it is intended to measure. An instrument may produce consistent scores but may be measuring something different, or may be producing biased readings. The two concepts are linked, though. For example, a scale that always gives the same weight regardless of what it is weighing has high reliability but no validity. A calibrated scale with a large measurement error has low reliability and also low validity. Reliability is thus considered a *sine qua non* condition for validity and therefore the first step in the evaluation of the validity of a questionnaire should be an assessment of its reliability. For the reasons presented above, claiming that an instrument is reliable does not mean that it is valid and proof of validity must be produced.

Validity is usually evaluated for two main types. Proof of **construct validity** is obtained by showing that a questionnaire fits the underlying theory. Construct validity has two components, content validity and convergent validity. **Content validity** is a thorough examination of the content of the questionnaire to evaluate whether the relevant facets of the construct's domain are accounted for in the questionnaire. For example, our artificial happiness questionnaire does not have content validity because it obviously does not cover many aspects of happiness.

Contrarily to content validity, which has no empirical testing, the proof of convergent validity is based on actual data. **Convergent validity** is the degree to which a questionnaire correlates with other measures that theory predicted it would. Conversely, a questionnaire should not correlate with other measures that are measuring something else. This is called **discriminant validity**.

Convergent validity is assessed in two ways. One way is by showing that each dimension of the questionnaire is correlated with some other dimensions and is uncorrelated with others, as postulated by the theory. This, of course, does not apply to questionnaires developed with factor analysis unless an oblique rotation has been applied, because otherwise all dimensions are uncorrelated.

The other way is by analyzing the correlation of the questionnaire scores and subscales with selected external measures as predicted by the theory. For example, in our happiness questionnaire we would expect the physical fitness component to correlate with measures like number of days practicing sports, number of weekly hours working out, or number of weekly pieces of fresh fruit eaten. The analysis of such studies is usually based on the determination of correlation coefficients between the questionnaire scores and the external measures.

The other type of validity is **criterion validity**, the degree to which a questionnaire measures the same as a validated measure of the same construct. Studies conducted to evaluate criterion validity require the simultaneous administration of a set of already validated questionnaires that will serve as the gold standard to which the instrument will be compared. Thus, this is called **concurrent validity**. Again, analysis is usually based on the determination of correlation coefficients between the questionnaire and the instruments used as reference.

In addition to reliability and validity, questionnaires should also have proof of sensitivity and generalizability. **Sensitivity** is the ability of the instrument to change its score in response to clinically important differences in the state or condition of a subject. Studies on the sensitivity of an instrument are often conducted within clinical drug trials and try to demonstrate that changes in the clinical state of patients are reflected in the scores of the instrument. **Generalizability** is shown by applying the instrument in different settings, by different health care professionals, in different populations and in different cultures.

11

Experimental studies

11.1 The purpose of experimental studies

The main feature of experimental studies, which distinguishes them from observational studies, is that an intervention is applied to the subjects, observing later in time whether a response is seen. The ultimate goal of experimental research is the demonstration of a causal relationship between an intervention and a response, although not all experimental studies have the ability to show causality.

As we have already seen, in order to establish causality it is necessary to demonstrate an association between an intervention and a response, an order factor whereby the intervention precedes the response, and the absence of plausible alternative explanations for the observed response (Figure 11.1).

We saw earlier that analytical studies allow the investigation of associations, and that some observational designs allow us to establish the order factor. It is not very difficult to establish these two conditions but, as we discussed earlier, analytical studies hardly establish a causal relationship because they are rarely able to exclude the possibility of alternative explanations for the observed response. As we also have seen before, this is because of the possibility of confounding.

In order to establish causality beyond reasonable doubt it is necessary that the study design, the procedures for the conduct of the study, and the methodology of data analysis allow the researcher to establish the simultaneous presence, in a single study, of the three required conditions for the presumption of causality.

Therefore, the greatest concern of experimental studies aimed at the demonstration of causality resides on the exclusion of explanations for the observed response, other than the intervention that was applied. Consequently, as we will see, the particularities of the design and conduct of experimental studies are essentially related to the need of eliminating, or at least controlling, what are commonly called the external factors to the experiment. Surely it is not possible to exclude the possibility of contamination of the experiment by some external factors, but the

Biostatistics Decoded, First Edition. A. Gouveia Oliveira.
© 2013 John Wiley & Sons, Ltd. Published 2013 by John Wiley & Sons, Ltd.

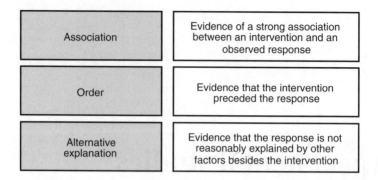

Figure 11.1   Conditions for the presumption of causality.

more these factors are eliminated, the more likely it is that the study findings correspond to reality, that is, that the conclusions are valid.

In the next sections we will focus on the methodology of **clinical trials**, the most common experimental studies in clinical research. The great importance of clinical trials is not so much the fact that they are a powerful method of scientific research, but mainly that they are firmly established as the paradigm for the evaluation of new products for the pharmaceutical market. Additionally, the evidence produced by clinical trials has been increasingly adopted as the major criterion for many decisions related to health care management and financing, both by government and by private institutions. Some examples are decisions on co-payment of drugs by national health services, health management organizations and health insurance, or the inclusion of drugs on hospital formularies.

Basically, a clinical trial aims to establish and quantify the effect of a drug or other therapeutic intervention on selected indicators of activity or severity of a disease and on the well-being of the patient. In a clinical trial, these indicators are typically clinical but almost always have some degree of relationship to the patients' perception of a real benefit to their health or well-being. It is also a major objective of clinical trials to determine the safety and tolerability of therapeutic interventions.

In general, clinical trials are primarily focused on the issue of the efficacy of treatments and secondarily on the issue of safety, although some clinical trials are concerned only with safety. Of course, efficacy and safety are just two of the many dimensions that can be considered in the evaluation of a treatment. Other dimensions are evaluated by specific disciplines that will be only mentioned because they are outside the scope of this book.

A discipline with great affinities with clinical trials is **outcomes research**. This discipline seeks to develop metrics related to treatment goals as these are perceived by patients, and designs clinical trials that evaluate treatment effects in terms of those metrics. In a way, outcomes research is more focused on the issue of treatment effectiveness than on the issue of efficacy.

Pharmacovigilance is the discipline dedicated to the evaluation of the safety of drugs, after they have been introduced into the pharmaceutical market. **Pharmacoepidemiology** is dedicated to the evaluation of the use of marketed drugs by health care providers and of their effect under real-world conditions. **Pharmacoeconomics** is dedicated to the economic evaluation of treatments, taking into account as many factors as possible, including treatment costs, patient monitoring cost, and costs of the treatment of adverse events, as well as the personal and societal gains obtained by improving the health status.

The subject of treatment evaluation is thus extremely wide and cannot be covered in its entirety in this book. Therefore, we will focus the discussion on the fundamentals of the most critical aspects of the methodology of clinical trials. To begin with we must have, as with any other investigational study, a clear formulation of the study objectives, a definition of the population, a sampling methodology, and a study design. We will begin by discussing these aspects.

11.2   The clinical trial population

As discussed earlier in this book, each clinical trial must have a conceptual definition of the patient population for which the treatment being tested will eventually be indicated, as well as an operational definition of the same population. The operational definition includes the criteria for the diagnosis of the condition under study, the criteria for the differential diagnosis, and possibly the criteria for disease activity or severity. This set of rules that allow the researcher to decide, in each case and without ambiguity, whether a patient does or does not belong to the study population, is called the **inclusion criteria** of the clinical trial.

There are particular groups of individuals who, despite belonging to the previously defined clinical trial population, should not participate in the research for reasons of a diverse nature. The criteria that identify those individuals ineligible for the clinical trial are called **exclusion criteria**. In most clinical trials, there are at least five types of exclusion criteria: (1) criteria related to the possible teratogenicity of the study drugs, and women who are pregnant and women of childbearing potential who are not using a contraceptive method of proven efficacy, are often excluded; (2) criteria related to the possibility of intolerance to the product, and patients with a history of intolerance or hypersensitivity to the drugs or to the ingredients used in their formulation, are excluded, as well as women who are breastfeeding; (3) criteria related to an inability to comply with the procedures of the clinical trial, such as the coexistence of mental illness or limited mobility; (4) criteria related to the coexistence of severe disease with a life expectancy shorter than the duration of the trial, such as advanced cancer or severe heart failure; (5) criteria related to failure of organs involved in the metabolism and excretion of drugs, such as the liver and kidneys.

In addition to these, several other exclusion criteria often exist, depending on the characteristics of the treatment. For example, patients may be excluded because they are taking medication that may interact with the drugs being tested and that cannot be stopped, or because they have other diseases or conditions that may interfere with the evaluation of the response. As a rule, all patient characteristics that define a population in which it may be considered that participation in the clinical trial may expose the subjects to a risk outweighing the expected benefits must be part of the exclusion criteria.

The sampling method typically used in clinical trials is consecutive sampling. Patients are enrolled in the clinical trial by the order they attend the trial site, provided that they verify all the inclusion criteria, none of the exclusion criteria, and have consented in writing to participate in the trial after they have been fully informed of the research objectives, of all the clinical trial procedures, of any foreseeable risks to their health, well-being, and privacy, of existing treatment options, and of the applicable laws and regulations.

As we have seen, this type of consecutive sampling is not probability sampling and this opens the possibility of biasing the estimates of treatment effects. The bias eventually introduced by convenience sampling is particularly worrisome when the aim of the clinical trial is to estimate the population effect of a treatment, but when the aim is to estimate the difference in efficacy between two or more treatments, it is widely assumed that differences between treatments are largely independent of the segments of the patient population that are sampled, and therefore estimates of treatment differences are generally considered essentially valid even if obtained from convenience samples.

One strategy often used to try to minimize the selection bias eventually introduced by convenience sampling is to conduct clinical trials in many sites simultaneously, which allows the enrollment of patients from a diversity of settings. This trials are called **multicenter clinical trials**. Multicenter trials have the added benefit of shortening the time for the enrollment of patients needed for the study. In these trials, sampling is therefore stratified by study site since patients are selected consecutively in each center. However, in this case the purpose of stratification is to increase patient accrual rate and to cover a wider population; it is neither to increase the precision of the estimates nor to decrease sample size, as in the stratified sampling method that we discussed in descriptive studies. We will see later that some clinical trials use stratification with a different purpose.

Having obtained a definition of the patient population and the sampling method, before we proceed to study design it is appropriate to define the clinical and/or laboratory variables that will allow us to measure the response to a treatment, the so-called **efficacy criteria**.

11.3 The efficacy criteria

Ideally, the efficacy criteria should consist of clinical parameters that are easily measurable (preferably by non-invasive methods), objective, reliable, and perceived by the patients as a real benefit to their health and well-being. The efficacy criteria may be directly related to the objective of the treatment (e.g., cardiovascular mortality in patients with hypercholesterolemia receiving a lipid-lowering drug) or may be **surrogate criteria**.

Surrogate criteria are sometimes used in situations where the variables that are directly related to the objectives of treatment are very difficult to obtain because it would take a very long period of observation. In such situations, it is common to use a variable highly correlated with the objective of treatment and to assume that a change in that surrogate variable is accompanied by a change in the unobserved variable that truly measures the objective of treatment. For example, the objective of the treatment for a chronic disease like primary biliary cirrhosis is the prolongation of survival, but because a trial aimed at this goal would require patients to be observed for decades, it is acceptable to use as a surrogate criterion the normalization of indicators of hepatic inflammatory activity.

Either type of efficacy criteria can be represented by a single variable, for example, the serum glycosylated hemoglobin in a diabetes trial, or by an efficacy criterion defined by a set of variables. The latter are called **composite endpoints**. For example, a composite endpoint can be a major cardiovascular event defined as a binary variable taking the value 1 in the event of non-fatal myocardial infarction, non-fatal stroke, or death from ischemic heart disease or stroke.

There are particular aspects related to efficacy criteria that should be discussed. First, it is essential to ensure that the measurements are valid and reliable. This applies particularly to the case of efficacy variables that are scores of clinical questionnaires. It is essential to ensure that these questionnaires have been formally validated and, if a translation of a validated questionnaire is to be used, the translation should have been performed according to the correct methodology and subsequently subjected to formal evaluation for validity. If measurements are to be performed by two or more observers, then the reliability of those measurements should be assessed using the analytical methods described in the previous chapter.

Second, it is important to consider the issue of **multiplicity of the efficacy criteria**. As we saw earlier, when several comparisons are made on the same data the probability of a type I error (the alpha error or the false positive rate of the statistical test) increases. This is called **inflation of the alpha error**. We have seen that the Bonferroni correction gives us the approximate value of inflation of the alpha error resulting from multiple comparisons. Therefore, if a clinical trial has, say, six efficacy criteria and the statistical significance is set at the 5% level for each comparison, then there is a probability of approximately $6 \times 5\% = 30\%$ that one of the comparisons will yield a statistically significant difference even though the treatment is ineffective.

Several statistical methods are available to control for inflation of the alpha error caused by multiple testing, and later on we will cover some of them. A common practice has been to classify the efficacy criteria as primary or secondary. **Primary efficacy criteria** are kept to a minimum, preferably just one or two key parameters and appropriate methods to control the alpha error being used. Additional variables perhaps representing complementary gains in patient health and well-being that are evaluated simultaneously during the clinical trial are classified as **secondary**

efficacy criteria and no adjustment for multiplicity is made in the statistical tests on these criteria. Therefore, if statistical differences are seen only in secondary but not in primary efficacy criteria, this cannot be taken as firm evidence of treatment efficacy because those differences could have originated from multiple testing. Thus, secondary endpoints are intended only for the documentation of possible additional beneficial effects of a treatment under a strictly exploratory approach.

Finally, at least one of the primary efficacy criteria must be binary, interval scaled, or an event. This is because it is of major importance in clinical trials to estimate the magnitude of treatment effects through the construction of confidence intervals.

11.4 Non-comparative clinical trials

The simplest experimental design we can conceive for a clinical trial consists of a single sample of patients to whom a treatment is applied and, after allowance is made for an adequate time span for the treatment to produce its effect, an evaluation of the efficacy criteria is performed. From the sample proportion of the patients achieving the pre-specified efficacy criteria, interval estimates are constructed for the efficacy rate in the patient population receiving that treatment. This design is shown in Figure 11.2a. If the efficacy criteria are interval variables, it is usually appropriate to take a measurement of those criteria just before the prescription of the treatment, that is, a **baseline evaluation**, and obtain interval estimates for the average intra-individual difference between the final and the baseline measurement. A statistical test for paired samples can also be done, and if the result shows a statistically significant difference between

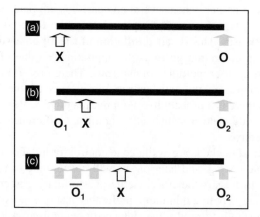

Figure 11.2   Non-comparative designs with a single end of trial observation (a), with baseline and final observations (b and c), with repeated measures at baseline (c). O represents a measurement of the efficacy variables, X the intervention.

the two observations, this would mean that the state of the patient changed after administration of the treatment. This design is shown in Figure 11.2b. The question, with both designs, is whether non-comparative clinical trials are sufficient to demonstrate a causal relationship between a treatment and an outcome. Let us check which conditions they demonstrate for the presumption of causality.

We can say that they demonstrate the order factor, because the intervention preceded the final observation. However, non-comparative trials do not show the existence of an association between the intervention and the outcome. As we know, in order to demonstrate an association we would need to compare the mean value of an efficacy variable in the population exposed to a treatment to the mean value of that variable in the population of patients in which the treatment was not administered. In other words, in order to demonstrate an association between a treatment and a response we need to have a control group. In the next chapter we will discuss the design of **controlled clinical trials**.

In addition to not demonstrating an association between a clinical response and an intervention, non-comparative clinical trials also do not rule out many alternative explanations for the observed response. For example, the simple effect of time on the natural history of disease might be a reasonable explanation for a change in the state of the patients. As we will see later on, there are quite a large number of plausible explanations for a change in the state of the patient other than the administered treatment. This is another reason for having a control group, because if a number of factors besides treatment may influence the response, in a controlled clinical trial their influence is exercised similarly in the intervention and control groups. The effect of the treatment itself is thus isolated.

An example of a possible alternate explanation for an observed change in efficacy criteria not due to the treatment intervention is a statistical phenomenon known as **regression to the mean**. This may happen in non-comparative trials evaluating the change from baseline in an efficacy variable.

Figure 11.3 illustrates how this phenomenon works. In situations where the primary efficacy variable shows considerable variation over time, if inclusion criteria determine that only patients with a value of this variable above a relatively high threshold are included, there is a strong possibility that the study population will contain a large proportion of subjects in which the average value of this variable over time is substantially lower than the chosen threshold. Therefore, it is very likely that a second observation done at a later time will produce a value that is closer to the mean value of that variable for that subject. Thus, at the end of the trial there is a high probability that a difference from the baseline value is observed, even if a treatment is completely inactive.

For example, consider a clinical trial of a substance intended to lower blood pressure. If we define as an inclusion criterion a value for a casual reading of the diastolic blood pressure greater than 90 mmHg, there is a reasonable probability of including subjects in whom the average diastolic pressure is less than 90 mmHg but that on a random reading happened to have a value greater than the threshold for inclusion in the trial. Regardless of whether or not a treatment is administered, there is a high probability that, on a second measurement later in time, a value closer to

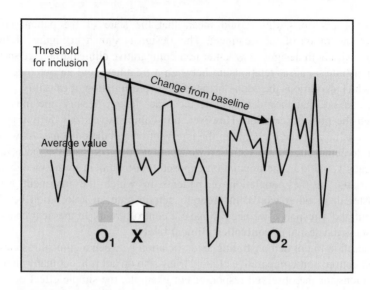

Figure 11.3   Illustration of the phenomenon of regression to the mean. The graph represents the variation over time of the efficacy variable in a given subject. O represents a measurement of the efficacy variables, X the intervention.

the average value for that subject, and therefore lower than the subject's baseline value, will be obtained. Hence the name regression to the mean.

Thus, in any non-comparative trial in which the value of the efficacy variable, or another attribute closely related to it, is used as an inclusion criterion, there is the possibility that any observed difference between the final and the baseline values are due only to regression to the mean. A conclusion is that non-comparative clinical trials where subjects are included on the basis of some pre-specified value of the efficacy criterion need a confirmation that the measured values are close to the average value of each subject and are not occasional outliers. If several measurements are made a few days apart, and the average value of those readings is above the threshold for inclusion, this will confer some protection against the possibility of regression to the mean. Figure 11.2c illustrates this design schematically.

At this point in the discussion it is natural to wonder if non-comparative trials are useful. Actually, as descriptive studies, they are very helpful in many ways. They may indicate that a treatment is ineffective if no response is seen after its administration; they allow identification of the adverse events of a treatment, which is important for the definition of the safety profile of a treatment; they may establish proof-of-concept that the effect of a treatment corresponds to the one predicted by theory; they may help identify subgroups of patients more prone to toxicity or less likely to respond to the treatment; and they provide data that will help plan further larger-scale comparative trials. Very often, the initial evaluation of a new treatment is made with non-comparative trials.

11.5 Controlled clinical trials

A clinical trial with a control group has the potential of allowing the establishment of the three conditions for the presumption of causality. This design demonstrates the order factor, because the intervention precedes the evaluation of the response; demonstrates an association between the intervention and the response, because now there is a comparison group that was not exposed to the intervention; and eliminates alternative explanations for an observed response because the comparison group allows us to control for the influence of many external factors. Because these factors act similarly in both groups, their influence on the response is canceled out and the effect of the intervention is therefore isolated.

Figure 11.4 illustrates schematically the design of a controlled clinical trial where O_1 and O_2 represent the observations of the efficacy criteria in each group and X the intervention. There may be only one measurement of the efficacy criteria in the last observation, or there may be a baseline measurement as well. This study design has, as its main characteristics, not just two or more groups, but also a method for the allocation of the subjects to the groups.

In most instances, the control group is observed at the same time as the intervention group, and they are called **concurrent controls**, but in special situations the clinical trial may use **historical controls**, that is, the comparison is done with patients who have in the past been included in a similar clinical trial. Clearly, the use of historical controls is limited when evidence of causality is sought, because the patient populations might have somewhat different characteristics and this alone could explain differences in the outcomes eventually observed.

Naturally, if we want to exclude the possibility of the observed response being due to factors other than the treatment, the first requirement is that the control group should have similar characteristics to the intervention group. One way to achieve this would be by using **matched controls**, that is, by selecting for the control group patients with characteristics similar to those who were included in the intervention group. In principle, this method would yield comparable groups. That would be only in principle, though, because in reality it is impractical to match individuals for more than a few attributes, and even with only a small

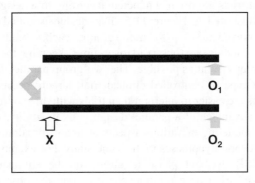

Figure 11.4   Diagram of the controlled clinical trial.

number of matching attributes it could take a long time until all the controls were found. The benefits of matching as a method for forming identical groups are actually small, because the method only ensures that the groups are comparable with respect to a small group of features.

Another possibility is simply to assign subjects to groups at random. If patients are randomly assigned to groups, not only will their characteristics tend to be similarly distributed in both samples, but more importantly there is the guarantee that the populations from which the samples were obtained are identical on all characteristics, not just on the known important factors but for all the factors too, both known and unknown, important or unimportant. Therefore, the random allocation of patients to the study groups eliminates the possibility that differences in the outcomes between treatments are due to differences in the characteristics of the populations from which the groups were sampled. The purpose of randomization is not to obtain study groups with identical distributions of patient attributes, which would be impossible due to sampling variation; rather, the purpose is to guarantee that the null hypothesis of no difference in population means of the efficacy criteria between study groups is true at baseline and, in this manner, to ascertain that a difference in population means at the end of the trial must be a consequence of the treatment, if contamination from external factors can be excluded.

If patient enrollment in the trial is consecutive and patients are randomized to the study groups, this should eliminate any interference from the researcher in the formation of the samples of patients. In other words, consecutive enrollment and randomization eliminate **selection bias**. Obviously, the absolute equality of the populations under study can only be guaranteed if the sampling procedures and randomization are performed without fault.

The design just described is known as the **randomized controlled clinical trial**, which is widely considered as the gold standard for the evaluation of treatments.

11.6 Classical designs

Controlled clinical trials are often conducted according to a design called **parallel group trial**, as illustrated in Figure 11.5. This design is called parallel because patients are randomized into two or more groups, each submitted to a different intervention, and patients are observed over time, keeping the trial conditions identical for all groups as far as possible. The two-group parallel design is certainly the most frequent type of controlled clinical trial, largely because it is easier to conduct and results are easier to interpret than trials with three or more groups.

An important advantage of a parallel design is its flexibility, which allows, on the one hand, the modeling of various aspects of treatment administration and, on the other, to test several hypotheses in the same study. For example, a clinical trial may have an initial washout phase to allow for the elimination of previous treatments, followed by a dose titration phase to find the dose that best fits the individual patient, and then by the treatment phase where one or more treatments in

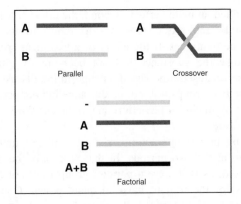

Figure 11.5   Classical designs of controlled clinical trials.

different doses and schedules of administration may be compared. Treatment effects can be compared at the end of the treatment phase to determine the relative efficacy of the treatments and again after a further period of observation to determine the rate of recurrence of symptoms, long-term efficacy, or long-term safety.

Another design is the **crossover trial**, also illustrated in Figure 11.5. In its simpler form, patients are randomized to one of two treatment groups and observed for a period of time. At the end of this period the efficacy criteria are evaluated and the treatments are withdrawn. After a period of time considered sufficient to allow for the effects of treatments to wear off completely, the patients are then administered the treatment that the other group received, again being observed to determine the response. Thus, after the first treatment, patients are crossed over to the alternate treatment.

The advantage of this design over the parallel group trial is that each patient serves as his or her own control. Therefore, the variability of the efficacy variable is much less than in parallel group trials because the variability between individuals is eliminated. Thus, for the same sample size the crossover design has greater statistical power than the parallel group trial and, therefore, requires a significantly lower number of patients to demonstrate the same effect. (In the worst case scenario, a crossover trial requires half the sample size of an equivalent parallel trial with the same power and alpha error, but if the outcomes of the treatments are correlated the sample size is further reduced by a factor equal to one minus the correlation coefficient between the two outcomes. That is, $n = (1 - r) \times N/2$ where N is the sample size required for a parallel group trial.) However, the other side of the coin is that this design has limitations in its application that diminish its usefulness.

In order that both treatments of a crossover trial can be compared, it is necessary that at the beginning of the second period the patients are in the same state as they were at the beginning of the first period. For that to happen, two conditions are necessary. First, the disease has to be stable over time, showing no tendency for progressive improvement, worsening, or seasonal variations, otherwise the mean of the efficacy variables will be different in the two periods, regardless of the received treatment. This is called an **order effect**. Second, the effects of the treatments have to be fully reversible and the disease has to return to its initial state after discontinuation of the treatment. If the effect of a treatment continues to the next period, either because the substance has not been totally eliminated from the organism or because it caused a persistent change in the disease, the difference in efficacy between the two groups will not be the same in both periods. That is, there will be a **treatment by period interaction** due to the carryover of effects from the first to the second period.

If a treatment by period interaction exists, it will not make much sense to combine the data from the two periods. In that case only the data from the first period should be used, as if it were a parallel group trial, but then there may be insufficient patients for the trial to have adequate power. Consequently, for the statistical analysis of a crossover trial, the procedure is always to look first for a treatment by period interaction. If an interaction is not present, then the analysis consists of a comparison of the treatment effects adjusting for a period effect. If a period effect is identified, it should be explained and taken into account in the interpretation of the results. There are several methods available for the analysis of crossover trials, from standard t-tests to more complex methods based on anova, and the particular method selected in each case will depend on a number of factors, including the characteristics of the problem, the specific design of the trial, and the balance of the number of subjects across the arms of the trial.

For example, for a simple two-group, two-period crossover trial of treatments A and B with a single measurement of the outcome variable at the end of each period, the analysis is as follows. For each group, we take the difference in the outcome variable in each subject between the first and second period and compute their mean and variance. Call D_{AB} and D_{BA} the means of the within-subjects differences in the first and second groups, respectively. Under the null hypothesis of no difference between treatments A and B, D_{AB} and D_{BA} should be equal apart from random variation. This null hypothesis may be tested with a standard two-sample t-test, by dividing $D_{AB} - D_{BA}$ by an estimate of the standard error of the difference obtained from the combined sample variances and their degrees of freedom. A statistically significant difference is evidence of a treatment effect adjusted for the order of administration of the treatments.

For the test of an order effect, the null hypothesis is that D_{AB} and D_{BA} are equal but opposite in sign. Accordingly, this hypothesis is tested with a standard t-test, by dividing $D_{AB} - (-D_{BA})$ by the estimate of the standard error.

For the test of the treatment by period interaction we must sum in each subject the outcome variables in the two periods and compute their means and variances. Each sum represents thus the treatment effect plus the period effect for each subject. Call S_{AB} and S_{BA} the means of the within-subjects sums in the first and second group, respectively. If there is no interaction S_{AB} and S_{BA} should be equal, apart from random variation and between-subjects variation. If there is a treatment by period interaction these quantities should be different.

Therefore, we test the null hypothesis of no interaction with a standard t-test by dividing $S_{AB} - S_{BA}$ by the estimate of the standard error of the difference obtained from the combined sample variance of the sums and their degrees of freedom.

It should be noted that the interaction test has small power and a non-significant result may not provide enough security of an absence of carryover effects. Because of the assumptions of crossover trials their application is restricted, but they are still used in selected clinical conditions that are of a chronic nature, are relatively stable over time, and when treatments are not expected to act on the underlying pathologic process, for example, epilepsy or asthma. More complex crossover designs that include multiple periods and multiple treatments are frequently used for dose determination studies of investigational drugs.

A third design found in the literature, although infrequently, is the **2 × 2 factorial design**, which is represented schematically in Figure 11.5. This design may be used for the investigation of the efficacy of two different treatments as well as of their combined administration.

A 2 × 2 factorial design consists of four parallel groups. One of the groups is administered treatment A, another treatment B, another the association of treatments A and B, and the fourth is the control group which receives neither treatment A nor B. The economy in the sample size is obtained because the determination of the efficacy of treatment A is based on a comparison of the outcomes of all the subjects that received treatment A with the outcomes of those that did not receive it. That is, the outcomes observed in the groups A and $A + B$ combined are compared to the outcomes of the group B and control group combined. This can be done because, if the effects of treatment A and B are independent, then both the difference between group A and the control group, and the difference between group $A + B$ and group B, estimate the effect of treatment A. If sample sizes are equal, a combined estimate of the effect of treatment A may be obtained by averaging the two differences. For treatment B the procedure is the same: groups B and $A + B$ combined are compared to group A and the control group combined. The factorial design has the same sample size requirement of a single parallel trial, with the same power and alpha error, comparing the least effective treatment to the control group. Therefore, with a factorial design two trials are conducted at the same cost, in terms of sample size, as a single trial.

However, if there is an interaction between A and B, this means that the effect of one treatment is modified by the simultaneous administration of the other treatment. Interaction may manifest itself in two ways. The simultaneous administration of the other treatment may change the magnitude of the effect compared to no treatment, which is called a **quantitative interaction**, or may reverse the effect, making it worse than no treatment, which is called a **qualitative interaction**. In either case, it is no longer possible to estimate the isolated effect of treatment A by combining groups A and $A + B$, nor the isolated effect of treatment B by combining groups B and $A + B$. However, a small quantitative interaction may be tolerated and an estimate of the average treatment effect may still be acceptable, although it must be taken into account that some patients will experience a little less benefit, and others a little more benefit, from the treatment than the estimate produced by the trial.

Therefore, whenever a significant interaction is identified, the trial must be analyzed as a parallel trial with three arms and a control, that is, by comparing each of the treatment arms to the control. In this event, the problem is that the sample sizes may not afford enough power for a demonstration of efficacy. In addition, the same considerations apply regarding the lack of power of interaction tests as discussed above for crossover trials.

11.7 The control group

We have been talking about the control group as a sample of individuals from the same population as the intervention group that is submitted to exactly the same trial procedures with the sole exception of not receiving the treatment under investigation.

However, if individuals in the control group do not receive any treatment, in the eventuality that a difference in efficacy between the two groups is observed, an alternative explanation for the observed difference could be that its cause was the fact that some patients had the perception of receiving a treatment while others had the perception that they were left untreated. This could have introduced a bias, called **participant's bias**, because patients in the treated group might have greater expectations of an improvement to their health than those receiving no treatment. This bias is caused by the **placebo effect**, which manifests itself in an unpredictable, often beneficial, but sometimes detrimental, effect for the patient.

In order to control for participant's bias, all groups in a controlled clinical trial must be given a treatment. If the aim of a trial is to compare the efficacy of a treatment against no treatment, then the control group should receive a placebo. In drug trials, a placebo is a pharmacologically inert substance, usually consisting of the excipients used in the formulation of the investigational product but without the active substance.

If the treatment under investigation is inactive and its effect is due only to a placebo effect, then if the control group receives a placebo treatment the outcomes observed in the two groups will be similar and their difference will be zero. If the treatment does have efficacy, then the observed outcomes in the treatment group represent the sum of the effect of the treatment with its placebo effect, and the difference to the placebo control group will estimate the effect of the treatment itself.

In practice it is relatively rare to administer only a placebo to the control group. For ethical reasons, it is not acceptable to deny treatment to patients under any justification, including for the purposes of a clinical trial. Thus, a placebo is reserved for clinical situations for which there is no treatment with proven efficacy and for selected situations in which, although an approved treatment exists, no harm will be done to the patients by withholding treatment for a clinical trial of short duration. In all other cases, the control group should receive the best treatment available and the treatment under investigation is to be compared to the standard of care. A control group that receives the approved treatment is called an **active control**.

A placebo is also acceptable in clinical trials evaluating an add-on treatment, that is, a treatment that is to be used in addition to the standard of care. Then, the placebo group receives the standard treatment plus the placebo add-on. In all cases of placebo-controlled trials the patient must be fully informed of the chances of being assigned to a placebo group.

11.8 Blinding

The administration of a placebo or an active treatment to controls does not discard entirely the possibility of attributing a difference in efficacy between groups to differences in the attitude of patients, and possibly also of researchers, regarding the results of treatment. The assessment of response can be influenced by knowledge of the substance being administered, systematically favoring the evaluation of one of the treatments. This is called **evaluator's bias**.

In addition, depending on their prior beliefs regarding the benefits and risks of the experimental treatment, investigators might adopt different patterns of care for patients in the intervention and in the control groups, such as systematically providing encouragement to stay on the trial, using supplemental interventions of care or treatment, or having different criteria for withdrawing patients from the trial to one of the groups, thereby compromising the comparability of the groups during the conduct of the trial.

Therefore, it is most important to hide from the patient and the investigator which treatment is being administered. This requires that the test treatment and the comparator, whether it is an active substance or a placebo, have exactly the same look and feel and that there are no identifying elements in any of the treatments.

This design is called **double-blind** and its purpose is to control for both participant's bias and evaluator's bias, therefore affording comparability of the groups throughout the trial. However, sometimes it is not possible to hide from the investigator the nature of the treatment being administered, for example, because one of the treatments is associated with characteristic adverse reactions. In this case there may be two options: either the efficacy criteria are evaluated by a third person who has no knowledge of the treatment being administered to the patient nor of any manifestations that could indicate the treatment, called a **triple-blind** trial; or the trial is conducted as **single-blind**, that is, only the patient has no knowledge of the treatment being administered. The latter option may, for the reasons presented above, open the possibility for an alternative explanation of the observed effect, particularly when the efficacy variables are not objective measures. When the trial cannot be single-blind, for example, when comparing a medication with surgery, the design is called an **open trial**.

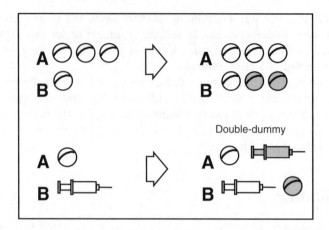

Figure 11.6   Uses of placebo for maintaining double-blinding. The placebo is represented in gray.

Sometimes a clinical trial compares drugs that are administered in different schedules, and in this case a placebo is used to allow the trial to be conducted as double-blind. For example, one medication is administered as one tablet once a day while the comparator is administered as one tablet three times daily. In this situation it is possible to conduct a double-blind trial by using placebo tablets to complete the three daily administrations, as illustrated at the top of Figure 11.6.

To the same end, when both treatments have completely different formulations, for example, tablets versus injections, it is possible to conduct a double-blind trial using the technique of the **double-dummy**. As shown at the bottom of Figure 11.6, this method consists of administering, in each group, a placebo identical to the treatment of the other group.

11.9 Randomization

Another feature of comparative studies that needs further discussion is the method of randomization of subjects. In principle, any process that distributes individuals in a perfectly random fashion is acceptable, including the tossing of a coin. However, this method of **simple randomization** has the inconvenience of not controlling the number of individuals that are included in each treatment group. Groups of similar size have several advantages, including obtaining the maximum power for the statistical tests, the possibility of using statistical methods that require equal numbers of observations in each group, as in some types of anova, and the relaxation of some assumptions of statistical tests, such as the equality of variances in Student's *t*-test.

The method of **block randomization** ensures that all trial arms will have approximately the same number of subjects, not only at the end of recruitment, but also at any time during the trial. This is convenient because, due to the usual difficulty in recruiting patients for a clinical trial, the study may end before the

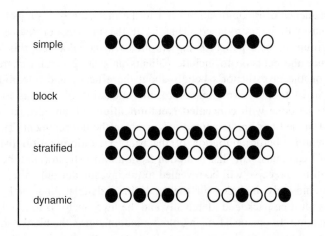

Figure 11.7   Randomization methods used in controlled clinical trials.

planned number of patients has been reached. In this method, before the trial begins subjects are randomized in sets of predefined size, called blocks, in a manner such that, in each block, the subjects are distributed evenly between treatment groups (Figure 11.7).

For example, with two treatment arms *A* and *B* the blocks are permutations of *A* and *B* with half the subjects allocated to each treatment in each block. This method ensures that, no matter what the final sample size is, the maximum difference in the number of subjects between groups will be no greater than half the size of the block.

The method of **randomized permuted blocks** goes even further in the attempt to achieve an even balance in the allocation of subjects to treatment arms. In this method the blocks are formed in a way that ensures that the samples will be evenly balanced not only in the number of subjects, but also in the order of inclusion in the trial. For example, if by chance the blocks always started with the sequence *AA* or *ABA*, the subjects in group *A* would systematically be included in the trial at an earlier time than subjects in group *B*. This technique prevents that from occurring.

Randomization need not be balanced among groups. Sometimes **unbalanced randomization** is selected, allocating different numbers of subjects to the treatment arms. This method may be selected for a number of reasons, for example, in placebo-controlled trials in order to give the trial subjects a greater probability of being randomized to the active treatment group. Sometimes unbalanced randomization is adopted because one of the treatments is more difficult to administer or more expensive, but perhaps the most common justification for using unbalanced randomization is because a larger sample is sought in some of the treatment arms in order to obtain better estimates of the incidence of adverse events.

It is of paramount importance that randomization is done without errors, otherwise the trial arms can no longer be considered as samples from the same population and, consequently, differences in outcome between treatments may not be caused solely by the treatment. In addition, the investigator must not be able to know the sequence of randomization prior to the inclusion of a subject, otherwise selection bias will be introduced. That is, if the investigator knows to which treatment group the next patient being enrolled will be allocated, this knowledge may influence the decision to include patients presenting certain characteristics, depending on the investigator's beliefs about the efficacy and tolerability of the treatments in each study arm. Therefore, the allocation of patients to treatment groups must be done with **concealed randomization** and an adequate procedure should be designed and implemented to guarantee the concealment. This is often done by creating, before initiation of the trial, a **randomization list** with the sequence of treatments. This list is hidden from the investigator and the treatment that a patient is to receive will be revealed to the investigator only after the patient has been definitively included in the trial. In a double-blind trial, the study medications must be packaged identically and the investigator is told only the code of the medication. In open trials using block randomization the blocks should be of varying size, otherwise the investigator will easily realize the size of the block and will be able to discover the treatment to be given to the last subject in a block.

A simple method of concealed randomization is to prepare envelopes numbered in the order of inclusion in the trial, each envelope containing a card indicating the treatment to be assigned to the patient. After a patient is definitely included in the trial, the envelope with the lowest number is opened and the treatment group is revealed. The procedure is illustrated in Figure 11.8.

In industry-sponsored pharmaceutical clinical trials, concealed randomization is more often achieved with the utilization of interactive voice response systems (IRVSs). In such systems, the medication code is given to the investigator by telephone, usong an automatic response system.

Another method of randomization, which may be combined with block randomization, is **stratified randomization**. This method may be used when one wants to ensure that the treatment groups will have identical distributions of one or more key attributes. These attributes are important because the values of the efficacy variables are different in the subsets of patients defined by those attributes. Thus, these attributes add variation to the efficacy criteria and, consequently, contribute to a decrease in the power of the study. In stratified clinical trials, the differences between group means in the efficacy criteria are computed separately for each stratum and then combined into a single estimate of the difference in efficacy between treatments, thus maintaining the power of the study.

For example, suppose we want to stratify a clinical trial by the presence of diabetes, because we know that the values of the efficacy criteria have different means in the diabetic and non-diabetic populations. Ideally, as in any comparative study, we want to make sure that at the conclusion of the trial the number of subjects in each treatment group is approximately equal within each stratum. This is achieved by creating separate randomization lists, one for each stratum. As each subject is included in the trial, the stratum to which the patient belongs is determined and the randomization list pertaining to that stratum is used. Randomization on multicenter clinical trials should always be stratified by study site.

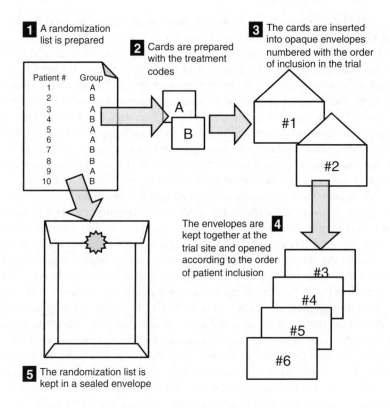

Figure 11.8   Concealed randomization using opaque envelopes.

A clinical trial can be stratified by several attributes, but then the number of strata and randomization lists will increase geometrically with the number of stratification variables. In practice, it is difficult to manage a trial with more than two strata, in addition to the trial site strata, and so this method should be considered only for attributes that have great influence on the response to treatment, and when the trial is so small that a difference between groups in only two or three individuals with that attribute can result in a significant increase in the variance of the efficacy criterion. Still, in this situation it may be more rewarding to increase the sample size than conduct a stratified trial.

Randomized stratification may also be used when it is desired to estimate treatment efficacy separately for each stratum as well as for the whole population. In this case the sample size is computed separately for each stratum. It is important to note that combining across the strata the differences in the efficacy criteria between groups to obtain a single estimate of the difference between treatments is only possible if the difference in efficacy between groups is equal across the strata. In other words, there can be no interaction of treatment by stratum. If an interaction exists, especially if it is qualitative, the treatments should be compared separately in each stratum but a combined estimate is no longer adequate.

A method of randomization known as **minimization** can be used to obtain a similar distribution of a large number of attributes between the study groups. This method determines, for each individual included in the trial, the treatment group to which the individual should be allocated in order to minimize differences in the distribution between groups of the stratification attributes. The method differs from matching because only the marginal totals are kept identical between the groups and not every individual has a control with identical characteristics. For example, if the matching attributes are gender and diabetes, the final sample of a matched study will have the same number of, say, diabetic women in both groups. With minimization, what is granted is that the groups will have the same number of women and the same number of diabetics in both groups. In the condition where both attributes are independent, the effect is identical to matching.

Finally, it is worth referring to the methods of **dynamic randomization** that adjust the randomization ratio of subjects into groups according to the results that are being observed throughout the study (Figure 11.7). In a trial with dynamic randomization the outcomes of the treatment groups are compared at predetermined time points during the trial and the randomization ratio is unbalanced in favor of the study arm showing better results up to that time point. For example, if the initial randomization ratio between groups A and B is $1:1$, that is, equal probability of allocation to the treatment groups, and a mid-trial evaluation of the results shows better responses for treatment B, the allocation ratio is changed to $1:2$. If a subsequent analysis continues to show better results for treatment B the ratio is increased to 1:3, otherwise it is changed to 1:1 again. The process is repeated several times and randomization rates are adjusted to promote the inclusion of a larger number of patients in the treatment group showing better results.

The purpose of dynamic randomization is thus an attempt to offer the more effective treatment to a larger number of subjects in the study population. For example, if with treatment A remission of the disease is achieved in 60% of the cases and with treatment B in only 30%, in a trial of 200 patients with a $1:1$ randomization ratio we would expect to observe 90 remissions (60 in group A and 30 in group B). However, if the randomization ratio had been $3:1$ in favor of group A, we would expect to observe, in the same 200 patients, 105 remissions (90 in group A and 15 in group B). Because we do not know at the outset of the trial which treatments are more effective, dynamic randomization seeks to maximize the number of responses in the trial population as information is being gathered during the trial.

As we have already seen, the power of a statistical test is at its maximum when the samples are of equal size. Therefore, when dynamic randomization is used the sample size must be adjusted to maintain the power of the trial.

11.10 The size of a clinical trial

From the previous discussion we can appreciate that it will be difficult to properly organize a randomization procedure without previously defining the total number of subjects required for a clinical trial. There are of course other important reasons for the determination of the sample size during the planning phase of a trial. It would be unethical to start an investigation involving patients without even having an estimate of the probability of success, that is, of reaching a useful conclusion.

Accordingly, for the calculation of the sample size we begin by deciding on the probability that a difference between treatments will be demonstrated if a difference truly exists. We call this probability the power of the trial, or true positive rate, or $1 - \beta$. Typically, the power adopted for clinical trials involving human subjects lies between 80 and 90%, but can be higher, and it is not uncommon for clinical trials to be powered to 99%.

Then we need to decide on the proportion of patients in each group. We already know that equal sample sizes afford the greatest power, but sometimes there are reasons that militate in favor of the treatment groups having different sizes.

If the efficacy variable is interval scaled, we need to obtain information about its variance in the target population. This information can usually be obtained from the literature on clinical trials conducted in a similar population, from databases of previous clinical trials, or, in the absence of this information, from a small number of observations of a group of patients.

The next decision is much more difficult. We must specify the difference between treatments that the trial will be able to identify with the previously defined power. The general approach is to adopt the criterion of the **minimal clinically important difference** (MCID), which is the difference in efficacy between treatments that represents a gain in health and well-being of the patient that justifies a decision for switching from the standard treatment to a new treatment.

Obviously, the definition of the MCID is not easy and carries great subjectivity. For this reason, a discussion with several experts experienced in the disease being researched, in the values and expectations of the patients, and in the treatments that are to be tested, is required to find a value that meets consensus.

More formal methods of reaching a value for the MCID are available. In **anchor-based methods** a meaningful external measure, the anchor, is used to provide a yardstick against which a difference in the efficacy variable may be compared. For example, one could use as anchor the patient's global impression of change and adopt as the MCID the mean difference in the efficacy variable between patient statements of 'no change' and 'somewhat improved.' This would require a longitudinal study of a cohort of patients evaluated before and after an intervention, but the MCID can also be assessed in cross-sectional studies. For example, one could use as anchor the performance status, a validated scale of well-being of cancer patients, and estimate the mean difference in the efficacy variable between two scores on the scale from a sample of patients. Several methodologies have been proposed for the determination of the MCID by anchor-based methods.

In **distribution-based methods** the efficacy variable must be interval scaled and the MCID is defined as a quantity that is orders of magnitude above the variability of the efficacy variable. One of the most commonly used measures is the **effect size**, which is the difference d in the mean value of the efficacy variable expressed as a proportion of its standard deviation s, that is, effect size $= d/s$. It is generally agreed that an effect size of 0.2 is indicative of a small change, 0.5 of a moderate change, and 0.8 or more of a large change. So for example if the primary efficacy variable has a standard deviation of, say, 12 mg/dL the MCID would be $0.2 \times 12 = 2.4$ mg/dL. This means that a difference of 0.2 standard deviations from the mean, corresponding to a difference of 2.4 mg/dL between treatments, represents an important difference.

One problem with this measure is its dependence on the variability of the efficacy variable. Therefore, if the variance is very small the MCID will also be very small and may not reflect a clinical improvement. For example, glycosylated hemoglobin in diabetic patients has a standard deviation of about 0.9 and an effect size of 0.2 represents a decrease of only 0.18, which is far from being clinically relevant. One way around this difficulty is to consider the MCID as a quantity that is orders of magnitude above measurement error. The **standard error of the measurement** can be estimated by the product of the variance of the primary efficacy variable and one minus the reliability of the measurements. Reliability may refer to the intraclass correlation coefficient or to the internal consistency of a scale. Values of change above 1.96 standard errors of measurement have been suggested as the MCID. Other approaches to distribution-based methods have been proposed.

Clearly, each of the above methods has pitfalls and limitations and often anchor-based, distribution-based, and expert-based methods are employed concurrently in the search for a consensus definition of the MCID.

The next decision regards the significance level that will be adopted. General practice is, as we know, to adopt the two-sided significance level of 0.05. However, often clinical trials seek to show that the difference in the primary efficacy variable between the test treatment and the control treatment is greater than the MCID. The null and alternate hypotheses are thus H_0: $\mu_T - \mu_C \leq$ MCID and H_A: $\mu_T - \mu_C >$ MCID. Accordingly, a difference between treatments is more logically tested with a one-sided test, but since it is always possible that the test treatment performs worse than the control treatment, even if the control is a placebo, if a one-sided test is used the significance level should be set at 2.5%.

If the clinical trial has more than one primary efficacy variable the significance level will have to be adjusted to account for multiple comparisons. In this case, sample sizes must be computed for each primary variable and the largest number should be selected. In the next chapter we will discuss several methods for the analysis of multiple endpoints and we will see how the selected method influences the sample size calculation. The same considerations apply to clinical trials with more than two treatment arms.

In Section 7.3 we saw how to calculate sample sizes for binary and interval efficacy variables, and the same calculations are used for clinical trials. However, the primary efficacy criterion in many clinical trials is time to event and in this case sample size calculations cannot be done with the formula used for the comparison of two proportions. This is because the proportion of subjects experiencing the event depends on the duration of the observation period and so it makes little sense to assume a given proportion of patients with the events in the control group. For example, consider a trial where the primary outcome is survival and the treatment is expected to decrease the mortality rate to 70% of that of the controls. To compute the sample size with given power and significance level using the method described previously for the comparison of two proportions we would need to define the proportion of survivors in the control group; however, this proportion decreases over time. Therefore, unless we establish a fixed observation time for all subjects, it makes no sense to define that proportion.

The solution is to compute the number of events required to afford the pre-specified power for the trial. This number can be estimated from the MCID established for the trial, which is the minimum difference in the event rates between two treatment groups that is considered clinically important. This difference is expressed as a ratio which, in the context of time-to-event studies, is called the hazard ratio.

If all subjects are observed until the event occurs, which is uncommon, the number of subjects needed is equal to the number of events required. However, if some of the subjects have censored observations, as is usually the case, the number of subjects necessary must be greater than the number of events required. From the number of events required and from estimates of the probability of the event in the controls at the end of the follow-up period, we can derive the total number of subjects to be enrolled. It should be noted that the power of the trial depends entirely on the number of observed events and not on the number of subjects enrolled. For this reason these trials are called **event-driven trials**.

Two popular methods for computing the number of events that need to be observed in a trial to detect a difference at the α significance level with power $1 - \beta$ are the Freedman method and the Schoenfeld method. For a two-tailed test with equal groups the **Freedman method** calculates the number of events E as

$$E = \left(z_{\alpha/2} + z_\beta\right)^2 \times \left(\frac{HR + 1}{HR - 1}\right)^2$$

and the **Schoenfeld method** as

$$E = \left(z_{\alpha/2} + z_\beta\right)^2 \times \left(\frac{2}{\ln(HR)}\right)^2$$

where HR is the hazard ratio of the event in the test treatment to the control treatment.

For example, suppose we are planning a clinical trial to compare the survivor functions of subjects submitted to one of two treatments where the MCID was defined as a 30% decrease in mortality; that is, it is expected that the hazard ratio of the test treatment to the controls is 0.70 or less. We plan to observe the patients for a period of three years. The null hypothesis will be tested with a two-sided logrank test with $\alpha = 0.05$ and we want the trial to have

80% power to detect that difference. From the table of the normal distribution we obtain the value for $z_{\alpha/2} = 1.96$ and $z_\beta = 0.84$. Using the Freedman method the number of events we need to observe is

$$E = (1.96 + 0.84)^2 \times \left(\frac{0.70 + 1}{0.70 - 1}\right)^2 = 252$$

With the Schoenfeld method the number of events is

$$E = (1.96 + 0.84)^2 \times \left(\frac{2}{\ln(0.70)}\right)^2 = 247$$

If all patients were observed up to the occurrence of the event, that number would be the total sample size (126 patients per treatment group). As we expect that a number of patients will be censored, the total number of patients required is equal to the total number of events divided by the probability of the event at the end of the follow-up period:

$$N = \frac{E}{P_E}$$

An estimate of P_E can be obtained from Kaplan–Meier estimates of the survivor function in the control group at time t, which in this example is three years. If we denote by $S_c(t)$ the cumulative probability of survival at time t in the controls, under the proportional hazards assumption the survival probability in the treatment group will be $S_t(t) = S_c(t)^{HR}$. If we average these estimates and subtract the result from one we will get the estimated probability of the event by the end of the follow-up.

Suppose we knew that the cumulative probability of survival at three years with the control treatment was 0.35. Then, the probability of survival in the test group is expected to be $0.35^{0.70} = 0.48$. The average is $(0.35 + 0.48)/2 = 0.415$. Thus the probability of the event is $1 - 0.415 = 0.585$. Therefore, we will need to enroll $252/0.585 = 431$ patients. This number must be rounded up to an even number, 432. In conclusion, we will need 216 patients in each group in order to observe 252 events.

Often a clinical trial does not establish a fixed follow-up time for the subjects. Rather, patients are followed up for as long as possible until termination of the study. Since patients are accrued progressively into the trial, some patients will be observed for a longer period. For example, if a trial has an accrual period of six months and patients are to be followed for a minimum of three years, then at the conclusion of the study, three and a half years after the first patient was included, on average the patients will have been observed for three years plus half the accrual period, that is, for three years and three months. During this extra follow-up time more events may have occurred, so we should take this into account when computing the number of patients needed for the trial.

In order to account for an accrual period and for the additional events occurring in the follow-up period, the Freedman method computes the number of subjects required as shown above, but the probability of the event is estimated from the survivor function at time t plus half the length of the accrual period. Suppose the accrual period for our clinical trial was six months. Say that the Kaplan–Meier estimate of the survival probability at 39 months in the control group is 32%. Then, the probability of survival in the test group is expected to be $0.32^{0.70} = 0.45$. The average is $(0.32 + 0.45)/2 = 0.385$. Thus the probability of the event is $1 - 0.385 = 0.615$. Therefore, we would need to enroll $252/0.615 = 410$ patients in order to observe 252 events.

An alternate method, **Simpson's rule**, considers the probability of the event at three time points to compute the number of patients required: at the end of the minimum follow-up period, at the end of that period plus half the accrual period, and at the end of that period plus the total accrual period. In our example, we would need to estimate the probability of the event at 36 months, 39 months, and 42 months.

Suppose that the cumulative probabilities of survival in the controls at those time points are 35%, 32%, and 28%. We calculate the probability of survival in the test group as 48%, 45%, and 41%. The probability of the event is computed as

$$P_E = 1 - \frac{(0.35 + 0.48)/2 + 4 \times (0.32 + 0.45)/2 + (0.28 + 0.41)/2}{6}$$

The result is 0.617. Therefore, we would need $252/0.617 = 408$ patients, 204 in each group.

It is also convenient to account for a proportion of patients that will withdraw from the trial or that will be lost to follow-up, so-called **patient attrition**. The final number of patients will be the computed number divided by one minus the proportion lost. For example, if the computed number was 410 and it is expected that 10% of the patients will be lost to follow-up, the total sample size should be $410/0.90 = 456$ or 228 patients per treatment arm.

11.11 Non-inferiority clinical trials

Until now we have always assumed that the objective of a clinical trial was to demonstrate the superiority of one treatment over another. Indeed, most clinical trials are conducted with this aim and they are planned and designed to reject the null hypothesis H_0: $\mu_1 = \mu_2$ and thus accept the alternative hypothesis H_A: $\mu_1 \neq \mu_2$.

However, clinical trials can be planned with different objectives and different H_0 and H_A. To better understand the reason for these designs and their applications, it is convenient to review briefly the development process of new treatments.

After a laboratory process of selection of **new molecular entities** with the potential for being used clinically, these substances enter a period of clinical experimentation and evaluation and acquire the status of **investigational new drug**s (INDs). This period is divided into four phases of clinical development. In phase I, or the dose determination phase, clinical trials are conducted on small numbers of subjects in order to determine the appropriate dosage of the INDs. In phase II the INDs are tested for the indications suggested by the theory in small to medium-sized clinical trials, which may be comparative or non-comparative. The principal aims of phase II clinical trials include the further definition of the dosage and an initial assessment of the efficacy and safety. Phase II trials are mainly geared to the identification of INDs which do not have enough activity to be used as medications. Those INDs that show some activity and acceptable toxicity in phase II trials are subsequently tested in phase III, the confirmatory phase, in randomized controlled clinical trials versus the best available therapy. If the results of at least two well-designed phase III clinical trials are conclusive of the efficacy of the IND and its safety is at such a level that the benefits expected from its administration outweigh the risks due to toxicity, and after an approval process conducted by the authorities, the drug will eventually be marketed. Phase IV, the post-marketing phase, is intended for the evaluation of the efficacy and safety of the drug on large samples of patients and to identify potential additional indications for a medicine. These clinical trials can be comparative or non-comparative and seek to reproduce the conditions in the real world where the drug is used.

Although it is desirable that new drugs achieve significant gains in health, sometimes a new drug may not have greater efficacy than the standard treatment but may nevertheless represent progress in therapy. For example, the new drug may have a more comfortable route or schedule of administration, may have a more favorable safety profile, may be from a new chemical class with interesting properties, or may represent a new paradigm for the treatment of a disease, among others.

Because of this, the concept of **equivalence trial** was developed. This design differs from the superiority trial because its objective is to demonstrate that the difference in efficacy between treatments is no greater than a predefined margin. This difference defines an **equivalence margin** and evidence in favor of the equivalence of two treatments is obtained if the 95% confidence limits for the difference between treatments in the primary efficacy variable do not exceed the equivalence margin. Figure 11.9 illustrates on the left the rationale of equivalence testing with examples of various possible outcomes and the respective conclusions to be drawn.

In practice, the issue is not whether two treatments are equivalent, but whether a new treatment is not inferior to the reference treatment, at least not by a clinically important difference. These are the **non-inferiority trials**.

The **non-inferiority margin** is admittedly difficult to establish, perhaps even more difficult than defining a MCID for superiority trials, and, as a rule, the non-inferiority margin is much smaller than the MCID. The non-inferiority margin should correspond to the largest loss of effect of the reference therapy that would be

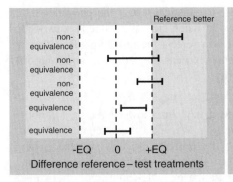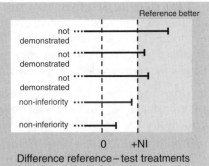

Figure 11.9   Interpretation of the results of equivalence (left) and non-inferiority (right) trials depicted as two-sided 95% confidence intervals (left) and one-sided 97.5% confidence intervals (right). EQ is the equivalence margin and NI the non-inferiority margin.

clinically acceptable. A commonly used approach is to estimate conservatively the effect of the reference therapy from previously published clinical trials, as the lower bound of the 95% confidence interval of the difference to a placebo in the primary efficacy variable, and to decide what proportion of the effect of the reference therapy should be retained by the new treatment. If several placebo-controlled clinical trials of the reference treatment have been published, a combined estimate of the 95% confidence interval can be obtained using meta-analysis, which will be covered in a forthcoming chapter. For example, suppose that the 95% confidence interval of the difference in the primary efficacy variable between the reference treatment and the placebo is 12 to 35, smaller values indicating lesser efficacy. Then, 12 is taken as a measure of the effect of the reference treatment. Suppose that we decided that no more than 50% of the effect should be lost to the new treatment. Thus, the non-inferiority margin would be set at $0.5 \times 12 = 6$. Usually no more than 50% of the effect of the reference treatment should be lost to the new treatment but, depending on the specific clinical problem, smaller proportions may be selected.

One way of establishing non-inferiority is by computing the one-sided 97.5% confidence interval for the difference between treatments. If the upper or lower bound of the confidence interval, depending on the direction of the difference of interest, is less than the non-inferiority margin, the test treatment will be considered as non-inferior to the reference treatment. Figure 11.9 illustrates, on the right, different results and the corresponding conclusions.

Non-inferiority can also be established with statistical tests. The hypothesis to be tested is thus one-sided, and denoting by μ_C and μ_T the population means of the primary efficacy variable in the control and in the test groups, and by d the non-inferiority margin, the null and alternative hypotheses can be formulated as H_0: $\mu_C - \mu_T \leq d$ and H_A: $\mu_C - \mu_T > d$ (or as H_0: $\mu_C - \mu_T \geq d$ and H_A: $\mu_C - \mu_T < d$, depending on the direction of what is considered inferiority).

For testing the null hypothesis we use a statistical test at the one-sided 0.025 significance level which, incidentally, has the same critical z-value as a two-sided test at the 0.05 significance level. If the test rejects the null hypothesis that the difference in efficacy between treatments exceeds the adopted non-inferiority margin, then we may declare non-inferiority.

For example, suppose we want to test a new lipid-lowering drug for non-inferiority against an already approved drug. We select the percentage change from baseline in LDL cholesterol as the primary efficacy variable. After reviewing the literature on clinical trials of the reference drug we conclude that the average percentage change from baseline in LDL cholesterol after three months of treatment with that drug is 42% with standard deviation 14%. We choose as non-inferiority margin a difference less than 0.2 standard deviations, that is, $0.2 \times 14\% = 2.8\%$.

We now compute the sample size required to afford 80% power at the one-sided significance level of 0.025 to detect a difference of 2.8% in the percentage change from baseline between the two treatments. In the table of the normal distribution we look for the values of $z_{0.025}$ and $z_{0.20}$, which are 1.96 and 0.84. We compute the sample size as $2 \times (1.96 + 0.84)^2 \times 14^2$ divided by 2.8^2, or 392 patients per treatment group.

Note that the sample size calculation for binary data is different in non-inferiority and in superiority trials because in the latter the difference was zero under the null hypothesis and non-zero under the alternative hypothesis, while in non-inferiority trials under both the null and alternative hypotheses there is a non-zero difference. Therefore, the variance of the proportions is assumed constant for both hypotheses and the calculation is

$$n = \frac{2(z_\alpha - z_\beta)^2 \pi(1 - \pi)}{\delta^2}$$

Now, suppose that the trial enrolled 380 patients for the reference treatment and 382 patients for the test treatment. Also assume that the mean and standard deviation of the percentage change from baseline of LDL cholesterol was $46.3 \pm 16.8\%$ in the control group and $45.9 \pm 16.4\%$ in the test group. The null hypothesis that we want to reject is that the difference between control and test treatments is at least 2.8%, which we can represent by H_0: $\mu_C - \mu_T \geq 2.8\%$. If we succeed in rejecting this hypothesis we will accept the alternative hypothesis H_A: $\mu_C - \mu_T < 2.8\%$ and declare non-inferiority. We will test the null hypothesis with Student's t-test

$$t_{\alpha, n_1 + n_2 - 2} = \frac{(m_C - m_T) - d}{\sqrt{\frac{s^2}{n_C} + \frac{s^2}{n_T}}}$$

where s^2 is the combined estimate of the common variance.

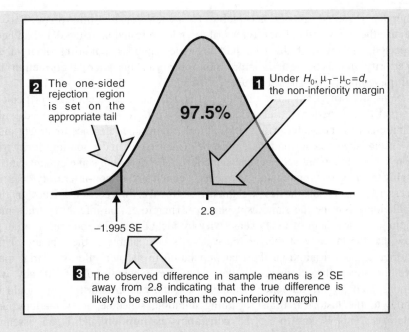

Figure 11.10   Testing for non-inferiority.

In our example, the combined estimate of the variance is $(379 \times 16.8^2 + 381 \times 16.4^2)/(380 + 382 - 2)$ or 275.58. The estimate of the standard error of the difference in sample means is the square root of $275.58/380 + 275.58/382$ or 1.203. The division of $46.3 - 45.9 - 2.8 = -2.4$ by the standard error yields $t = -1.995$.

On Student's t distribution with 760 degrees of freedom the limit of the one-sided 2.5% rejection region is -1.963, so the observed difference is within the rejection region. The null hypothesis is rejected and non-inferiority of the test treatment may be declared. Figure 11.10 illustrates the null hypothesis, the rejection region, and the position of the observed difference.

The one-sided 97.5% confidence limit of the difference between the reference and the test treatments is $(46.3 - 45.9) + 1.963 \times 1.203 = 2.76$. Therefore, the upper bound of the confidence interval is less than the non-inferiority margin.

If non-inferiority is established, the new treatment can be tested for superiority against the reference treatment. As the testing for superiority will be done only if the non-inferiority test is significant, no correction for multiple comparisons is required because a treatment that is superior must also be non-inferior. However, the reverse is not true: that is, if in a clinical trial designed for superiority the test treatment fails to show greater efficacy than the control, it is generally regarded as not acceptable to switch for non-inferiority. This is because the hypothesis of non-inferiority would be tested after the data had been inspected. However, if the trial had been designed as a non-inferiority trial, superiority could be tested before non-inferiority but then a correction for multiple comparisons should be used.

Non-inferiority trials have a number of issues that make them more difficult to interpret than superiority trials. One difficulty we have already mentioned is the definition of the non-inferiority margin. Furthermore, the reference treatment must have established efficacy and there must be evidence that in the non-inferiority trial the reference treatment achieves the same level of efficacy. In some circumstances, notably in trials of symptomatic treatments, a conclusion of non-inferiority does not necessarily mean that the test treatment is effective. Indeed, its efficacy in that population might be the same as a placebo. Therefore, non-inferiority trials may require an assessment of **assay sensitivity** by including a placebo arm in order to show that the efficacy of the reference drug is comparable to the efficacy of the historical trials. In addition, the trial population in the non-inferiority trial must have the same characteristics as the population where proof of efficacy was demonstrated for the reference treatment, and the non-inferiority trial should be similar to the historical trials regarding duration, schedule of evaluations, and endpoints. This is known as the **constancy assumption** and is an essential requirement for non-inferiority trials. In situations where this assumption may not be allowed, the trial design should include a demonstration of assay sensitivity. Furthermore, protocol deviations, patient withdrawals, and drop-outs should be minimal in non-inferiority trials because their occurrence tends to decrease differences between treatments.

Another type of clinical trial where one-sided significance tests are used is the **phase II futility trial**. This design was developed with the aim of identifying early in the drug development process those molecules whose efficacy is not large enough to make them interesting medicines, thus saving resources for other more promising molecules. These trials are often single arm, and historical controls are used for the definition of the efficacy expected with the standard therapy. However, it is safer to include a **calibration group**, that is, a sample of patients receiving the reference treatment to assess whether the efficacy hypothesized from the historical controls holds. A **futility threshold** is defined as the efficacy that the new molecule must demonstrate in order to be considered a clinically important improvement on the standard treatment. The trial is powered to test the null hypothesis that the efficacy of the new molecule is greater than the futility margin. If the null hypothesis is rejected, the conclusion is that the new molecule does not possess enough efficacy and it is thus futile to proceed to a phase III trial. The null hypothesis is tested one-sided with a one-sample statistical test. As we do not want to conclude too readily that the development of a new medicine should be stopped, the one-sided alpha error is usually set at a value of 0.10 or 0.15. On the other hand, we do not want to miss the identification of those molecules that are ineffective, so we set a small beta error, usually at 0.15 or 0.20. Thus, in this design the false negative rate now corresponds to the alpha error and the false positive rate to the beta error.

In problems where we have one sample mean m and the question is whether a population mean μ could be equal to a specified value μ_0 we use **one-sample tests**. The null and alternative hypotheses for a two-sided test are thus H_0: $\mu = \mu_0$ and H_A: $\mu \neq \mu_0$. The one-sample t-test is simply

$$t = \frac{m - \mu_0}{\sqrt{s/n}}$$

For proportions, using the normal approximation to the binomial distribution, if p is the sample proportion and π_0 is the specified value of the population proportion, the test is

$$z = \frac{p - \pi_0}{\sqrt{\pi_0(1 - \pi_0)/n}}$$

In the case of futility trials μ_0 and π_0 correspond to the futility threshold and the null hypothesis is formulated one-sided. For example, suppose that from historical data of clinical trials we have established that in a given disease the standard treatment achieves a response rate of 35% and that a clinically meaningful gain in efficacy would be an increase of no less than 10% in the response rate. The futility threshold would thus be defined as a response rate of 45%. Therefore, the null and alternative hypotheses are H_0: $\pi \geq 45\%$ and H_A: $\pi < 45\%$. We wish to conduct a phase II futility trial of a new molecule and choose a one-sided alpha error of 0.10 and a power of 0.85. The sample size for a one-sided, one-sample test that a proportion equals a given value is computed using the normal approximation as

$$n = \frac{\left[z_\alpha \sqrt{\pi_0(1 - \pi_0)} + z_\beta \sqrt{\pi(1 - \pi)}\right]^2}{(\pi - \pi_0)^2}$$

and the sample size for a one-sided, one-sample test of a mean is

$$n = \frac{\left[(z_\alpha + z_\beta)\sigma\right]^2}{(\mu - \mu_0)^2}$$

In this example $z_\alpha = 1.28$, $z_\beta = 1.04$, and the sample size is

$$n = \frac{\left[1.28 \times \sqrt{0.45(1 - 0.45)} + 1.04\sqrt{0.35(1 - 0.35)}\right]^2}{(0.35 - 0.45)^2} = 129$$

Suppose that we observed a response rate of 39% in 129 patients who were administered a new treatment. The one-sample test is

$$z = \frac{0.39 - 0.45}{\sqrt{0.45 \times (1 - 0.45)/129}} = -1.37$$

The *p*-value is 0.085 and we reject the null hypothesis, thus concluding with 90% confidence that it is futile to take the new molecule to a phase III trial.

11.12 Adaptive clinical trials

In section 11.9, when discussing the topic of dynamic randomization, we pointed out the convenience of using the information that is collected during the course of a clinical trial for the optimization of the study design. Dynamic randomization addressed an ethical issue, but other design modifications could be introduced that might have a positive impact not only on the ethics of the research, but also on the success of the research. This is the motivation behind the concept of **adaptive clinical trials**. In adaptive trials one or more analyses of the data are done while the trial is being conducted. These analyses of partial trial data are called **interim analyses**. The results of the interim analyses are interpreted by an independent **Data Monitoring Board** (DMB) that may propose a number of design modifications. Figure 11.11 shows some examples of opportunities for optimization of a clinical trial comparing two different doses of the test treatment with the standard treatment.

Some of the most common modifications are as follows. (A) If the initial estimates of the variance of the efficacy variable or of the treatment effect prove not to be correct during the course of the trial, a sample size reestimation can be done, increasing or decreasing the number of patients, or, in event-driven trials, increasing

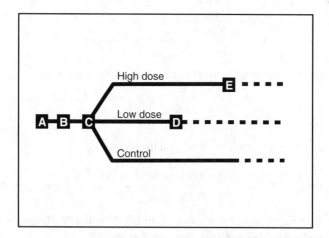

Figure 11.11   Opportunities for optimization of a clinical trial design: (a) sample size recalculation; (b) patient enrichment; (c) dynamic randomization; (d) premature discontinuation of a treatment arm; (e) premature termination of the trial.

or shortening the length of the trial. (B) The selection of patients could be modified in the course of the trial if there is a suggestion that a particular subgroup of patients has a different response to the test treatment, and further evidence regarding that subgroup can be obtained with the **enrichment** of the sample with patients belonging to that subgroup. (C) Randomization could be unbalanced if there is a suggestion that one treatment arm might be superior, in order to offer patients a greater probability of receiving the treatment with better outcomes. (D) One or more treatment arms could be discontinued prematurely if the probability of achieving a significant difference from the controls is very small. (E) The entire trial could be terminated prematurely if there is accumulated evidence that the test treatment has greater efficacy, or that it is futile to continue the trial to its normal conclusion.

These interim analyses of the clinical trial data would of course increase the risk of a type I error and, as we have already seen, the alpha error would increase with the number of interim analyses.

Several times we have addressed the problem of inflation of the alpha error when multiple comparisons are done on the same data. For example, consider that three interim analyses were done in a clinical trial comparing two treatments. In all, there will be four 'looks' at the data: the three interim analyses plus the final analysis. If the significance level was set at 5% for each analysis, the probability of obtaining a statistically significant difference just by chance in one or more analyses can be calculated as one minus the probability of observing no statistically significant difference in the four analyses. That is, $1 - 0.95^4 = 18.5\%$. Therefore, if at the end of the trial a p-value of 0.05 is obtained, 0.05 is just the **nominal p-value** and the true alpha error is actually 0.185. The more comparisons are made, the greater is the inflation of the alpha error. At the first interim analysis the nominal p-value corresponds to the true p-value, at the second interim analysis the alpha error increases to $1 - 0.95^2 = 0.0975$, and at the third it increases to $1 - 0.95^3 = 0.143$.

One way of overcoming this problem of the progressive increase in the alpha error due to multiple 'looks' at the data could be to use the Bonferroni correction we mentioned in Section 7.4. Thus, for the example of three interim analyses plus a final analysis, the Bonferroni correction would yield a nominal p-value of $0.05/4 = 0.0125$ for each analysis and the overall alpha error of 0.05 would be distributed evenly by the four planned analyses. Thus, the true alpha error at the second interim analysis is $1 - 0.9875^2 = 0.025$, at the third it is $1 - 0.9875^3 = 0.037$, and at the final analysis $1 - 0.9875^4 = 0.049$.

Inflation of the alpha error is not exactly as shown above because the sample analyzed in an interim analysis is not the same as the sample in the final analysis (it is a subsample of it) and also because the analyses are sequential, not in a random order. These features of the multiple comparison procedure call for special methods for correcting the inflation of the alpha error, generally called **group sequential plans**.

11.13 Group sequential plans

Earlier work in this area led to the proposal of the so-called **Pocock boundaries** where the alpha error is distributed evenly between the planned analyses but with more accurate values than the simple Bonferroni correction. So, for example, for the three interim analyses and the final analysis the nominal p-value is 0.0182, somewhat larger than with the Bonferroni correction. The Pocock boundaries are tabulated as a function of the number of interim analyses and the significance level. The method assumes that the interim analyses are done at constant increments of the trial data. The boundaries define a stopping rule for the trial whereby, if the nominal p-value obtained at an interim analysis is less than the boundary, then a difference between treatments should be declared and the trial stopped. Of course, other considerations are involved in the decision to stop a trial, such as the consistency of the findings, the coverage of the patient population, the risks incurred by the patients, budgetary aspects, and so on. If a decision is made to stop the trial, the reported alpha error must be the true alpha error, not the nominal p-value.

Two concerns have been expressed regarding the Pocock boundaries. One is that they may lead to an early termination of the trial when sample sizes are still small and a large treatment difference seen early in a trial might not be maintained at subsequent analyses with a larger sample size. The other is that a trial may be declared inconclusive despite a nominal p-value obtained in the final analysis that is much smaller than the traditional 0.05. For example, a p-value of 0.02 in the final analysis of a trial with three interim analyses does not allow one to reject the null hypothesis of no difference between treatments. This is awkward because, had the trial been conducted without interim analyses, the treatment effect would have been considered different at a high level of statistical significance.

To avoid these problems, a different boundary was proposed, called the **Haybittle–Peto boundary**. This boundary assigns a nominal p-value of 0.001 to each interim analysis and of 0.05 to the final analysis (or the remaining p-value for the final analysis, which is called the Bonferroni fix of the Haybittle–Peto boundary), no matter how many interim analyses are performed. Of course, this boundary is very conservative and a trial will be stopped early only if there is a very large treatment effect.

Another type of stopping boundary is the **O'Brien–Fleming boundary**. This method combines the idea of the Pocock boundary of distributing the alpha error evenly by a number of predefined 'looks' at the data, with the idea of the Haybittle–Peto boundary of reserving a large portion of the alpha error for the final analysis. In this method, a difference between treatments is declared at each interim analysis if the result of the test statistic multiplied by n/N, where n is the nth 'look' and N the total number of 'looks' at the data, exceeds a value $P(N, \alpha)$. The value of $P(N, \alpha)$ is a function of N and of the alpha error set for the trial, and can be found in appropriate tables. For example, for a two-arm trial with three interim analyses and a final analysis ($N = 4$), with an alpha error of 0.05, the tabulated coefficient $P(4, 0.05)$ is 4.170. Suppose that in the third interim analysis ($n = 3$) a total of

60 patients in each group had completed the trial and that 52 responses had been observed in the test treatment and 40 in the standard treatment. A chi-square test would then yield a result of 6.708 with 1 degree of freedom. The quantity $6.708 \times 3/4 = 5.031$ exceeds the rejection limit of 4.170 and superiority of the test treatment may be declared. If a table of the values of $P(N, \alpha)$ is not available, an approximate value is the chi-square statistic with 1 degree of freedom (3.84 for $\alpha = 0.05$, 6.63 for $\alpha = 0.01$).

As mentioned above, with the O'Brien–Fleming boundary a trial may be stopped in its early phases only if the difference between treatments is very large, and most of the alpha error is saved for the final analysis. For example, with three interim analyses and one final analysis the nominal p-values at each 'look' have to be < 0.00004 for the first 'look,' < 0.004 for the second, < 0.02 for the third, and < 0.04 for the final one.

The methods just presented are the most commonly used ones among a variety of methods proposed for the definition of stopping boundaries. Figure 11.12 summarizes the differences between the three methods, showing the respective boundaries for a trial with three interim analyses and one final analysis.

All three methods have the inconvenience that the number of interim analyses needs to be specified before the beginning of the trial and the number of patients accrued between two consecutive 'looks' is assumed to be constant. Therefore, when planning a trial, the number of interim analyses must be specified and, in addition, the sample size determination must account for the loss of power due to the several analyses (the loss of power exists in group sequential plans because multiple 'looks' increase the probability of the beta error). Tables are available with

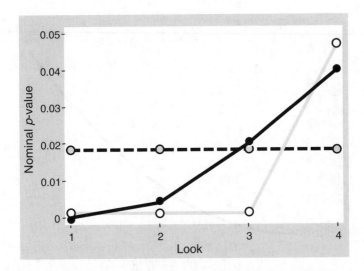

Figure 11.12   Nominal p-values for different methods of defining stopping boundaries for sequential analyses of clinical trials: Pocock (gray circles); Haybittle–Peto (white circles); and O'Brien-Fleming (black circles).

correction factors that must multiply the sample size computed with the standard methods. The adjustment is significant for Pocock's boundaries but small for O'Brien–Fleming boundaries. For example, if four 'looks' are planned, to maintain power with Pocock's boundaries the sample size has to be increased by about 20%, and with O'Brien–Fleming boundaries only by about 2%. The planned sample size is then divided by the number of 'looks' to obtain the number of patients with complete efficacy data that must be accrued before each interim analysis.

11.14 The alpha spending function

The idea of distributing the pre-specified alpha error among the several 'looks' at the data was further developed and led to introduction of the **Lan–DeMets alpha spending function**. This new idea was to define the stopping boundaries as a continuous function of the amount of information accrued during the trial. Thus, the alpha spending function allocates some of the total pre-specified alpha error to each fraction of the total information of the trial. Therefore, each interim analysis spends a portion of the total allowable alpha error for the trial. Figure 11.13 shows an example of an alpha spending function. At the first interim analysis (look 1) the trial will be stopped for efficacy if the nominal alpha error is less than 0.008. Therefore, about 0.008 of the total alpha error is spent in look 1. In the second interim analysis the cumulative alpha error defined by the alpha spending function is 0.015 and the alpha error allocated to the second look is $0.015 - 0.008 = 0.007$. For the final analysis the allocated alpha error is 0.035 because the remaining 0.015 ($= 0.008 + 0.007$) has already been spent in the two interim analyses.

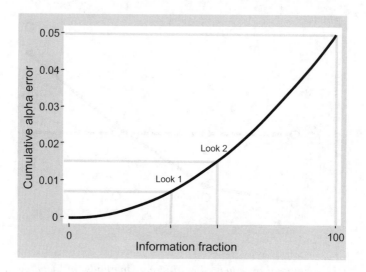

Figure 11.13   Example of an alpha spending function illustrating two interim analyses and the alpha error spent between the two analyses.

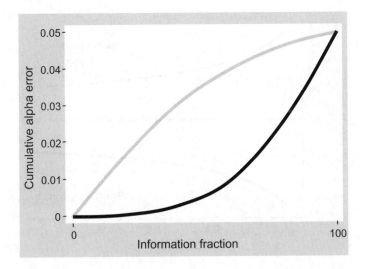

Figure 11.14 Types of alpha spending functions: Pocock type (gray line) and O'Brien–Fleming type (black line).

In the Lan–DeMets group sequential method, an alpha spending function is defined before the beginning of the trial. As we have seen, this spending function allocates some of the total alpha error to each interim analysis according to the fraction of the total information of the trial that has been accrued up to the time of the analysis. For trials with binary or interval efficacy variables, the information fraction is approximately the observed complete cases divided by the expected maximum sample size; for event-driven trials it is approximately the observed events divided by the expected total number of events. The spending function can be defined in many ways but very often either a Pocock type or an O'Brien–Fleming type of function is adopted (Figure 11.14). The first is an aggressive approach used when a large treatment effect is anticipated and a large amount of the alpha error is spent in the earlier phases of the trial in an attempt to stop it after a small number of patients. The second is a conservative approach, allocating most of the alpha error to the final analysis.

The alpha function determines, in turn, the boundaries for the stopping rule. At each interim analysis a test statistic of the comparison between treatment arms is computed from the observations accrued since the previous look at the data. This result is combined with the test statistics obtained at each of the previous analyses, taking into account the increment of information between analyses, to obtain the test statistic for that interim analysis. If this test statistic exceeds the boundary defined for that point in time, termination of the trial may be considered. Figure 11.15 illustrates an example of an alpha spending function with O'Brien–Fleming-type boundaries showing the test statistic obtained at sequential interim analyses. At the sixth interim analysis, with 80% of the trial information accrued, the test statistic crossed the boundary and the trial was stopped.

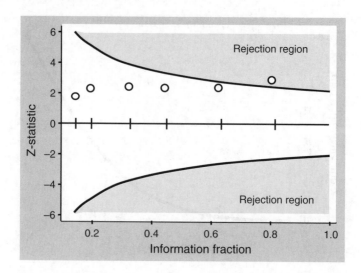

Figure 11.15   Illustration of a sequence of interim analyses in a clinical trial using the alpha spending method with O'Brien–Fleming-type boundaries. The open circles represent the value of the z statistic at each interim analysis. At the sixth interim analysis the z statistic crossed the boundary and a recommendation for early stopping was issued.

The advantage of the Lan–DeMets alpha spending function is that neither the number nor the timing of the interim analyses need to be pre-specified, and the flexibility afforded by this method is at the cost of a minimal loss in statistical power. What needs to be pre-specified, though, and cannot be changed during the trial, is the alpha spending function.

The group sequential approach can be applied to the definition of stopping rules for non-inferiority and futility as well. An alternative approach to monitoring a trial for efficacy or futility is **stochastic curtailment**, where the decision to stop the trial is based on an assessment of conditional power. **Conditional power** is the conditional probability of a significant result at the end of the trial given the data observed thus far. Inasmuch as a trial may be stopped when superiority of one treatment over another is shown, making futile the continuation of the trial, so can a trial also be stopped if the probability of reaching a statistically significant difference in the final analysis is below a predefined threshold (typically 10 or 15%), given the results observed in an interim analysis. Conditional power is usually computed under different hypotheses, often under the original assumptions of both treatment effect and variance of the primary efficacy variable, under the original assumption of the treatment effect but with the empirical estimate of variance, under the empirical estimate of both the treatment effect and variance, and under the null hypothesis of no difference between treatments.

11.15 The clinical trial protocol

As with any research project, the protocol is an essential tool for planning and conducting a clinical trial. The study protocol is structured in several sections, according to a fairly uniform methodology.

The protocol begins with the *title page*, identifying the authors, the sponsor of the trial, the participating centers, and expected dates for the beginning and end of the study. The following section is a *synopsis* of the trial, where the most important methodological aspects (objectives, trial design, efficacy criteria, number of patients, main inclusion and exclusion criteria, study treatments, and statistical methods) are described concisely. The protocol itself begins with a review of the treatment of the disease in question, a report of previous experience with the treatments that will be compared, and a justification of the scientific and clinical relevance of the clinical trial, the choice of the comparator if any, followed by the presentation of the general objectives.

The main body of the protocol begins with the formulation of the specific objectives of the study and the identification of the experimental design which will be adopted. The section on *study treatments* describes the medicines to be tested, their formulations, presentation, dose, and schedule of administration. Next comes a general definition of the *study population* followed by the detailed presentation of the inclusion and exclusion criteria. The primary and secondary efficacy variables and criteria are described in the section on *evaluation methodology*, along with a description of the methods used for their measurement. The methodology and procedures of *randomization* are described in detail in their own section. If *blinding* is to be used, a section should describe the methods and procedures to achieve and maintain blinding throughout the trial.

The section called *study plan* describes the flow of the trial, that is, what the trial periods are (washout, placebo run-in, dose titration phase, treatment phase, follow-up, long-term extension, and so on), the schedule of patient visits, all procedures performed at each observation of the patient, and all the examinations that the patient will perform.

A further section describes the *concomitant treatments*, that is, all the treatments, pharmacological or not, that may or may not be administered to the patient during the trial. If during the clinical trial a patient requires one of the treatments that are not allowed, and if there is no suitable and allowed substitute for that treatment, the patient will usually have to be discontinued from the trial. The section on *patient discontinuations* describes the circumstances that should lead to the discontinuation of a patient and the procedures to be undertaken in that event. Usual causes of discontinuation include, in addition to the administration of medication that is not allowed, poor patient compliance to the study medication, delays or failure to show up at the study visits, adverse events requiring discontinuation of the study medication, intercurrent disease, the patient's desire not to continue in the trial, and in general any reason where the investigator considers continued participation in the trial is not in the best interest of the patient. Another section describes how *adverse events* will be recorded and how they will be managed. A section on the *ethical aspects* of the research should discuss any identifiable ethical issues raised by the clinical trial and the procedure for obtaining informed consent from the patients.

The **patient information** describes the clinical trial, its objectives, and risks in a manner that is intelligible to the patient. This document must be delivered to the patient prior to study entry and must be accompanied by a verbal explanation of its content. The document must clearly present the justification and objectives of the clinical trial, all the procedures that will be performed, and, in particular, the eventual risks and discomforts caused to patients because of their participation in the trial. It should explain in what ways the trial procedures deviate from normal clinical practice and what treatment modalities are available for that clinical condition. The document should inform about the destination and uses of the data that will be collected, and what dispositions are in place to protect the confidentiality of the information and the privacy of the patient. Finally, it must be clearly explained to the patient that he or she is entirely free to accept or decline participation in the trial and that, having agreed to participate, he or she may at any time and without justification terminate participation in the trial. The **informed consent form** is a document that the patient signs, confirming that he or she has been informed and has understood the procedures of the trial.

The protocol ends with two technical sections that detail the statistical methodology. In one of the sections the *planned sample size* must be justified, presenting the assumptions made regarding the expected treatment effect and the power of the trial. If interim analyses are planned the adopted methodology should be presented. The other section is the *statistical analysis plan*. This section includes the precise definition of the study variables, the formulation of the research hypotheses, all the analyses that will be performed, the subject's eligibility criteria for each analysis, the statistical tests that will be used or the terms that will be included in regression models, the secondary analyses, and so on. Each of these aspects will be developed in the next chapter, where the main aspects of the general methodology for the analysis of clinical trials will be discussed.

11.16 The data record

In addition to the study protocol, before the beginning of the trial it is necessary to design the forms where the data will be recorded. These forms are grouped in a file for each individual to be included in the trial, usually called the **Case Report of Form** and abbreviated to CRF.

The CRF is intended not only for the recording of the data necessary to investigate the problem under study, but also to document all the relevant information related to the patient's participation in the trial, all observed clinical findings, all coexisting diseases and any illnesses occurring during the trial period, any adverse events, and, in general, all the data related to the health status of the patient.

The CRF is typically structured in several sections. The first sections are for the identification data of the subject, the patient demographics, and verification of the inclusion criteria and exclusion criteria. The following section is for recording data that characterizes the disease under study and includes data on the characteristics of the disease (start date, major manifestations, complications, and so on) and, if applicable, on the prior exposure to risk factors or aggravating factors (e.g., smoking, alcohol).

The characterization of the general clinical condition of the patient is recorded in the following sections and includes data regarding past medical history with the specification of previous diseases, surgical interventions, and medications, with the respective dates. Every disease present at the time of inclusion must be recorded.

All the findings on the physical examination should be recorded in a separate section for future reference, because it is important to check whether any findings observed during the trial were already present at baseline or appeared during the study period.

The following sections are repeated for each scheduled visit during the trial and are intended to record the values of the efficacy criteria, vital signs (temperature, heart rate, and blood pressure), patient compliance to the study medication, and any concomitant medication prescribed for other clinical conditions of the patient.

In any clinical trial, it is always necessary to monitor and analyze the toxicity of the investigational products being used. Safety and tolerability are evaluated through clinical laboratory safety parameters and the recording of adverse events. The **clinical laboratory safety parameters** consist of a panel of tests that monitor possible effects of the research products on those organ systems most often involved in drug toxicity – hematopoietic, liver, and kidney. **Adverse events** include all unfavorable events for the patient occurring from entry to the study to its end, regardless of their relationship to the study medication. Thus, not only are adverse symptoms and signs mentioned or observed in a patient (e.g., headache, eczema) considered, but also all intercurrent illnesses, deaths from any cause, hospitalizations, visits to an emergency room, voluntary or accidental overdose, and also traumas, accidents, and, in general, all events in the medical sphere occurring during the study. This means that the divorce or death of family members, for example, despite being unpleasant events, are generally not considered adverse events.

Finally, additional sections are used to record treatment allocation, compliance to the medication, deviations from the study protocol, and causes of patient discontinuation from the trial.

12

The analysis of experimental studies

12.1 General analysis plan

We saw in Section 11.1 that the goal of many experimental studies is the establishment of a cause–effect relationship between an intervention and an observable response. In the context of clinical trials, the aim is to assign a therapeutic response to a particular experimental medicine or therapeutic regimen. We have mentioned several times that in order to establish this relationship it is necessary to produce evidence, in the same study, for an order factor and for an association between the intervention and the response, and that plausible alternative explanations for the observed response can be excluded with reasonable confidence.

It turns out that the data analysis itself opens up many opportunities for alternative explanations for the results. The choice of the statistical methods, the exclusion of individuals or observations from the analysis, and multiple comparisons are examples of procedures that, when inadequately performed, can increase the likelihood of false positive results.

For this reason, a large part of the full statistical analysis of clinical trials is concerned with the exclusion of the possibility that the obtained results are fortuitous. The analysis of clinical trials is highly standardized and uniform. International standards published by the International Conference on Harmonization (ICH), which have been adopted by many countries, regulate the methodology of the analysis of clinical trials. These standards determine the procedures to be used in data preparation and cleansing, in the statistical methods, and in the presentation of the trial results.

Basically, the data analysis of a clinical trial is processed in seven sequential stages. First of all, it is necessary to clean the data. Understandably, a trial never runs exactly as predicted. On some occasions the procedures set out in the protocol

Biostatistics Decoded, First Edition. A. Gouveia Oliveira.
© 2013 John Wiley & Sons, Ltd. Published 2013 by John Wiley & Sons, Ltd.

are not fully met, such as when a patient takes a medication that is not allowed. Sometimes efficacy data is missing, which happens in patients who abandon the trial before the last scheduled observation. These situations must be analyzed and fully resolved before moving on to the next phase.

When there are problems with the data of some subjects, or if they did not comply with the trial protocol as planned, or if they did not have all of the eligibility criteria for the trial, there is the question of whether or not these cases should be included in the analysis. It is therefore necessary to decide which individuals are eligible for analysis, as well as which of their observations contain valid data.

After the set of data that will be analyzed is identified, the next step is the preparation of the data for statistical analysis. For example, it may be necessary to transform variables, dichotomize some variables, create dummy variables, create interaction terms, and so on.

The fourth phase corresponds to the main statistical analysis of the primary and secondary study variables, which must be performed strictly as described in the statistical analysis plan in the study protocol. Any deviation from the plan must be fully documented and justified.

After the definitive results of the statistical analysis have been obtained, it is necessary to check the robustness of the findings. While cleansing the data and selecting the population valid for analysis it is necessary to make several decisions. Consequently, it is important to verify that the results obtained in the main analysis are not particularly sensitive to the decisions that have been made. We do this through several secondary analyses.

In the sixth phase, additional analyses of a purely exploratory nature are done, with the aim of obtaining further information about the treatments. Typically, these analyses attempt to identify prognostic factors of therapeutic response and patient attributes that define subgroups of subjects with a different response to treatment.

Finally, in the seventh phase, the safety and tolerability of the treatments are evaluated. Data analysis of safety data is mostly descriptive.

In the following sections we will discuss each of these phases in greater detail. The subject of clinical trials is very extensive and in many aspects controversial, so it cannot be discussed here in full. In the following sections the discussion will focus on the main aspects in which an inappropriate statistical methodology may eventually contaminate the study by factors external to the experience, thus compromising the assignment of causality of the observed effects to the test treatment.

12.2 Data preparation

Protocol deviations represent the main indicators of the quality of conduct of a trial, yet it is not common to see in the literature a clinical trial report that includes a description of their number and type.

It is customary to classify **protocol deviations** into three different types. Minor deviations are small deviations from the protocol that in all likelihood will not affect the results, for example, a delay of a few days in a study visit or the omission of one or two administrations of the treatment. Major deviations are not avoidable and eventually affect the final results of the trial. Examples of major deviations are deaths, patient drop-outs, and discontinuations due to intercurrent diseases, adverse events, or therapeutic inefficacy. Naturally, as these deviations are unavoidable, they are by themselves not indicators of the quality of conduct of the trial. However, if they are sizeable in number then they may reflect an inadequate design of the clinical trial, perhaps because its duration is too long, or because it included patients who had little chance of complying fully with the procedures of the trial. The third type is **protocol violations**. Examples include the administration of medications that are not allowed during the trial, the administration of the wrong treatment or the wrong dose of the study medication, failure to meet the schedule of study visits, or the omission of patient visits. These violations almost certainly have an influence on the trial results and possibly reflect faulty administration of the trial or poor cooperation between the investigator and the trial subject.

After all the protocol deviations have been identified, it is necessary to decide which assessments are valid for analysis. It is usually considered that the study visits following a protocol violation are not valid for analysis as their data does not reflect the treatment effect. Consequently, all efficacy data pertaining to those visits are ignored and treated as missing data.

The handling of missing data is thus the next issue that arises in the analysis of clinical trials. The main causes of missing data are drop-outs and premature discontinuations from the study. In these cases there is no efficacy data after the last time the patient was observed.

The most commonly used method for dealing with missing data is to replace the missing values with the values observed in the last valid assessment. This method is called the **LOCF (Last Observation Carried Forward)**. With this method one assumes that the value of the last valid observation is the best result obtained with the treatment and the values that will be compared between treatment groups are thus those of the last valid observation. It is generally agreed that the LOCF method is acceptable as long as it not applied to more than about 10% of the patients.

In clinical trials for diseases characterized by a progressive decline, such as degenerative diseases of the central nervous system, the LOCF method is not suitable. In these situations, the sooner a patient abandons the study, the better the patient's state and therefore the value of the efficacy criterion does not reflect the best result obtained with the treatment. A method frequently used in these situations is interpolation by fitting a least squares line to the observed values over time, where the missing values are estimated from the slope of the line and the time intervals between visits.

12.3 Study populations

After the study data have been prepared, apparently we should be able to proceed to test the differences between treatments. However, before that we will need to decide upon the study population to be analyzed. This means that, contrary to what would

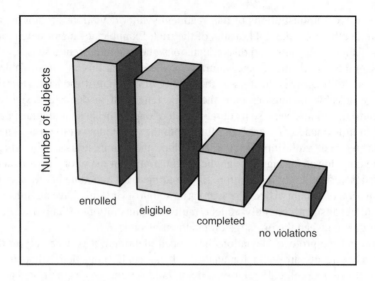

Figure 12.1   Several populations that can be identified in a clinical trial.

logically be assumed, the analysis does not necessarily focus on all the subjects included in the trial.

The explanation for this is illustrated schematically in Figure 12.1. Within the total number of subjects enrolled in a clinical trial, some may not meet all the inclusion and exclusion criteria. Moreover, from the total number of eligible individuals, only a proportion will have completed the trial and, among these, only a few did so with no protocol violations.

There is, therefore, the matter of deciding which of these trial populations will be analyzed. The question is not an easy one because arguments for and against each population can easily be produced. For example, we could say that the best population for analysis would be that of all eligible individuals who completed the trial with no protocol violations, because the responses observed in these patients do reflect the effect of the treatment when administered during the period of time necessary for its action and according to the proposed does and schedule in the patient population for which the treatment would be indicated. On the other hand, we could argue that selecting that particular study population would exclude from the analysis many individuals in whom the treatment response was unsatisfactory and who, for that reason, have abandoned the trial, leaving us to compare only those patients who were having some benefit with the treatments. This, of course, would significantly bias the results.

Intuition tells us that the primary efficacy analysis of clinical trials should be conducted in the study population which completed the trial with no protocol violations. Although this seems logical, the problem with this approach is that the exclusion of subjects from the analysis will invalidate the purpose of randomization. Remember that randomization gives us the assurance that the distribution of patient characteristics at the beginning of the trial is identical in the populations from where the groups were sampled. Now, if non-compliant patients are excluded from the analysis, then we can no longer assume the equality of populations at baseline. The systematic exclusion of particular types of patients will introduce a bias with an effect equivalent to a selection bias and, consequently, any differences between treatments at the end of the trial may be due to the differences between the patient populations, not to a treatment effect. Therefore, we had gone through all the trouble of randomizing patients to treatment groups to make sure that their characteristics had identical distributions in the population, only to ruin that effort by excluding patients from the groups after they have been included. So how can we analyze the data if some patients have abandoned the trial before efficacy data could be obtained from them?

A solution to this dilemma was found in the principle of the **intention to treat**. Basically, this principle states that what is being evaluated in a clinical trial is not the pharmacological activity of a substance, but a decision between two or more distinct therapeutic options. In other words, according to this principle, when we conduct a clinical trial comparing treatments A and B, what will be compared are the results obtained with the decision to opt for treatment A versus the results obtained with the decision to opt for treatment B. If you look back at the figures in the previous chapter showing diagrams of clinical trial designs you will notice that the intervention is represented by an arrow pointing to a specific moment in time. This arrow represents the moment when a decision is made to prescribe one or another treatment and the outcome of that decision is what is actually compared in the intention to treat approach.

In practical terms, the intention to treat principle states that all subjects included in a clinical trial must be analyzed. Granted, this method will most likely underestimate the actual treatment effects, because patients abandon a trial before the full effect of the treatment is expressed, but this is of much less importance than introducing bias into the trial.

Therefore, the analysis of the population which completed the study protocol without violations estimates the efficacy of the treatment itself when administered under ideal conditions. It is said that this method adopts an explanatory approach of treatment effects.

On the other hand, the intention to treat analysis takes a pragmatic approach, that is, it estimates the outcome of a therapeutic decision in near real-world conditions. Therefore, the treatment effects estimated with the intention to treat approach represent treatment effectiveness rather than efficacy. This approach is convenient for supporting decision making in actual clinical practice because what matters from the clinical perspective is the result actually obtained with a treatment decision, not the expected outcome in ideal patients and conditions of administration.

Accordingly, under the intention to treat principle one should not exclude dropouts, treatment discontinuations, and errors or omissions in the administration of treatments, because these events happen in the real world and in the same proportion in which they occur in the study population. The mean values of the efficacy criteria effectively observed in this population are, in short, the true therapeutic activity of the treatment minus the poor results observed in patients who do not complete a course of treatment as prescribed. This quantity is estimated without bias in the intention to treat population.

Nowadays it is widely accepted that the primary efficacy analysis of a superiority clinical trial must be based on the intention to treat population, but the definition of this population is not consensual. Some authors consider that all randomized patients must be included in the intention to treat population, but most accept that the intention to treat population consists only of those randomized subjects who were eligible for the trial. However, the exclusion of subjects not eligible according to the inclusion and exclusion criteria is acceptable only if that decision is clearly independent of the administered treatment and the outcome.

This definition of intention to treat, however, has raised some objections from the community of clinicians who found it difficult to understand, for example, the logic of including in the population for analysis those patients who abandoned the trial before even taking a dose of the study medication. Of course, if we think in terms of comparing therapeutic decisions this aspect raises no concern since in the real world some patients will also not take the prescribed medication.

As a compromise, statisticians and clinicians have somewhat relaxed the definition of the intention to treat population so that trial results would be easier to understand and better accepted by clinicians, without appreciable impact on the results. One consequence of this broader concept of the intention to treat is that, when reporting the results of a clinical trial, it is not sufficient to mention that the analysis was done on the intention to treat population – it is necessary to define

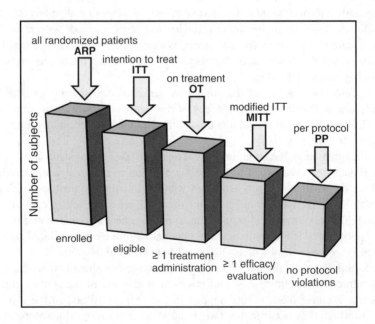

Figure 12.2   Study populations eligible for analysis.

what is understood by intention to treat. Figure 12.2 illustrates several populations for efficacy analysis of a clinical trial, together with the acronyms and their common designations.

Perhaps the most common definition is of all eligible patients who received at least one administration of the study treatment and who have some efficacy data; that is, they remained in the study at least until the first visit in which efficacy data was obtained, which is usually called the **modified intention to treat population**. It is generally accepted that the bias caused by the exclusion of patients from analysis will not be significant as long as they do not exceed 5% of the total eligible patients. If this happens, one should select the **on treatment population** (also called the **all patients treated population**), which includes all eligible patients who received at least one administration of the study treatment and may or may not have efficacy data. If this population is still less than 95% of the eligible patients, the analysis should be performed on the intention to treat population.

International standards concerning the analysis of clinical trials for the registration of new medicines require that the efficacy analysis always includes, in addition to the intention to treat population according to one of the above definitions, an analysis of the **all randomized patients population**, that is, all randomized patients, whether or not they were eligible for the trial, and an analysis of the population which completed the trial without protocol violations. The latter is called the **per protocol population**. The rationale for doing analyses in different populations will be explained later, when we consider secondary analyses of clinical trials.

12.4 Primary efficacy analysis

Once the population for the primary efficacy analysis has been identified, the next step is to define the measure of efficacy that will be compared between groups. In clinical trials analyzing the time to an event, the efficacy variable is naturally the occurrence of the event and the time from baseline to the event or till censoring occurred. In many trials the efficacy variable is a binary variable encoding the success or failure of a treatment according to a previously agreed definition of success, and patient discontinuations are considered failures. When the efficacy criterion is ordinal or interval, the usual practice is to use the **last observed value**, that is, the value obtained at the last valid observation of the efficacy variable. With interval-scaled efficacy criteria, and with some ordinal-scaled criteria that are assumed to be measuring an interval-scaled underlying attribute, the efficacy variable is often defined as the **change from baseline**, that is, the difference from the last valid observation to the baseline value of the efficacy criterion in each subject. If the efficacy criterion is measured in a ratio scale, then the **percentage change from baseline** can be used, that is, the change from baseline divided by the baseline value.

An analysis using one of the latter two measures has greater power than a comparison of the last observed value because the variability between subjects is eliminated, thereby substantially reducing the variance of the efficacy criterion.

However, they have the drawback of making the meaning of the difference between treatments sometimes more difficult to interpret. In addition some precautions must be taken in the analysis when the efficacy variable is the change from baseline.

In fact, as we saw in Section 11.4, due to the phenomenon of regression to the mean we expect that individuals who have at baseline a more extreme value of the efficacy variable will also have a greater change from baseline at a later assessment. Thus, the group with the higher mean value of the efficacy variable at baszeline is likely to be favored in the comparison of changes from baseline.

For this reason, whenever we use the difference from baseline as the efficacy variable, we need to adjust the analysis by the baseline value. The proper method is, as we already know, a multiple regression analysis in which the dependent variable is the change from baseline and the independent variables are one or more dummy variables coding for the treatment groups and the baseline value of the efficacy variable. This model, adjusting for the baseline value of the dependent variable, is called **analysis of covariance**, sometimes abbreviated ANCOVA.

The next aspect to be considered is whether or not it is a multicenter clinical trial. In this case, the analysis should be stratified by study center as this will afford a bit more power to the trial by removing the variability of the efficacy variable between centers. However, in order to do a stratified analysis, differences between treatments are required to be uniform across study centers. In the section on meta-analysis, the reason for this requirement will be explained in greater detail.

Therefore, in order to verify that a stratified analysis can be performed, first we must test the homogeneity of the differences between groups across the centers with a test of the treatment by center interaction. This analysis may be done with anova or with multiple regression if the efficacy variable is interval, or with logistic regression if it is binary. It is customary to use as the significance level for the interaction test a value of $p < 0.10$ or $p < 0.15$ due to the low power of the interaction test.

If the interaction test is negative, the stratified analysis is done with a multiple regression or a logistic regression model including dummy variables encoding the study centers. If the test is significant this means that the difference between treatments is variable from one center to another and it is not appropriate to average the treament with a stratified analysis. Thus, the distribution of patients by center should be ignored, by pooling all patients from each study arm into a single group.

If the trial was stratified by patient attributes, the analysis should also be stratified but, as we have just seen, only in the case where no interaction exists between treatments and strata. For this analysis we use a regression model with dummy variables encoding the treatments, the centers, and the strata, and interaction terms that are the product of treatment with strata. If the interaction is not significant this means that treatment effects are the same across all strata and may therefore be averaged to obtain a common estimate of the treatment effect, so the analysis is done with stratification. If the interaction is significant, it is not appropriate to pool the data from all patients, ignoring the strata. Therefore, the efficacy analysis must be done and reported separately for each stratum.

The data analysis in the presence of treatment by center and of treatment by stratum interactions is different because strata are considered a fixed effect and centers a random effect. A variable is a **fixed effect** if its values or levels are purposively selected for a study because they are the only levels we are interested in. Therefore, no generalization to a population beyond the selected levels of that variable is intended and we may or may not be interested in drawing conclusions about that effect, for example, by making comparisons between levels. A variable is a **random effect** if its levels are a random sample of all the levels existing in a population, we want to generalize our results to all the levels that could have been selected, and we have no intention of drawing conclusions about that effect. Accordingly, if the trial centers had been purposively selected to represent specific health care settings and we wished to draw conclusions about differences in the treatment effect between those settings, then the centers would be a fixed effect. This is not the case in multicenter clinical trials, where several centers are included in the analysis because we want to control for the extent to which they account for the variance of the efficacy variable.

The primary efficacy analysis may be adjusted by several prognostic variables. The rationale for this is that, although in theory randomization will ensure that the treatment groups are samples from populations with identical distribution of all patient attributes, in practice this may not be the case because clinical trials are based on convenience samples, not on probability samples. Therefore, as a safeguard for accidental imbalances in the distribution of important prognostic variables across study groups that might result in confounding, one may adjust the analysis by those variables.

12.5 Analysis of multiple endpoints

Some diseases require the assessment of multiple primary efficacy criteria because the objective of treatment is the improvement of several manifestations of the disease. In such cases, evidence of efficacy is obtained only if significant differences are shown for all primary efficacy criteria. For example, in Alzheimer's disease, a chronic progressive degenerative disease of the central nervous system, the evaluation of treatment efficacy requires the simultaneous demonstration of a decrease in the loss of cognitive function and a patient impression of a positive change in health. In order to maintain the overall alpha error at the 5% level, we can divide 0.05 by the number of primary efficacy variables and allocate equal amounts of the alpha error to each comparison. For example, in a trial on Alzheimer's disease, we can divide the alpha error by the two comparisons and allocate 0.025 to each comparison. With this procedure, the probability of a false positive test under the null hypothesis of no difference between treatments is one minus the probability that the two tests are non-significant, that is, $1 - 0.975^2 = 0.0494$, a value that does not exceed 0.05. Therefore, this procedure, known as the **Bonferroni correction**, guarantees control of the type I error at a given level alpha in a situation of multiple comparisons. Technically, the overall alpha error is called the **family-wise error rate** (FWER), that is, the probability of one or more false positives (or type I errors) when multiple hypotheses tests are done. The Bonferroni correction is a simple and effective method for controlling the FWER and, because it guarantees that the alpha error is not exceeded, this method is said to provide **strong control** of the FWER. Naturally, when this method is used, the estimates of treatment differences must be presented as adjusted 95% confidence intervals. In this example, the adjusted 95% confidence interval corresponds to the 97.5% confidence interval.

In other situations, the primary efficacy criteria will be measuring a number of distinct features of a disease, and in this case differences are not required to be shown in all primary criteria. For example, a clinical trial of a disease-modifying agent for rheumatoid arthritis, in addition to the primary objective of treatment, which is a clinically important improvement in the disease symptoms and signs assessed with a standardized scoring system, may attempt to demonstrate additional benefits to the patients, such as a decrease in structural joint damage, an improvement in a score of health-related quality of life, a decrease in the duration of morning joint stiffness, and an improvement in a score of activity of daily living and function.

In this case, with five primary efficacy variables and using the Bonferroni correction, each null hypothesis would be rejected only if the nominal p-value were less than 0.01. This means that we were requiring the same strength of evidence for a conclusion of superiority in the primary objective of treatment and in each of the other variables, which might not be reasonable.

Therefore, we could define a testing strategy whereby we would allocate some of the type I error to the test of the primary objective of treatment, and the remainder to the set of four tests on the other efficacy variables. For example, the most important comparison could be tested at a nominal p-value of 0.025 and each of the other four comparisons at $0.025/4 = 0.006\,25$. The FWER would be $1 - 0.975 \times 0.993\,75^4 = 0.049$. We can allocate the alpha error any way that seems better. For example, we could allocate 0.035 to the test of the primary treatment objective and 0.015 to the set of the remaining four tests. The FWER would be $1 - 0.965 \times 0.996\,254 = 0.049$.

The Bonferroni correction is rather conservative and in a situation like this it may not be the most efficient method for controlling the alpha error. A number of less conservative methods providing strong control of the FWER have been proposed, but the two most commonly used are the Holm–Bonferroni and the Hochberg procedures.

In the **Holm–Bonferroni procedure** we begin by performing all the statistical tests without adjustment for multiplicity. The nominal p-values are evaluated sequentially in ascending order at decreasing significance levels defined by the allocated FWER divided by the number of performed comparisons minus the number of previously tested hypotheses. The procedure stops the first time we fail to reject a null hypothesis and the remaining hypotheses are not rejected.

In the **Hochberg procedure** we go from the largest to the smallest nominal p-value and evaluate each one sequentially at increasing significance levels defined by the allocated FWER divided by the number of hypothesis tested thus far. The procedure stops when we reject a null hypothesis, and then all the remaining hypothesis are also rejected.

For example, say that in the rheumatoid arthritis trial the set of tests of the four variables had been allocated 0.025 of the type I error. Suppose that the results of the comparison between treatments were: structural joint damage $p = 0.004$, quality of life $p = 0.02$, morning joint stiffness $p = 0.007$, and activities of daily living $p = 0.05$.

The Holm–Bonferroni procedure is as follows. The smallest p-value is 0.004 and, being less than $0.025/4 = 0.006\,25$, we reject the null hypothesis. The next smallest p-value is 0.007, which is less than $0.025/3 = 0.0083$, and so we reject this null hypothesis as well. The next smallest p-value is 0.02, which is larger than $0.025/2 = 0.0125$. Therefore the procedure stops and we do not reject this null hypothesis or the remaining hypotheses. Accordingly, we have evidence of a statistically significant treatment effect for structural joint damage and duration of morning stiffness, but not for quality of life and activities of daily living.

In the Hochberg procedure we start with the largest p-value, 0.05. Since this value is greater than 0.025 we do not reject the null hypothesis. The next largest p-value is 0.02. As this value is larger than $0.025/2 = 0.0125$ we also do not reject the null hypothesis and proceed to the next largest p-value, 0.007. This value is smaller than $0.025/3 = 0.0083$. Consequently, the procedure stops and we reject this null hypothesis as well as all the remaining hypotheses. The conclusions of the analysis are the same as above.

These methods are called **closed testing procedures**. A closed testing procedure is a general method for performing multiple comparisons while providing strong control of the overall type I error. Both methods are less conservative than the Bonferroni correction. The Hochberg procedure is the most powerful of all but has the assumption that the p-values are independent, while the Holm–Bonferroni procedure makes no assumptions. The downside of closed testing procedures is that adjusted confidence intervals cannot be computed.

The rheumatoid arthritis trial could also be analyzed with a different strategy. We could consider that the comparison of the four less important efficacy variables would be of interest only if we had first obtained evidence of a treatment effect on the primary objective of the treatment. Thus, we could design a hierarchy of analyses, starting with the testing of the more important outcomes and proceeding to the secondary outcomes only if a significant difference had been shown.

This strategy is known as a **gatekeeping procedure**. The basic idea is to group the study hypotheses into families of hypotheses and to define the sequence in which the families will be tested. Within each family of hypotheses a closed testing procedure is used to keep the FWER at the preset alpha level. Each family serves as a gatekeeper to the next family in the hierarchy, allowing the analysis to continue into the next family only if significant differences have been shown.

For example, in the rheumatoid arthritis trial we could have defined three families of hypotheses. The primary efficacy criterion, the score on a measure of signs, and symptoms of the disease would be included in the first family, structural joint damage and duration of morning stiffness in the second family, and quality of life and activities of daily living in the third family. Then, we define a testing strategy and let us suppose that we have chosen to test each family if and only if all the hypotheses of the preceding family had been rejected.

We first test the primary efficacy variable at the 0.05 significance level. Suppose that the treatment effect was significant with $p = 0.01$. Since we have shown that the treatment improves the primary objective of the treatment, it now makes sense to look for additional benefits for the patient. We now proceed to test the second family of hypotheses. This family will be tested with a closed testing procedure providing strong control of the FWER at the 0.05 level, such as the Holm–Bonferroni procedure. If both hypotheses are rejected, then the next family will be analyzed with the same procedure at a FWER of 0.05. Otherwise, the next family will not be analyzed and the untested null hypotheses will automatically be accepted.

The gatekeeping procedure just described is called **serial gatekeeping** because each family of null hypotheses will be tested only if all null hypotheses in the previous family are rejected. This is illustrated in Figure 12.3. An alternative gatekeeping procedure is **parallel gatekeeping**. In this case, each family of null hypotheses will be tested if at least one null hypothesis in the previous family has

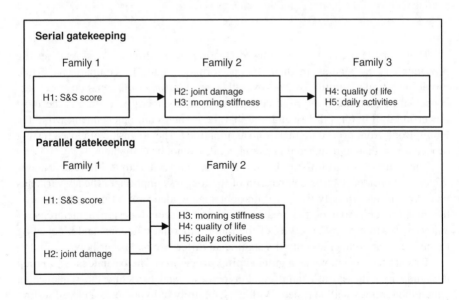

Figure 12.3   Serial and parallel gatekeeping strategies.

been rejected (Figure 12.3), but then a multiplicity adjustment has to be introduced to account for the previously analyzed families.

For example, suppose that the objective of the treatment of rheumatoid arthritis was either an improvement in the score of symptoms and signs or a decrease in structural joint damage. In that case, we could have designed a parallel gatekeeping strategy with two families of hypotheses, the first family including the two treatment objectives and the second family the three secondary variables. If one of the null hypotheses in the first gatekeeper family is rejected, then the three null hypotheses in the second family are tested at a FWER of 0.025. Otherwise, all null hypotheses are accepted.

Tree gatekeeping is a more complex approach that combines serial and parallel gatekeeping and also accounts for logical constraints among the multiple analyses.

Needless to say, all multiplicity adjustment methods mentioned in this section must be fully defined in the study protocol before any data is collected and may not be changed at analysis time.

12.6 Secondary analyses

The main efficacy analysis should be followed by a number of additional analyses. Some of these analyses are aimed at evaluating the quality of the conduct of the clinical trial, others at evaluating the robustness of the results of the main analysis, and still others seek supplemental information regarding the pathophysiology of the disease, the mechanism of action of the treatment and the patient attributes associated with greater or lesser effectiveness of the treatment.

The adequacy of the randomization procedure is usually evaluated with the comparison of the baseline characteristics of the subjects between treatment groups. If heterogeneity in the distribution of the patient attributes between groups is detected, then the adequacy of the randomization procedure may be questioned. Clearly, with a correctly executed randomization procedure, we expect 1 out of 20 comparisons of baseline characteristics between treatment groups to result in a statistically significant difference, and the number of false positive tests to be even greater. However, if the observed differences are systematic, for example, if the patient characteristics are consistently worse in one of the groups, then one can question the adequacy of randomization, and therefore the validity of the conclusions of the study.

Another secondary analysis consists of replication of the main analysis in different study populations. Typically, the main analysis is repeated in the all randomized patients population and in the per protocol population. It is commonly accepted that the per protocol population should be used only when it comprises at least 80% of the intention to treat population, otherwise the data may be so much biased that any results are useless. The purpose of this analysis is to show that the study conclusions do not depend on the selected study population. It is expected, therefore, that the results of this analysis do not contradict those of the main analysis.

A third analysis, called **post-stratification**, consists of the comparison of the efficacy criteria between groups, in the primary efficacy analysis population, adjusting for important patient attributes. Important attributes are those that are, or that may be, associated with both the efficacy criteria and the treatment groups. We saw earlier that variables with these characteristics are called confounders.

The purpose of this analysis is to verify that any differences between groups in the distribution of important patient attributes do not affect the results of the main efficacy analysis. If the treatment differences shown in the main analysis disappear when controlling for important patient characteristics, this may be an indication that the observed differences in the main analysis could be due to confounding and, again, this would suggest an inadequate randomization procedure and compromise the validity of the conclusions.

We have seen in previous sections that several decisions about the methodology of analysis must be made during the formulation of the statistical analysis plan of a clinical trial. Therefore, when interpreting the results of the efficacy analysis of a confirmatory clinical trial we must ask ourselves whether those decisions may have had an influence on the results. Accordingly, **sensitivity analyses** may be performed in order to test the robustness of the results by repeating the main analysis using alternative statistical methods or models, making different assumptions on the data, and applying different methods of imputation of missing values. Again, the results of these analyses should not contradict the conclusions of the main analysis.

The clinical trial data may also be used to perform additional analyses of an exploratory nature with the purpose of identifying patient factors that predict a greater response to the treatment. The analysis of **prognostic factors of treatment response** is performed only in the group of patients who received the test treatment. This information may provide useful insights into the mechanism of action of the treatment.

With the purpose of identifying subgroups of patients with a better response to the test treatment, **subgroup analyses** are often performed. Subgroup analyses seek to identify patient variables associated with a greater response to treatment. However, it is most inappropriate to analyze differences between treatments separately for several subgroups defined by patient variables, because one can almost always find some patient attribute in which a large treatment effect is seen in one of the levels and no difference is seen is the others. Subgroup analyses must be done by comparing the difference between treatments across levels of a patient variable and the correct procedure is to test the interaction of each patient variable with the treatment groups. A statistically significant interaction would suggest that the treatment effect is different across levels of the patient variable. Subgroup analyses are purely exploratory analyses and their results cannot be used for claims of treatment effectiveness on a particular subpopulation of patients.

12.7 Safety analysis

To conclude, a few words are in order on the analysis of safety and tolerability. This analysis is performed on the **safety population**. Unlike the primary efficacy population, safety is consensually defined as consisting of all subjects enrolled in the trial, whether eligible or not, who have received at least one administration of the study medication. In contrast, there is no consensus on the methods for the analysis of safety data. Therefore, safety data is usually presented only descriptively.

For clinical laboratory safety data, the usual method is to prepare several summary tables conveying different perspectives of the data. For example, the mean value and standard deviation of each laboratory safety variable observed at each trial visit where a safety assessment was made allow us to evaluate whether there has been a significant variation in the average values of these parameters. A more sensitive analysis, but more complicated to perform, consists of comparing the mean value of each parameter at baseline to the average of the most extreme values observed in the subsequent evaluations.

These presentations of data may suggest the existence of a treatment effect on one or more laboratory parameters, but they are not informative of the clinical significance of this effect. For example a decrease of, say, 5% in the average leukocyte count does not clarify whether there has been a slight reduction in the number of leukocytes in most individuals, or if there has been a large reduction in a small number of individuals, and the clinical implications are completely different.

Thus, it is convenient to count the number of subjects whose values observed in laboratory parameters exceeded, in either direction, a specified value. Generally, the reference value for each parameter in the specific laboratory where the test was performed is used for this purpose. Typically, the total counts of individuals that exceeded at least once this reference value in either direction are presented. Additionally, the total discounting those subjects who at baseline already exceeded the reference value should be presented, that is, the number of **treatment-emergent laboratory changes**.

Safety data also includes the so-called vital signs (heart rate, systolic and diastolic blood pressure, body temperature, and possibly body weight). The analysis and presentation of this data are identical to those of laboratory safety parameters.

Adverse events are presented in frequency tables counting the number of patients who presented each observed event. In oncology trials, the frequency of each type of adverse event is usually also presented per treatment cycle. Frequency tables are also constructed for the different types of adverse events (symptoms and signs, intercurrent illness, worsening of a preexisting condition) and their severity (serious and significant adverse events). **Serious adverse events** are those which result in death or place patients at high risk of death; which result in hospitalization or prolongation of hospitalization; which result in significant, persistent, or permanent disability of a bodily function or structure, physical activity or quality of life, or in which a medical or surgical intervention was required to prevent that disability; and which resulted in a congenital malformation in a newborn child from a patient exposed to the study treatments. Significant adverse events are non-serious adverse events leading to the permanent discontinuation of the treatment and the premature termination of the trial for a patient.

Finally, it is customary to display a frequency table of adverse events disaggregated by degree of severity (mild, moderate, and severe) and, in each severity level, by causal relationship to the study treatment (not related, related) in the investigator's opinion.

Due to the problem of multiple comparisons, usually in none of these tables are statistical tests presented for the comparison between treatment groups.

13

Meta-analysis of clinical trials

13.1 Purpose of meta-analysis

Meta-analysis is a methodology for integrating information by combining the results of several clinical trials. The term meta-analysis literally means the analysis of analyses. Meta-analysis can be defined as the analysis and synthesis of the results of independent studies by a systematic, explicit, and quantitative method that takes into account all available information and which is based on a solid theoretical statistical framework.

The most common application of meta-analysis of clinical trials is the combination of results of inconclusive clinical trials with the purpose of increasing the sample size and thereby conferring increased power to the statistical testing of treatment effects. However, other important uses of meta-analysis include situations where published clinical trials have provided contradictory results, or when a more accurate estimate of the treatment effect needs to be obtained, or for the analysis of particular subgroups of patients for whom the results of clinical trials have not yet provided conclusive evidence. In essence, meta-analysis is a methodology that seeks to draw the information needed to support clinical decisions from reports of clinical trials found in scientific journals.

As with any other scientific project, adequate planning and the careful preparation of a study protocol are essential. The most important steps in the planning and conduct of a meta-analysis can be defined as follows: (1) a clear definition of the research problem and a precise formulation of the objectives of the study; (2) a definition of the inclusion and exclusion criteria of the clinical trials; (3) a methodology for finding clinical trials; (4) a definition of the efficacy criteria; (5) extraction of data form published clinical trials; (6) data analysis; (7) sensitivity analyses; (8) secondary analyses; and (9) the presentation and interpretation of the results.

Biostatistics Decoded, First Edition. A. Gouveia Oliveira.
© 2013 John Wiley & Sons, Ltd. Published 2013 by John Wiley & Sons, Ltd.

Methods for finding clinical trials include a search of bibliographic databases, looking for papers cited by other papers already found (**search by reference**), a search of public registries of clinical trials, inquiries to authors of published clinical trials about their eventual knowledge of other authors involved in clinical trials on the same treatments, and inquiries to pharmaceutical companies which market the medicine in question regarding the existence of any unpublished trials. Clinical trial results published only as abstracts of presentations at conferences are not usually considered.

Data is extracted from the retrieved reports, preferably by at least two independent raters, and any discrepancies between raters should be solved by consensus.

The analysis methodology consists basically of the calculation of the average treatment effect across clinical trials, that is, the average difference in efficacy between treatment groups. In a meta-analysis that is not a simple average; it is a weighted average takes into account a number of characteristics of the clinical trials that we will discuss later. The first decision to be made in the analysis phase is the definition of a measure of treatment effect that might be combined across the various clinical trials.

13.2 Measures of treatment effect

In the case of binary efficacy variables, results are normally reported as proportions, for example, the response rate or the proportion improved in each treatment arm. The most commonly used measures of the treatment effect are the relative risk (the proportion of responses in one treatment group divided by the proportion of responses in the other group) and the odds ratio (the ratio between the proportion of responses divided by the proportion of non-responses in one group and the proportion of responses divided by the proportion of non-responses in the other group). These measures are identical to those used in epidemiology and have a number of interesting mathematical properties, but are less easily interpretable from the clinical standpoint than the **risk difference** (the arithmetic difference between the proportions of responses in the two groups), since the latter reflects the net gain that can be expected in terms of successful outcomes. Moreover, the relative risk and the odds ratio cannot be determined when there are no responses in the comparison group. Nevertheless, relative risks and odds ratios express differences between proportions better than the risk difference and are the preferred measure in meta-analyses.

Besides these, other measures of treatment effect that can be used are the **risk reduction** (one minus the reciprocal of the relative risk) and the **number needed to treat** (NNT) to obtain an additional response (the reciprocal of the risk difference). However, both these measures are difficult to treat mathematically and are usually reserved for transforming the results of the meta-analysis into measures that clinicians consider to have a more straightforward interpretation. For example, if the risk difference between two treatments is, say, 10 percentage points, the NNT will be $1/0.10 = 100$. This means that if 100 patients are treated with the test treatment, we expect one more successful outcome than if the patients had been treated with the control treatment.

In the case of interval-scaled efficacy variables, one measurement of the treatment effect is, naturally, the difference in means between treatments. However, this measure can be used only if all clinical trials have adopted the same efficacy criterion and it has been measured on the same scale.

However, in a meta-analysis the various clinical trials may not always have adopted the same efficacy criterion or it may not always have been measured on the same scale. In such cases it is therefore necessary to standardize the scale of measurement of the efficacy variable before it can be combined across the various trials.

A popular measure of the treatment effect which is used when the efficacy criterion is quantitative and measured in an interval scale is the **effect size** of Glass, which consists of the difference between the averages of the efficacy variable in both groups, divided by the standard deviation of the efficacy variable in the control group. The effect size thus expresses the differences between treatments as a quantity that has the standard deviation as its unit. Thus, a clinical trial with an effect size of, for example, 0.4, has an effect that is half the effect size of another study with an effect size of 0.8.

The standardization of treatment effects by dividing by the standard deviation of the controls does not have much support when comparing two active treatments. In this situation it is preferable to use the **method of Cohen**, which defines the effect size as the difference between group means divided by the combined estimate of the standard deviation computed from the standard deviations of the two groups. This measure is often called the **standardized mean difference**.

For efficacy criteria measured on ordinal scales, the measure of the treatment effect relies heavily on the data provided by the authors of the publications. If the results have been expressed as means and standard deviations, and the scale has a large range of values, the usual practice is to analyze the data as if it was an interval scale. If the efficacy variable was measured on a scale with a small range of values and the authors have presented the distribution of individuals by each value of the scale, the usual procedure is to dichotomize the efficacy variable and to use one of the measures mentioned above for proportions.

In clinical trials where the efficacy criterion was the time to an event, the measurement of treatment effects is usually by the hazard ratio.

13.3 The inverse variance method

As mentioned above, meta-analysis is based on a weighted average of the treatment effects. The weighting is intended to differentiate the contribution of the several clinical trials to the average treatment effect, giving more weight to the estimates of treatment differences obtained in the clinical trials that have greater precision.

The weighting method is a general method of analysis that is used when we wish to test differences among several means. This method can be used both when we have individual data and when we only have aggregate data. Hence it is particularly suited to the case of meta-analysis where, in most situations, we have access only to aggregate data.

When the treatment effect is measured on an interval scale and the results observed in clinical trials are homogeneous, we may estimate the true value of the difference between treatments using a weighted average of the observed differences, the weights being the reciprocal of the squared standard errors.

This method, called the **inverse variance method**, gives us a weighted estimate of the true treatment effect, its confidence interval, and a test of homogeneity of treatment effects across trials. However, those estimates and their confidence intervals are valid only on the condition that treatment differences across clinical trial are homogeneous. In the next section we will discuss methods that are adequate when clinical trial results are heterogeneous.

The principle of weighting seems to make sense, but leaves two questions open. First, which weighting factor should be used? Second, does the use of a weighting factor improve the combined estimate of the difference between treatments?

In a meta-analysis, we are analyzing several variables (the differences between treatments observed in the clinical trials) whose values are differences between sample means. Each of these variables has a different variance (squared standard error) because the sample sizes are not equal in all clinical trials. We also know that the values of those variables correspond to differences between means of independent samples, and thus have a normal distribution if the samples sizes are large. As clinical trials almost always have at least 20 patients in each group, this assumption usually holds.

Under the null hypothesis of equality of the differences between treatments across all the clinical trials, that is, of **homogeneity** of the differences between treatments in the populations, all the observed treatment differences estimate the same quantity, which is the true value of the difference between treatments. We can test this null hypothesis by evaluating whether the spread of the values of the observed differences around the true treatment difference is within or larger than expected from normal sampling variation.

To this end, we might proceed as we usually do when we want to quantify the dispersion of values about their mean: that is, we would compute the squared difference of each observation to the true mean and sum all the results. But since the observed differences were obtained from clinical trials of different sample sizes and, therefore, have different variances, we must first standardize by dividing each squared difference by the respective variance (squared standard error). Accordingly, this is equivalent to a sum of squares weighted by the inverse variance and is obtained with

$$\text{wSSq} = \sum \frac{(d_i - \mu)^2}{\text{SE}_i^2}$$

where d_i is the observed difference between treatments, μ is the true difference between treatments and SE_i the standard error of the difference in the ith clinical trial.

Under the null hypothesis of homogeneity of treatment differences across clinical trials, the sum of the squared differences of each observed treatment effect to the true treatment effect divided by their respective variance has a chi-square distribution with degrees of freedom equal to the number of observations. We know this because, when we subtract the mean from a random variable with a normal distribution and divide the result by the standard deviation, we will get a standardized normal deviate, and the sum of n squared standardized normal deviates has a chi-square distribution with n degrees of freedom.

At this moment we cannot yet test the null hypothesis of homogeneity of the differences across clinical trials because we do not know the true value μ of the difference between treatments. We can, however, replace that value by the average of all the observed differences between treatments.

However, for the calculation of this average we will weight the observed differences by the same factor as before, i.e., the inverse variance. If treatment effects are homogeneous, this weighted mean is the best estimator of the true difference between treatments, because no other has lower variance. Therefore, we obtain the value of the mean treatment difference by summing the product of each observed difference by its weight and dividing the result by the sum of all weights. We may represent this by

$$D = \frac{\sum w_i d_i}{\sum w_i}$$

where $w_i = 1/SE^2_i$.

Finally, we need only to know the variance of each observed difference. We have seen already that the estimate of the standard error obtained from the observed data is quite accurate. Therefore, we may use the square of the standard error estimated from the data as a good approximation to the true value of the variance of the observed differences.

The quantity resulting from the sum of squared differences of the observed treatment effects to their weighted average, each one divided by the respective variance, now has a chi-square distribution with degrees of freedom equal to the number of clinical trials minus one, since we have used our data for the estimation of the true value of the difference between treatments, therefore losing 1 degree of freedom. High values of this result provided evidence against the null hypothesis of homogeneity of the treatment effect across trials.

If the null hypothesis is not rejected we have no evidence for the heterogeneity of treatment differences across clinical trials and we can obtain an estimate of the true value of the difference between treatments from the weighted average of the observed differences between treatments, as shown above. Still, under H_0, the variance of the weighted average is the reciprocal of the sum of the weights and its standard error is the square root of this quantity. This allows us to obtain confidence intervals using the normal distribution.

Thus, the 95% confidence limits are

$$D \pm 1.96\sqrt{1/\sum w_i}$$

The inverse variance method can be used whenever the measures of the treatment effect are interval scaled (difference between means, effect size, or standardized mean difference). It can also be used when the measure of the treatment effect is the odds ratio, the risk ratio, or the incidence rate ratio, applying the same method to the logarithmic transformation of those quantities.

However, the preferred method for binary efficacy variables in the absence of heterogeneity appears to be the **Mantel–Haenszel method**. For this test it is necessary to obtain the total of subjects in each treatment group and the total with a response to the treatment. The Mantel–Haenszel method is equivalent to a logistic regression where the dependent variable is the efficacy variable and the independent variables are a set of dummy variables encoding the different clinical trials and a binary variable encoding the treatment group. The exponential of the coefficient of this latter variable is approximately the meta-analytic estimate of the odds ratio.

13.4 The random effects model

The method discussed in the previous section assumes that each clinical trial represents a measurement of the true treatment difference and that the results obtained in the several clinical trials differ only because of sampling variation. As the error due to sampling variation is random, if we combine the differences between treatments that were observed in several clinical trials we will obtain unbiased estimates of the true treatment difference. This scenario corresponds to the so-called **fixed effects model**.

Sometimes, however, we suspect that such an assumption does not hold because of obvious differences in treatment effects across the clinical trials, that is, there is **heterogeneity** of the results. In this situation, the differences between clinical trial results cannot be explained solely by sampling variation. There must also exist an unexplained variation, presumably due to systematic differences between the characteristics of the trials, such as the eligible population, duration of treatment, local conditions of trial administration, and so forth. This variation manifests itself as random variation superimposed on sampling variance and, if it can be quantified, we can take it into account when we calculate the precision of the estimate of the true treatment effect.

This scenario corresponds to the so-called **random effects model**. This model is therefore appropriate in the presence of heterogeneity. Unlike the fixed effects model, which when wrongly assumed will produce erroneous estimates of the treatment effect, the wrong specification of the random effects model entails only a lesser accuracy of the estimated treatment effect. Accordingly, the random effects model will be less likely to demonstrate a statistically significant effect. In other words, when the wrong model is applied the fixed effects model is subject to bias, whereas the random effects model is too conservative.

The most commonly used method in the presence of heterogeneity is the **DerSimonian–Laird method**, which can be applied to any type of measurement of the treatment effect. This method produces an estimate of the variance between clinical trial results and the weighting factor is the reciprocal of the sum of the between-trials variance with the variance of each trial.

The homogeneity test described in the previous section largely determines the model that should be adopted. A significant result is indicative of heterogeneity and the random effects model and the DerSimonian–Laird method should be selected for an estimation of the true treatment effect. A non-significant result, however, does not confirm the hypothesis of homogeneity, even more so because the homogeneity test is notoriously underpowered. In the next section we will discuss additional methods for identifying heterogeneity.

13.5 Heterogeneity

The identification of heterogeneity is of considerable importance because when it exists and is disregarded it may lead to invalid conclusions, while if it does not exist but is considered it may lead to inconclusive results. The test of heterogeneity has low statistical power and a non-significant result by no means excludes the possibility of heterogeneity. For this reason, methods have been introduced for the visual identification of heterogeneity.

An often used method, applicable when the efficacy variable is binary, is the **L'Abbé plot**, which is a scatterplot of the proportion of responses observed in each treatment group for each clinical trial. Figure 13.1 shows examples of L'Abbé plots. The solid diagonal line represents the equality of treatment effects, the dashed line represents the meta-analytical estimate of the treatment effect, and each circle represents a clinical trial. In Figure 13.1, the measure of treatment effect is the risk difference.

In the absence of heterogeneity, it is expected that the circles corresponding to the clinical trials will be located along a straight line, as in Figure 13.1a, and heterogeneity is suggested if some trials depart considerably from that line, as in Figure 13.1b.

The L'Abbé plot also provides an indication of the best measure of treatment effect. In situations where the response rate is very low in one of the treatment groups, clinical trials can be heterogeneous with one measure of the effect (e.g., the risk difference) and homogeneous with another measure (e.g., relative risk), and the L'Abbé plot provides a visual indication of such a situation. For example, the dotted line in the graph of Figure 13.1b is the meta-analytical estimate of the treatment effect based on relative risks and, with this measure, the results appear to be homogeneous.

Another method for investigating heterogeneity is the **Galbraith plot**. This method consists of a scatterplot of the test statistic of the difference between

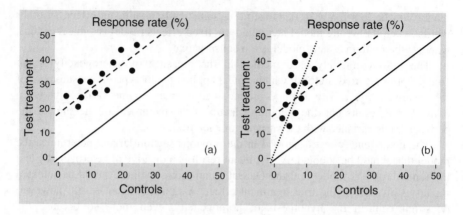

Figure 13.1   L'Abbé plot: solid line, line of equality; dashed line, meta-analytical estimate of the risk difference; dotted line, meta-analytical estimate of the relative risk; (a) homogeneity; and (b) heterogeneity of risk difference, but homogeneity of relative risk.

treatments in each trial (the treatment difference divided by its standard error) against a measure of the precision of the trial (the inverse of the standard error of the treatment difference). A least squares line through the origin will have a slope equal to the estimate of the treatment effect from a fixed effects meta-analysis, and the vertical distance of each trial to the regression line will give an indication of the contribution of the trial to the statistic of homogeneity (Figure 13.2). In the absence of heterogeneity we expect that 95% of the difference between treatments, in the observed clinical trials, are within two standard errors of the true treatment effect. This region is represented by the dashed lines in Figure 13.2. Clinical trials outside

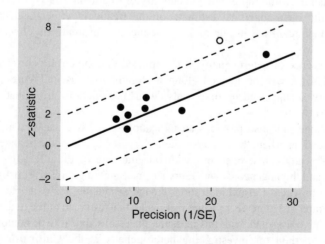

Figure 13.2   Galbraith plot for the detection of heterogeneity.

this region are trials that make a significant contribution to the statistic of homogeneity. The second largest trial (open circle) makes an important contribution to the statistic of homogeneity. In this situation, a meta-analysis with a random effects model would be more appropriate.

The investigation of the source of heterogeneity may provide interesting indications about what factors influence the treatment effect, which is analogous to a subgroup analysis in clinical trials. This investigation can be carried out visually with Galbraith plots, labeling each point on the graph according to the trial characteristics (e.g., type of trial, treatment duration, medication dose), or by using formal statistical methods. For example, we can extend the principle of the Galbraith plot and use linear regression to analyze the extent to which one or more factors may explain the heterogeneity of treatment effects.

13.6 Publication bias

Another issue that may compromise the validity of a meta-analysis is the possibility that the assumptions of the method are not met due to a particular type of clinical trial having systematically been excluded from publication. Both the fixed effects and the random effects models are based on the assumption that the observed clinical trials represent a random sample from a population of clinical trials, and if a particular type of clinical trial is systematically excluded from the sample, the results of the meta-analysis will almost certainly be biased.

Apparently, the most frequent reason for this **publication bias** is simply because the authors of a clinical trial are not as committed to publishing the results of a trial showing no significant differences as they are to 'positive' trials. Another important reason for publication bias is the greater likelihood of rejection of negative studies by editors of scientific journals.

In addition, other causes of bias have been noted. Some examples are citation bias (frequently cited trials have a greater probability of being selected for a meta-analysis), location bias (clinical trials conducted in developing countries have a lower probability of being published), English language bias (clinical trials published in English have a greater probability of being selected), multiple publication bias (clinical trials that have been published several times in different journals have a greater probability of being selected), and confirmation bias (the preferential rejection by reviewers of scientific journals of manuscripts presenting results that are contrary to their beliefs or to mainstream opinion).

The only way of minimizing publication bias is to ask potential authors of clinical trials about any unpublished results, a task that can be aided by querying public registries of planned and ongoing clinical trials. As this is a long and laborious undertaking, a number of statistical methods designed to detect the presence of publication bias have been developed.

One of the first approaches for detecting the eventuality of publication bias was the **funnel plot** (Figure 13.3). This method assumes that, since all clinical trials in a meta-analysis are estimating the same and unknown true difference between

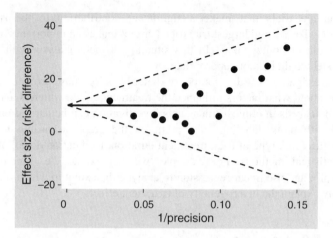

Figure 13.3   Funnel plot for the identification of publication bias: solid line, meta-analytical estimate of the treatment effect; dashed lines, 95% confidence limits.

treatments, the treatment effects observed in the individual trials should be distributed about the true treatment effect and their dispersion should be proportional to their variances.

In other words, in a scatterplot of the treatment effect versus the accuracy of the trials, it is expected that small-sized clinical trials will have widely dispersed treatment effects around the average effect and that the dispersion of effect sizes decreases as the precision (sample size or inverse variance of the treatment difference) of the trial increases. Hence the designation of funnel plot. The existence of publication bias is suspected by an asymmetry of the funnel plot, particularly by the absence of trials in the bottom right part of the graph (as shown in Figure 13.3), since that region corresponds to small-sized clinical trials with inconclusive results, precisely those clinical trials that are less likely to be considered for publication by scientific journals or even to be submitted for publication.

Because the interpretation of the funnel plot is essentially subjective, formal statistical tests have been proposed for the identification of publication bias. **Begg's test** is based on the observation that the presence of publication bias induces an asymmetry in the funnel plot and thus a correlation between the treatment effects and the precision of the trial will be evident (Figure 13.4a and b). Consequently, Begg's test assesses the significance of Kendall's correlation coefficient between treatment effects and their variances. In the presence of publication bias, this correlation will be statistically significant (Figure 13.4b).

Another test which seems to have greater power than the previous test is **Egger's test**. This test detects an asymmetry in the funnel plot through a linear regression of the treatment effect, standardized by division by its standard error, on the precision of the clinical trial (approximately equivalent to the sample size). The standardized treatment effect, as we saw above, corresponds to the statistic of the

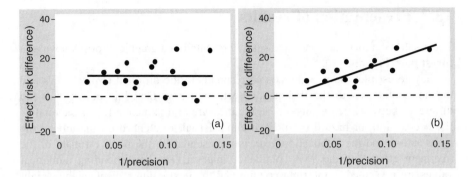

Figure 13.4   Begg's test: (a) absence of publication bias; (b) presence of publication bias.

test of a difference between treatments. Egger's test is based on the observation that, in the absence of publication bias, if a difference between treatments truly exists then small-sized trials will have low precision (nearing zero) and a small test statistic (no significant difference), whereas large clinical trials will have high precision and a large value of the test statistic (statistically significant difference). Consequently, a regression line of the test statistic on the precision of the trial should pass through the origin (Figure 13.5 a).

In situations of publication bias, one would expect an absence of clinical trials in the region of small test statistics and low precision (small inconclusive trials), but not in the region of large test statistics and low precision (small trials showing statistical differences between treatments). Consequently, the regression line will be shifted from the origin (Figure 13.5b). Egger's test consists of testing whether the intersection of the regression line is significantly different from zero. The results of these two tests should be interpreted with caution, because of their notorious low power.

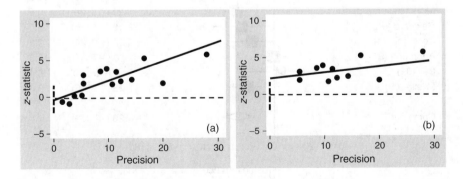

Figure 13.5   Egger's test: (a) absence of publication bias; (b) presence of publication bias.

13.7 Presentation of results

The results of a meta-analysis are usually presented in a graphical form known as a **forest plot** (Figure 13.6).

In a forest plot the vertical axis corresponds to the different clinical trials and the horizontal axis to the treatment effect, with a vertical line drawn at 0 (no difference between treatments) or 1 if the measure of treatment effect is the relative risk, odds ratio, or hazard ratio. Each clinical trial, referenced on the left by the main author and the publication year, is represented by the point estimate of the treatment effect and the 95% confidence interval. The corresponding numerical values are presented in the right column and the next column displays the weight given to each trial for the calculation of the meta-analytical estimate. The weighting is depicted visually by the size of the symbol of the point estimate of the difference between treatments, whose area is proportional to its weight. The meta-analytical estimate and the respective 95% confidence interval are represented by a diamond at the bottom of the graph.

From the above discussion of the methodology of meta-analysis it should be clear that a number of aspects must be considered in the interpretation of the results. When clinical trials have identical designs, the same eligibility criteria, and the same efficacy criteria, and their results are clearly homogeneous, then the fixed effects model is likely to be adequate and the meta-analytical estimates and their confidence intervals will be reliable, provided that publication bias can be assumed

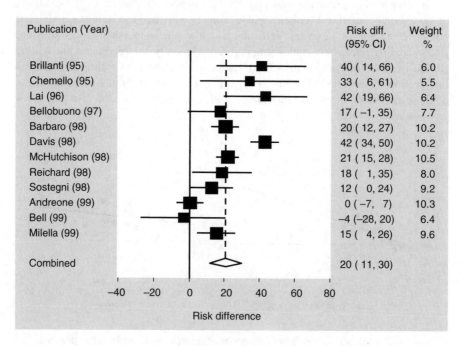

Figure 13.6   Forest plot presenting the results of a meta-analysis.

to be absent or minimal. Unfortunately these conditions are not often met and more commonly a meta-analysis is performed in situations where clinical trials have different eligibility criteria, study designs, medication doses, length of observation, or efficacy variables, and the treatment effects are heterogeneous. A random effects model may be used in these cases, but this method assumes that the observed clinical trials are a representative sample of all the trials that can be conducted for that therapeutic condition and that treatment, and this is, of course, a very strong assumption. In both cases the possibility of publication bias is worrisome and no effort should be spared in the search for published and unpublished trials.

The robustness of the results may be evaluated with sensitivity analyses, by repeating the meta-analysis for different subsets of clinical trials (e.g., only for large trials, or only for high-quality trials) and using the alternative model. The results of sensitivity analyses should not contradict the conclusions of the main analysis, otherwise this will be an indication that the assumptions of the method do not hold.

Despite its limitations, meta-analysis remains an extremely important methodology for the formation of an informed and objective opinion about the available information on existing treatment modalities.

Further reading

Aaker, D.A., Kumar, V., and Day, G.S. (2001) *Marketing Research*, 7th edition, John Wiley & Sons, Inc., New York.

Altman, D.G. (2001) *Practical Statistics for Medical Research*, Chapman and Hall, London.

Armitage, P., Berry, G., and Matthews, J.N.S. (2002) *Statistical Methods in Medical Research*, 4th edition, Blackwell Science, Oxford.

Bland, M. (2000) *An Introduction to Medical Statistics*, 3rd edition, Oxford University Press, Oxford.

Chow, S.-C. and Liu, J.-P. (2004) *Design and Analysis of Clinical Trials: Concepts and Methodologies*, 2nd edition, John Wiley & Sons, Inc., Hoboken, NJ.

Crocker, L. and Algina, J. (1986) *Introduction to Classical and Modern Test Theory*, Harcourt Brace Jovanovich College Publishers, Forth Worth, TX.

Egger, M., Smith, G.D., and Altman, D.G. (eds.) (2001) *Systematic Reviews in Health Care: Meta-Analysis in Context*, 2nd edition, BMJ Publishing Group, London.

Fleiss, J.L. (1999) *The Design and Analysis of Clinical Experiments*, John Wiley & Sons, Inc., New York.

Fleiss, J.L., Levin, B., and Paik, M.C. (2003) *Statistical Methods for Rates and Proportions*, 3rd edition, John Wiley & Sons, Inc., New York.

Hosmer, D.W. and Lemeshow, S. (2000) *Applied Logistic Regression*, 2nd edition, John Wiley & Sons, Inc., New York.

Hosmer, D.W., Lemeshow, S., and May, S. (2008) *Applied Survival Analysis: Regression Modeling of Time-to-Event Data*, 2nd edition, John Wiley & Sons, Inc., New York.

Pawitan, Y. (2001) *In All Likelihood: Statistical Modelling and Inference Using Likelihood*, Oxford Science Publications, Oxford.

Piantadosi, S. (2005) *Clinical Trials: A Methodological Perspective*, 2nd edition, John Wiley & Sons, Inc., New York.

Pocock, S.J. (1983) *Clinical Trials: A Practical Approach*, John Wiley & Sons, Inc., New York.

Rao, C.R., Miller, J.P., and Rao, D.C. (2008) *Handbook of Statistics: Epidemiology and Medical Statistics*, Elsevier, Oxford.

Rothman, K.J. (2002) *Epidemiology: An Introduction*, Oxford University Press, Oxford.

Rothman, K.J., Greenland, S., and Lash, T.L. (2008) *Modern Epidemiology*, 3rd edition, Lippincott Williams & Williams, Philadelphia.

Biostatistics Decoded, First Edition. A. Gouveia Oliveira.
© 2013 John Wiley & Sons, Ltd. Published 2013 by John Wiley & Sons, Ltd.

Siegel, S. and Castellan, N.J., Jr. (1988) *Nonparametric Statistics for the Behavioral Sciences*, 2nd edition, McGraw-Hill, New York.

Thompson, S.K. (2002) *Sampling*, 2nd edition, John Wiley & Sons, Inc., New York.

Tryfos, P. (1996) *Sampling Methods for Applied Research: Text and Cases*, John Wiley & Sons, Inc., New York.

Vittinghoff, E., Glidden, D.V., Shiboski, S.C., and McCulloch, C.E. (2004) *Regression Methods in Biostatistics: Linear, Logistic, Survival and Repeated Measures Models*, Springer, New York.

Index

Actuarial
  curve, 155
  method, 151–154
Adaptive clinical trial, 284
Add-on treatment, 267
Adjusted R^2, 185
Adjustment, 82
Adverse events, 293, 309
  serious, 309
  significant, 309
Alpha error, 112, 135
  inflation of, 257, 285
Alpha spending function, 288
Alternative hypothesis,
  108, 131
Analysis of covariance, 302
Analysis of variance
  in linear regression, 180, 183
  in test-retest reliability, 247
  test of difference between means,
  123–129
Analytical studies, 6, 87–91
Anchor-based method, 273
ANCOVA, 302
ANOVA, 123–129
Approximate relative risk, 102
Assay sensitivity, 282
Association studies, 6
Attributable
  fraction, 102
  fraction among the exposed, 103
  fraction in the population, 103
  risk, 102
Attribute, 11
  continuous, 17

discrete, 17
  multivalued, 16
Average, 15

Backward elimination, 193
Bartlett method, 239
Begg's test, 320
Beta error, 135
Bias
  citation, 319
  confirmation, 319
  English language, 319
  evaluator's, 267
  location, 319
  multiple publication, 319
  participant's, 266
  publication, 319
  selection, 256, 262
Binomial distribution, 54
Binomial test, 115
  one-sample, 283
Biostatistics, 1–4
Blinding
  double, 267
  double-dummy, 268
  open, 267
  single, 267
  triple, 267
Bonferroni correction, 141, 257, 303
Boundary
  Bonferroni fix, 286
  Haybittle–Peto, 286
  O'Brien–Fleming, 286–288
  Pocock, 286
Breslow method, 223

Biostatistics Decoded, First Edition. A. Gouveia Oliveira.
© 2013 John Wiley & Sons, Ltd. Published 2013 by John Wiley & Sons, Ltd.

Calibration group, 282
Case report form, 292
Case series, 5
Case, 77
Case-control study, 89
　matched, 90
　nested, 90
Case-crossover study, 90
Case-fatality rate, 79
Cause-effect relationship, 4–5, 253
Censored data, 152
　non-informative, 219
Census, 6
Central limit theorem, 31
Central tendency measures, 19–20
Centralized patient registry, 82
Centroid, 172
Change from baseline, 301
　percentage, 301
Chi-square
　distribution, 118, 209
　statistic, 118
　test with Yates' continuity correction, 122
　test, 116–121
Classical test theory, 240
Clinical case, 5
Clinical laboratory safety parameters,
　　293, 309
Clinical trial population, 297–301
　all patients treated, 301
　all randomized patients, 301
　modified intention to treat, 301
　on treatment, 301
　per protocol, 301
　safety, 308
Clinical trial, 254
　adaptive, 284–286
　baseline evaluation, 258
　composite endpoint, 257
　controlled, 261
　crossover, 263–265
　efficacy criteria, 256–258
　equivalence, 278
　event-driven, 275
　factorial 2x2, 265
　futility, 282
　inclusion/exclusion criteria, 255
　informed consent, 256, 292
　intention to treat, 299

　multicenter, 256
　non-comparative, 258–260
　non-inferiority, 277–284
　parallel group, 262
　phase I–IV, 278
　protocol, 291
　protocol deviations, 296
　protocol violations, 297
　randomized, 262
　sensitivity analyses, 308
　subgroup analysis, 308
　surrogate criteria, 257
Closed testing procedure, 305
Cluster sampling, 74
Coefficient
　of determination, 179
　of multiple determination, 185
Cohen's kappa, 248
Cohort study, 64, 89
　analytical, 88
　descriptive, 64
Combined sampling, 76
Communality, 235
Comparison
　of hazard functions, 159
　of paired samples, 147
　of several means, 123
　of two means, 93–98, 108–115
　of two proportions, 98–100, 115, 116–121
Composite endpoint, 257
Concordance, 245–251
　intraclass correlation coefficient, 246
　kappa, 248
　weighted kappa, 249
Concurrent validity, 252
Conditional power, 290
Confidence bands, 158
Confidence interval, 45
　exact, 56
　for cumulative survival, 158
　for incidence rate ratio, 166
　for means, 45–46
　for paired samples, 149
　for proportions, 55–58
　for the difference of two means, 93–98
　for the difference of two proportions, 98
　for the regression coefficient, 178
　in non-inferiority testing, 281
　one-sided, 279

Confounding, 83, 161, 187, 253
Consecutive random sampling, 66, 256
Constancy assumption, 282
Construct, 229
  validity, 251
Content validity, 251
Controls
  active, 267
  concurrent, 261
  historical, 261
  matched, 261
Convenience sampling, 65, 91
Convergence of distributions
  binomial to normal, 56
  Poisson to binomial, 165
  Poisson to normal, 165
  t to normal, 50
Convergent validity, 252
Cook–Weisberg test, 189
Correlation
  coefficient, 179
  interdimension matrix, 244
  intraclass coefficient, 246
  item-rest, 244
  item-test, 244
  item-to-own-dimension, 244
  matrix, 231
  Spearman's coefficient, 245
Covariance, 241
  matrix, 241
Cox model, 219–223
  assumptions, 223–225
  with competing risks, 227
  with multiple failure data, 227
  with time-varying covariates, 226
CRF, 292
Criterion validity, 252
Cronbach's alpha, 240–245
Cross-sectional study
  analytical, 88
  descriptive, 64
Crude rate, 82

Data monitoring board, 284
Degrees of freedom, 49, 121–122
Dependent variable, 169
DerSimonian–Laird method, 317
Descriptive studies, 5, 63–64, 77–79
Design effect, 76

Deviance, 209
  analysis of, 208
DfBeta, 189
Dimension, 230
Direct age-adjusted mortality rate, 83
Discriminant validity, 252
Discrimination, 212
Distribution
  binomial, 54
  chi-square, 118
  F, 128, 128
  hypergeometric, 68
  normal, 31–34
  Poisson, 163
  Student's t, 48
Distribution-based methods, 273
Double-dummy technique, 268
Dummy variables, 194
Duncan's test, 141

Effect size, 273, 313
  method of Cohen, 313
  method of Glass, 313
Egger's test, 320
Eigenvalue, 234
Encoding
  binary, 195
  by reference, 195
Enrichment, 285
Equation
  of the parabola, 200
  of the straight line, 171
Equivalence
  clinical trial, 278
  margin, 278
Error
  alpha, 112, 135
  beta, 135
  type I, 135
  type II, 135
Error of the estimate, 60
  maximum, 61
Event, 151
Event-count analysis, 163
Exclusion criteria, 255
Exogenous/endogenous variable, 227
Experimental studies, 6, 253–293
Explained variable, 169
Explanatory variable, 169

F
distribution, 125, 128
test in linear regression, 180–183
test in multiple regression, 186
Factor, 230
analysis, 230–236
loading, 232
rotation, 236–238
score, 238
standardized score, 238
Factor analysis, 229
common, 239
confirmatory, 239
exploratory, 239
principal, 239
False negative/positive rate, 135, 214
Family-wise error rate, 304
Finite population correction, 68
Fisher's exact test, 120, 145
Fixed effects, 303
Force of mortality, 220
Forest plot, 322
Forward elimination, 193
Freedman method, 275
Frequency
absolute, 18
cumulative, 19
distribution, 18
relative, 18
Funnel plot, 319
Futility
clinical trial, 282–284
threshold, 282
FWER, 304

Galbraith plot, 317
Gatekeeping
parallel, 305
serial, 305
tree, 306
Gaussian curve, 31
Generalizability, 252
Group sequential plan, 286–288

Hazard
adjusted ratio, 163
baseline rate, 220
cumulative, 224
function, 159, 220

rate, 220
ratio, 161, 222, 225
Heterogeneity, 316
Heteroscedasticity, 129
Histogram, 22
Hochberg procedure, 304
Holm–Bonferroni procedure, 304
Homogeneity, 314
Homoscedasticity, 129
Hypergeometric distribution,
68
Hypothesis
alternative, 108,131
null, 108, 131
one-sided, 133, 274, 279

Immediate effects model, 91
Incidence study, 78
Incidence, 78
density, 80, 164
rate ratio test, 167
rate ratio, 166
rate, 78
Inclusion criteria, 255
Independence
of random variables, 28
Independent variable, 169
Informed consent form, 292
Intention to treat, 299–301
modified, 301
Interaction, 196–199, 264, 271
qualitative, 265, 271
quantitative, 265
second order, 198
treatment by period, 264
Interactive voice response system,
270
Intercept, 171
Interim analysis, 284
Internal consistency, 240
reliability, 240
Interquartile range, 21
Inter-rater reliability, 248
Interventional studies, 6
Intraclass correlation coefficient, 246
Inverse variance method, 313–316
Investigational new drug, 278
Item list, 17
Item response theory, 240

Kaiser criterion, 234
Kaplan–Meier
  curve, 157
  method, 155–156
Kappa, 248
  Cohen's, 248
  Fleiss's, 251
  weighted, 249
Kruskal–Wallis test, 145

L'Abbé plot, 317
Lan-DeMets
  alpha spending function, 288
  group sequential method, 289
Last observation carried
  forward, 297
Latent variable, 238
Least squares method, 171–174
Leverage, 189
Likelihood, 205
  function, 205
  ratio test, 208
Likert scale, 17, 230
Limits, 21
LOCF, 297
Log likelihood, 207
  ratio, 209
  ratio test, 208
Logistic regression, 201–219
  assumptions, 207
  maximum likelihood method in,
    204–206
  stepwise, 212
Logit, 105, 202
Log–log plot, 224
Log-odds, 105, 202
Logrank test, 159–161, 226
  adjusted, 161–163

Mann–Whitney–Wilcoxon U test, 143–145
Mantel–Haenszel test, 212, 316
Matched case-control study, 90
Maximum likelihood
  estimate, 205
  method, 204
  partial, 223
MCID, 273
McNemar's test, 150
Mean deviation, 23

Mean square, 128
  due to regression, 182
  residual, 177
Mean squared error, 184
Mean, 15, 19
  inference from, 43
  population, 25, 41
  sample, 25, 41
  squared error, 184
  weighted, 173
Measures of dispersion, 20–21
Median, 19
Meta-analysis, 311
  fixed effects model, 316
  random effects model, 316
  sensitivity analyses, 323
  test for heterogeneity, 314
Minimal clinically important difference,
    273
  anchor-based methods, 273
  distribution-based methods, 273
Minimization, 272
Mode, 20
Modern test theory, 240
Morbidity index, 77
Mortality rate, 79
Multicollinearity, 191
Multiple comparisons, 139–141, 285,
    303–307
Multiple regression, 185–188
Multistage sampling, 74–77
Multivariate analysis, 185

Nested case-control study, 90
New molecular entity, 278
NNT, 312
Non-inferiority
  clinical trial, 277–284
  margin, 278
Non-parametric test, 143–145
Non-probability sampling, 65, 81, 91
Non-sampling error, 66
Normal curve, 31
Normal distribution, 29–34
  of sample means, 39
  parameters, 31
  properties of, 31–34
Null hypothesis, 108, 131
Null model, 208

Number at risk, 152
Number needed to treat, 312

Oblique rotation, 237
Observational studies, 6
Odds, 101
Odds-ratio, 101, 211
  adjusted, 211
One-sample statistical test, 283
One-sided tests, 132
Order effect, 264
Orthogonal rotation, 236
Outcomes research, 254
Overfitting, 192, 207

Paired t test, 147–149
Panel data, 147
Patient attrition, 277
Patient information, 292
Patient's global impression of change, 273
Performance status, 273
Person-years method, 80
Pharmacoeconomics, 255
Pharmacoepidemiology, 255
Pharmacovigilance, 255
Phases of clinical development of drugs, 278
Placebo, 266
  effect, 266
Point estimate, 59
Poisson distribution, 163
Poisson regression, 166
Population at risk, 77
Population standard deviation, 41
Population, 2, 4
  conceptual definition, 4, 255
  operational definition, 4, 255
Post-stratification, 308
Power, 136
Predictive value positive/negative, 215
Prevalence study, 77
Prevalence, 77
  rate, 77
Primary sampling unit, 66, 77
Probability density function, 32
Probability distribution, 26, 29
  of sample means, 38
  of sample proportions, 54
  of variance ratios, 125
Probability sampling, 65

Product-limit estimate, 157
Prognostic factor, 212
  of treatment response, 308
Projection of estimates, 84
Promax rotation, 237
Proportional hazards model, 219–223
Proportional stratified sampling, 72
Proportionality assumption, 223–225
Proportions
  confidence interval for, 57
  exact confidence limits for, 56
  inference from, 55–58
  variance of, 53
Prospective study, 64, 89
Protocol
  deviations, 296
  of clinical trial, 291
  structure, 291
  violations, 297
PSU, 66, 77
p-value, 111
  nominal, 285

Quadratic term, 200
Qualitative studies, 5
Quartiles, 21

R^2, 179, 185
  pseudo, 210
Ramsey RESET test, 189
Random effects, 303
Random variable, 26, 29
Randomization
  blocked, 268
  concealed, 270
  dynamic, 272
  list, 270
  minimization, 272
  permuted blocks, 269
  simple, 268
  stratified, 270
  unbalanced, 269
Range, 21
Regression
  assumptions, 178, 191
  coefficient test, 179
  coefficient, 174
  constant, 174
  curvilinear, 200

diagnostics, 188–191
  equality of slopes, 199
  line, 170
  linear, 169–184
  logistic, 201–219
  multiple, 185–188
  nonlinear, 199–201
  partial coefficient, 185
  polynomial, 200
  selection of variables, 192
  slope, 171
  stepwise, 193
  sum of squares due to, 182
  to the mean, 259
Rejection region, 109
Relative death rate, 161
Relative risk, 101
Reliability, 239
  internal consistency, 240–245
  test-retest, 245
Residual, 175
  analysis of, 188
  mean square, 177, 181
  plot, 189
Retrospective study, 64, 89
Risk
  attributable, 102
  difference, 312
  population at, 77
  ratio, 101
  reduction, 312
  stratification, 212
Robust, 98
ROC curve, 215
  area under the, 216
Root mean square error, 184

Sample size estimation, 59, 273
  for clinical trials, 273
  for differences in means, 136
  for differences in proportions, 138
  for hazard ratios, 274
  for means, 69
  for non-inferiority trials, 280
  for one-sample tests, 283
  for proportions, 60, 69
  for sampling without replacement, 69
  for stratified sampling, 74
  unbalanced, 139

Sample, 3
  mean, 25
  representativeness, 6
  standard deviation, 41
Sampling, 3, 6–13, 64–77
  cluster, 74
  combined, 76
  consecutive random, 66, 256
  convenience, 65, 91, 256
  distributions, 37–39
  fraction, 66
  frame, 66
  multistage, 74–77
  non-probability, 65, 81, 91
  probability, 65
  proportional stratified, 72
  simple random, 66
  stratified, 70
  stratified with optimal allocation, 73
  systematic random, 66
  variation, 38
  with replacement, 67
  without replacement, 67
Scale transformation
  logarithmic, 142
  rank, 145
  reciprocal, 142
  square-root, 142
Scales of measurement, 15–18
  binary, 15, 53
  categorical, 16
  counts, 17
  interval, 17
  nominal, 16
  ordinal, 16
  ratio, 17
Scatterplot, 169
Schoenfeld method, 275
Scree plot, 234
Screening instrument, 78
Search by reference, 312
Selection bias, 256, 262
Sensitivity, 136, 213
  of clinical questionnaires, 252
Set
  risk, 222
  testing, 217
  training, 216
  validation, 218

Shapiro–Wilk test, 189
Sign test, 150
Simple random sampling, 66
Simpson's rule, 277
Slope, 171
SMR, 85
Spearman's correlation coefficient, 245
Specific rate, 82
Specificity, 135, 213
Standard
    error of the measurement, 274
    normal deviate, 34, 40
    normal variable, 40
Standard deviation, 23
    population, 24
    sample, 24
Standard error
    of difference in means, 94, 97
    of difference in paired samples, 149
    of difference in proportions, 100
    of incidence density, 165
    of survival probability, 157
    of the incidence rate ratio, 166
    of the mean, 40–42
    of the regression coefficient, 175–177
    value of, 42
Standardization, 82–85
    direct, 82
    indirect, 84
Standardized mean difference, 313
Standardized mortality rate, 85
STATA, 183
Statistical table, 34
    of the binomial distribution, 58
    of the chi-square distribution, 122
    of the F distribution, 129
    of the normal distribution, 34–36
    of the t distribution, 51
Statistical tests
    one-sided, 132–135
    rationale, 107
    two-sided, 132
Stochastic curtailment, 290
Strata, 70
Stratified sampling, 70
    with optimal allocation, 73
Strong control of the FWER, 304
Student's t distribution, 48–51
Student's t test, 112–115

in non-inferiority trials, 280
one-sample, 283
one-sided, 133
Study design, 4–6
Subgroup analysis, 308
Sum of squares, 23
    between-groups, 127
    partition of, 127
    within-group, 127
Surrogate criterion, 257
Survival-time analysis, 152
Survivor function, 154, 157
Systematic random sampling, 66

t test, 112–115
    in crossover clinical trials, 264
    in non-inferiority trials, 280
    of the slope of a regression
        line, 179
    one-sample, 283
    one-sided, 133
Table
    cell, 117
    classification, 212
    contingency, 117, 248
    grand total, 117
    marginal total, 117
Tabulation, 18–19
Test
    false negative rate, 135, 214
    false positive rate, 135, 214
    sensitivity, 136, 213
    specificity, 135, 213
    true negative rate, 135, 213
    true positive rate, 136, 213
Test–retest, 245
Treatment-emergent, 308
True negative/positive rate, 135,
    136, 213
Tukey's test, 141
Type I error, 135
Type II error, 135

Uniform distribution, 26
Uniqueness, 235
Validity, 251
    concurrent, 252
    construct, 251
    content, 251

convergent, 252
criterion, 252
discriminant, 252
Variance, 23
  between-groups, 126
  of incidence density, 165
  of the logarithm of incidence density, 165
  population, 25, 42
  ratio, 125, 181
  residual, 125
  sample, 25, 42
  total, 124

within-group, 125
Varimax rotation, 236
Visual analogic scale, 17

Weighting, 314–316
Wilcoxon rank-sum test, 143–145
Wilcoxon signed-rank test, 149

Yates' continuity correction, 122

z-score, 238
z-test, 108–110